T0091715

Green
Communications
and Networking

Green

Communications
and Networking

Edited by
F. Richard Yu
Xi Zhang
Victor C.M. Leung

CRC Press
Taylor & Francis Group
Boca Raton London New York

CRC Press is an imprint of the
Taylor & Francis Group, an **informa** business

CRC Press
Taylor & Francis Group
6000 Broken Sound Parkway NW, Suite 300
Boca Raton, FL 33487-2742

First issued in paperback 2019

ISBN-13: 978-1-4398-9913-7 (hbk)
ISBN-13: 978-0-367-38054-0 (pbk)

Library of Congress Cataloging-in-Publication Data

Green communications and networking / editors, F. Richard Yu, Xi Zhang, Victor C.M. Leung.
   pages cm
   Includes bibliographical references and index.
   ISBN 978-1-4398-9913-7 (hardcover : alk. paper)
   1. Telecommunication--Energy conservation. 2. Computer networks--Energy conservation. 3. Telecommunication--Environmental aspects. 4. Greenhouse gas mitigation. 5. Sustainable engineering. I. Yu, F. Richard. II. Zhang, Xi. III. Leung, Victor Chung Ming, 1955-

TK5102.5.G733 2013
621.382028'6--dc23   2012026685

Contents

II   Green Wireline Communications and Networking

A Brief Journey through "Green Communications and Networking"

F. Richard Yu, Carleton University, Ottawa, ON, Canada
Xi Zhang, Texas A&M University, College Station, TX, USA
Victor C. M. Leung, The University of British Columbia, Vancouver, BC, Canada

Introduction

As concerns about climate change, rising fossil fuel prices and energy security increase, companies and governments around the world are committing great efforts to develop new technologies for the green strategies addressing climate change globally and facilitating low greenhouse gas (GHG) development. Currently, the GHG emissions produced by the Information and Communication Technology (ICT) industry alone are said to be equivalent to the GHG emissions of the entire aviation industry. It is estimated that one small computer server generates as much GHG as a sport utility vehicle (SUV). Furthermore, with the increasing demand for higher data rates, the energy consumption for the ICT industry is increasing by 16-20% per year, and the energy costs for mobile network operators can already be as high as half of their annual operating budgets. The role of ICT includes not only the emission reduction and energy savings in ICT products and services, but also enabling low carbon emissions in other industries, such as electric power smart grids. Indeed, networks are crucial technologies for a smart power grid, which monitors, protects and optimizes the operation of its interconnected elements from end to end, with two-way flow of electricity and information to create an automated and distributed energy delivery network.

The contributed articles in this book from the leading experts in this field cover different aspects of modeling, analysis, design, management, deployment, and optimization of algorithms, protocols, and architectures of green communications and networking. In particular, the topics include energy efficiency, resource management, relay techniques, cross-layer design and optimization, rate adaptation, graph-theoretic approach, router architecture, dynamic scheduling, smart grid communications, demand and response in smart grids, and wireless networks in the smart grid environment. A summary of all of the chapters is provided in the following sections.

PART I: Green Wireless Communications and Networking

As the first chapter of this book, *Chapter 1*, authored by *A. Attar, H. Li and V. C. M. Leung*, introduces a novel solution, named broadband wireless access with fiber-connected massively distributed antennas (BWA-FMDA) to deliver green last-mile access. The advantages of BWA-FMDA architecture are its flexibility of deployment, scalability of coverage from a few meters for indoor access to several kilometers for outdoor communications and superior performance in terms of throughput as well as power efficiency. The focus of this chapter is mainly on power saving capabilities of BWA-FMDA compared with state-of-the-art last mile access solutions. In particular they compare the power consumption model for several last-mile solutions and demonstrate that through integrating optical fibers with wireless access a more power efficient access solution can be envisioned which also enhances the network throughput.

Chapter 2, authored by *X. Zhang and W. Cheng*, develops a Demanding-Based Resources Trading (DBRT) model for green communications. They propose a mechanism to minimize the energy consumption of wireless networks without compromising the quality-of-service (QoS) for users. Applying the DBRT model, they develop a novel scheme – Wireless Networks Resources Trading, which characterizes the trading relationships among different wireless resources for a given number of QoS performance metrics. According to wireless networks resources trading relationships, different wireless resources can be consumed to meet the same set of QoS performance metrics. Therefore, to minimize the energy consumption for given QoS performance metrics, they can trade the other type of wireless networks resources for the energy resources while satisfying the demanded QoS performance metrics. Based on the developed wireless networks resources trading relationships, this chapter derives the optimal energy-bandwidth, energy-time, energy-space and energy-code wireless networks resources trading relationships for green wireless networks. Two example-case studies are also conducted to show how to use the available bandwidth or the acceptable delay bound to achieve the minimum energy consumption while guaranteeing the required QoS performance metrics in wireless networks.

Chapter 3, authored by *Y. Qi, F. Héliot, M. A. Imran and R. Tafazolli*, analyzes the relaying technique at link and system levels from both spectrum efficiency (Se) and energy efficiency (EE) perspectives. A thorough investigation will be provided for a variety of approaches at the relay node (RN) to forward information including amplify-and-forward (AF), decoding-and-forward (DF), compress-and-forward (CF). Advanced relaying schemes, where the conventional relaying schemes are combined in a hybrid manner to adapt to the variations of the channel states, are introduced and investigated. Furthermore, the relaying techniques are combined with retransmission protocols for packet oriented communication systems and a study from spectrum and energy efficiency perspectives is presented. Finally, this chapter also addresses the challenge of designing and positioning RNs in a state-of-the-art wireless cellular system, namely LTE system, coupled with practical power consumption models.

Chapter 4, authored by *T. Zhu, S. Xiao and C. Zhou*, introduces (i) energy-efficient hardware platforms, (ii) energy-efficient MAC, (iii) energy-efficient networking, and (iv) energy-efficient applications. In addition, motivated by the insufficiency of link-layer designs, the authors introduce cross-layer optimization in energy static low-duty-cycle wireless networks. The cross-layer design in energy dynamic low-duty-cycle wireless networks is also studied in this chapter.

Chapter 5, authored by *Z. Zhao, Z. Dou and Y. Shu*, studies energy-efficient rate adaptation in long-distance wireless mesh (LDmesh) networks. The authors propose an efficient probing algorithm to obtain the Frame Delivery Ratio (FDR)-Received Signal Strength Indicator (RSSI) envelope mapping for each bit rate. FDR-RSSI is linear and remains invariant for a period of time so that it can be used to facilitate rate selection. Moreover an energy-efficient rate selection approach is presented to leverage the path loss information based on channel reciprocity. In addition, this chapter provides a technique to detect the distortion of FDR-RSSI that arises from external WiFi interference. The simulation results show that the proposed schemes can improve link throughput efficiently with minimum energy consumption.

PART II: Green Wireline Communications and Networking

Chapter 6, authored by *F. Cuomo, A. Cianfrani and M. Polverini*, studies energy saving in the Internet. The authors present graph-theoretic solutions that can be adopted in an IP network for energy saving purposes. The common idea of these solutions is to reduce the number of links, e.g., router line cards, that are used in the network during the off-peak period. To this aim different properties of the graph that models the network are used. By controlling the impact of the removal of some links on the algebraic connectivity, the proposed scheme derives a list of links that can be switched off. It combines the algebraic connectivity also with the edge betweenness. This latter parameter allows to cut from the network graph links that are crossed only by few paths. The resulting graph algebraic connectivity is then used to control that the network remains connected and that its connection degree is above a suitable threshold.

Chapter 7, authored by *C. Hu, B. Liu, M. Zhang, B. Zhang and X. Wang*, studies the architectural design of energy-efficient wireline Internet nodes. The authors concentrate on the exploration of power/energy-saving mechanisms through the design of Internet transmission equipment, e.g., routers. By revisiting the characteristics of the Internet behaviors and the modular architecture of routers, this chapter suggests the approach for engineering energy-efficient Internet from three different perspectives and discusses the imposed technical challenges. To address the challenges and seize the energy-saving opportunities, a new conceptual router model/architecture as the guide to design and implement power efficient router, as well as the Internet, is pursued.

Chapter 8, authored by *M. P. Anastasopoulos, A. Tzanakaki and D. Simeonidou*, studies the converged optical network and IT infrastructures suitable to support cloud services. More specifically, the concept of Virtual Infrastructures (VIs), over one or more interconnected Physical Infrastructures (PIs) comprising both network and IT resources, is considered. Taking into account the energy consumption levels associated with the ICT today and the expansion of the Internet, energy efficient infrastructures with reduced CO_2 emissions become critical. To address this, a hybrid energy power supply system for the high energy consuming IT resources is adopted. In this system conventional and renewable energy sources are cooperating to produce the necessary power for the IT equipment to operate and support the required services. The reduction in CO_2 emissions is further increased by applying energy aware planning of VIs over the converged PI. To quantify the benefits of the proposed approach a Mixed Integer Linear Programming model suitable for planning VIs is proposed and developed. This model takes into account multi-period and multi-service considerations over an integrated hybrid-solar powered IT and optical network infrastructure and aims at minimizing the CO_2 emissions of the planned VIs. The modelling results indicate significant reduction of the overall CO_2 emissions that varies between 10-50% for different levels of demand requests.

Chapter 9, authored by *M. J. Neely*, presents a methodology for optimizing time averages in systems with variable length frames. Applications include energy and quality aware task scheduling in smart phones, cost effective energy management at computer servers, and more. The author considers energy-aware control for a computing system with two states: *active* and *idle*. In the active state, the controller chooses to perform a single task using one of multiple task processing modes. The controller then saves energy by choosing an amount of time for the system to be idle. These decisions affect processing time, energy expenditure, and an abstract *attribute vector* that can be used to model other criteria of interest (such as processing quality or distortion). The goal is to optimize time average system performance. The solution methodology of this chapter uses the theory of *optimization for renewal systems*.

PART III: Smart Grid Communications and Networking

Chapter 10, authored by *Z. Li, D. Ishchenko, F. Yang and Y. Ye*, reviews the recent development of utility communication networks, including the advanced metering infrastructure (AMI) and the supervisory control and data acquisition (SCADA). The standardizations of communication protocols in both AMI and SCADA systems are the major focus of this chapter. In addition, some potential grid management applications that are facilitated by the real-time communication mechanism and enable efficient grid operations are also discussed.

Chapter 11, authored by *Q. Dong, L. Yu and W. Song*, surveys the ongoing research through elaborating a representative number of Demand and Response (DR) methods in smart grid and discusses future directions. DR refers to the dynamic demand mechanisms to manage electricity demand in response to supply conditions, and is one of the most important functions of smart grids. DR offers several benefits, including reduction of peak demand, participant financial benefits, integration of renewable resources and provision of ancillary services. This chapter focuses on a classification that is based on the optimization objective. A representative number of DR methods have been stated, which belong to customer profit optimization category operation cost of electric utility reduction category and social welfare maximization category.

Chapter 12, authored by *S. Bu, F. R. Yu and P. X. Liu*, considers not only energy-efficient communications but also the dynamics of the smart grid in designing green wireless cellular networks. Specifically, the dynamic operation of cellular base stations depends on the traffic, real-time electricity price and the pollutant level associated with electricity generation. Coordinated multipoint (CoMP) is used to ensure acceptable service quality in the cells whose base stations have been shut down. The active base stations decide on which retailers to procure electricity from and how much electricity to procure. We formulate the system as a Stackelberg game, which has two levels: a cellular network level and a smart grid level. Simulation results show that the smart grid has significant impacts on green wireless cellular networks, and our proposed scheme can significantly reduce operational expenditure and CO_2 emissions in green wireless cellular networks.

Conclusion

A summary of the contributed chapters has been provided that will be helpful to follow the rest of this book. These chapters essentially feature some of the major advances in the research on green communications and networking for the next generation communications and networking systems. Therefore, the book will be useful to both researchers and practitioners in this area. The readers will find the rich set of references in each chapter particularly valuable.

About the Editors

F. Richard Yu is currently an associate professor in the Department of Systems and Computer Engineering, School of Information Technology, at Carleton University, Ottawa, ON, Canada. He received the Ph.D. degree in electrical engineering from the University of British Columbia, Vancouver, BC, Canada, in 2003. From 2002 to 2004, he was with Ericsson, Lund, Sweden, where he worked on research and development of third-generation cellular networks. From 2005 to 2006, he was with a startup company in California, where he worked on research and development in the areas of advanced wireless communication technologies and new standards. He joined the School of Information Technology and the Department of Systems and Computer Engineering, Carleton University, Ottawa, ON, Canada, in 2007. His research interests include cross-layer design, security, and quality-of-service provisioning in wireless networks.

He received the Carleton Research Achievement Award in 2012, the Ontario Early Researcher Award in 2011, the Excellent Contribution Award at IEEE/IFIP TrustCom 2010, the Leadership Opportunity Fund Award from Canada Foundation of Innovation in 2009 and the Best Paper Awards at IEEE/IFIP TrustCom 2009 and Int'l Conference on Networking 2005. His research interests include cross-layer design, security and QoS provisioning in wireless networks.

Dr. Yu is a senior member of the IEEE. He serves on the editorial boards of several journals, including *IEEE Transactions on Vehicular Technology, IEEE Communications Surveys & Tutorials, ACM/Springer Wireless Networks, EURASIP Journal on Wireless Communications Networking, Ad Hoc & Sensor Wireless Networks, Wiley Journal on Security and Communication Networks*, and *International Journal of Wireless Communications and Networking*, and a guest editor for *IEEE Systems Journal* for the special issue on Smart Grid Communications Systems. He has served on the Technical Program Committee (TPC) of numerous conferences, as the TPC Co-Chair of IEEE CCNC'13, INFOCOM-CCSES'2012, ICC-GCN'2012, VTC'2012S, Globecom'11, INFOCOM-GCN'2011, INFOCOM-CWCN'2010, IEEE IWCMC'2009, VTC'2008F and WiN-ITS'2007, as the Publication Chair of ICST QShine 2010, and the Co-Chair of ICUMT-CWCN'2009.

Xi Zhang received the B.S. and M.S. degrees from Xidian University, Xian, China, the M.S. degree from Lehigh University, Bethlehem, PA, all in electrical engineering and computer science, and the Ph.D. degree in electrical engineering and computer science (electrical engineering systems) from The University of Michigan, Ann Arbor.

He is currently an associate professor and the founding director of the Networking and Information Systems Laboratory, Department of Electrical and Computer Engineering, Texas A&M University, College Station. He was an assistant professor and the founding director of the Division of Computer Systems Engineering, Department of Electrical Engineering and Computer Science, Beijing Information Technology Engineering Institute, China, from 1984 to 1989. He was a research fellow with the School of Electrical Engineering, University of Technology, Sydney, Australia, and the Department of Electrical and Computer Engineering, James Cook University, Australia, under a fellowship from the Chinese National Commission of Education. He was with the Networks and Distributed Systems Research Department, AT&T Bell Laboratories, Murray Hill, NJ, and with AT&T Laboratories Research, Florham Park, NJ. He has published more than 200 research papers in the areas of wireless networks and communications systems, mobile computing, network protocol design and modeling, statistical communications, random signal processing, information theory and control theory and systems.

Dr. Zhang received the U.S. National Science Foundation CAREER Award in 2004 for his research in the areas of mobile wireless and multicast networking and systems. He is an IEEE Communications Society Distinguished Lecturer. He received the Best Paper Awards in the IEEE GLOBECOM 2007, IEEE GLOBECOM 2009, and IEEE WCNC 2010. He also received the TEES Select Young Faculty Award for Excellence in Research Performance from the Dwight Look College of Engineering at Texas A&M University, College Station, in 2006. He is currently serving or has served as an editor for the *IEEE Transactions on Communications*, an editor for the *IEEE Transactions on Wireless Communications*, an associate editor for the *IEEE Transactions on Vehicular Technology*, a guest editor for the *IEEE Journal on Selected Areas in Communications* for the special issue on Broadband Wireless Communications for High Speed Vehicles, a guest editor for the *IEEE Journal on Slected Areas in Communications* for the special issue on Wireless Video Transmissions, an associate editor for the *IEEE Communications Letters*, a guest editor for the *IEEE Communications Magazine* for the special issue on Advances in Cooperative Wireless Networking, a guest editor for the *IEEE Wireless Communications Magazine* for the special issue on Next Generation of CDMA Versus OFDMA for 4G Wireless Applications, an editor for the *John Wiley Journal on Wireless Communications and Mobile Computing*, an editor for the *Journal of Computer Systems, Networking, and Communications*, an associate editor for the *John Wiley Journal on Security and Communications Networks*, an area editor for the *Elsevier Journal on Computer Communications*, and a guest editor for the *John Wiley Journal on Wireless Communications and*

Mobile Computing for the special issue on next generation wireless communications and mobile computing. He has frequently served as a panelist on the U.S. National Science Foundation Research-Proposal Review Panels. He is serving or has served as the Technical Program Committee (TPC) chair for the IEEE GLOBECOM 2011, TPC area chair for the IEEE INFOCOM 2012, general co-chair for INFOCOM 2012 - Workshop on Communications and Control for Sustainable Energy Systems: Green Networking and Smart Grids, TPC co-chair for IEEE ICC 2012 - Workshop on Green Communications and Networking, general co-chair for IEEE INOFOCOM 2011 - Workshop on Green Communications and Networking, TPC co-chair for the IEEE ICDCS 2011 - Workshop on Data Center Performance, Panels/Demos/Posters Chairs for the ACM MobiCom 2011, TPC vice-chair for IEEE INFOCOM 2010, general chair for the ACM QShine 2010, TPC co-chair for IEEE INFOCOM 2009 Mini-Conference, TPC co-chair for IEEE GLOBECOM 2008 - Wireless Communications Symposium, TPC co-chair for the IEEE ICC 2008 - Information and Network Security Symposium, symposium chair for IEEE/ACM International Cross-Layer Optimized Wireless Networks Symposium 2006, 2007, and 2008, respectively, the TPC chair for IEEE/ACM IWCMC 2006, 2007, and 2008, respectively, the demo/poster chair for IEEE INFOCOM 2008, the student travel grants co-chair for IEEE INFOCOM 2007, the general chair for ACM QShine 2010, the panel co-chair for IEEE ICCCN 2007, the poster chair for IEEE/ACM MSWiM 2007 and IEEE QShine 2006, executive committee co-chair for QShine, the publicity chair for IEEE/ACM QShine 2007 and IEEE WirelessCom 2005, and a panelist on the Cross-Layer Optimized Wireless Networks and Multimedia Communications at IEEE ICCCN 2007 and WiFi-Hotspots/WLAN and QoS panel at IEEE QShine 2004. He has served as the TPC member for more than 100 IEEE/ACM conferences, including IEEE INFOCOM, IEEE GLOBECOM, IEEE ICC, IEEE WCNC, IEEE VTC, IEEE/ACM QShine, IEEE WoWMoM, IEEE ICCCN, etc.

Victor C. M. Leung received the B.A.Sc. (Hons.) degree in electrical engineering from the University of British Columbia (U.B.C.) in 1977, and was awarded the APEBC Gold Medal as the head of the graduating class in the Faculty of Applied Science. He attended graduate school at U.B.C. on a Natural Sciences and Engineering Research Council Postgraduate Scholarship and completed the Ph.D. degree in electrical engineering in 1981.

From 1981 to 1987, Dr. Leung was a senior member of technical staff at MPR Teltech Ltd., specializing in the planning, design and analysis of satellite communication systems. In 1988, he started his academic career at the Chinese University of Hong Kong, where he was a lecturer in the Department of Electronics. He returned to U.B.C. as a faculty member in 1989, and currently holds the positions of Professor and TELUS Mobility Research Chair in Advanced Telecommunications Engineering in the Department of Electrical and Computer Engineering. He is a member of the Institute for Computing, Information and Cognitive Systems at U.B.C. He also holds adjunct/guest faculty appointments at Jilin University, Beijing Jiaotong University, South China

University of Technology, the Hong Kong Polytechnic University and Beijing University of Posts and Telecommunications. Dr. Leung has co-authored more than 500 technical papers in international journals and conference proceedings, and several of these papers had been selected for best paper awards. His research interests are in the areas of architectural and protocol design, management algorithms and performance analysis for computer and telecommunication networks, with a current focus on wireless networks and mobile systems.

Dr. Leung is a registered professional engineer in the Province of British Columbia, Canada. He is a Fellow of IEEE, a Fellow of the Engineering Institute of Canada, and a Fellow of the Canadian Academy of Engineering. He is a Distinguished Lecturer of the IEEE Communications Society. He is serving on the editorial boards of the *IEEE Transactions on Computers, IEEE Wireless Communications Letters, Computer Communications,* the *Journal of Communications and Networks,* as well as several other journals. Previously, he has served on the editorial boards of the *IEEE Journal on Selected Areas in Communications Wireless Communications Series,* the *IEEE Transactions on Wireless Communications* and the *IEEE Transactions on Vehicular Technology.* He has guest-edited several journal special issues, and served on the technical program committee of numerous international conferences. He is a General Co-chair of GCSG Workshop at Infocom 2012, GCN Workshop at ICC 2012, CIT 2012, FutureTech 2012, CSA 2011. He is a TPC Co-chair of the MAC and Cross-layer Design track of IEEE WCNC 2012. He chaired the TPC of the wireless networking and cognitive radio track in IEEE VTC-fall 2008. He was the General Chair of AdhocNets 2010, WC 2010, QShine 2007, and Symposium Chair for Next Generation Mobile Networks in IWCMC 2006-2008. He was a General Co-chair of Chinacom 2011, MobiWorld and GCN Workshops at IEEE Infocom 2011, BodyNets 2010, CWCN Workshop at Infocom 2010, ASIT Workshop at IEEE Globecom 2010, MobiWorld Workshop at IEEE CCNC 2010, IEEE EUC 2009 and ACM MSWiM 2006, and a TPC Vice-chair of IEEE WCNC 2005. He is a recipient of an IEEE Vancouver Section Centennial Award.

Contributors

Muhammad Ali Imran
University of Surrey
Surrey, UK

Markos P. Anastasopoulos
Athens Information Technology
    Center
Peania Attikis, Greece

Alireza Attar
Department of Electrical and
    Computer Engineering
The University of British Columbia
Vancouver, Canada

Shengrong Bu
Department of Systems and
    Computer Engineering
Carleton University
Ottawa, Canada

Wenchi Cheng
Department of Electrical and
    Computer Engineering
Texas A&M University
College Station, TX, USA

Antonio Cianfrani
University of Rome Sapienza
Rome, Italy

Francesca Cuomo
University of Rome Sapienza
Rome, Italy

Qifen Dong
Zhejiang University of Technology
Hangzhou, China

Zhibin Dou
Tianjin University
Tianjin, China

Fabien Heliot
University of Surrey
Surrey, UK

Chengchen Hu
MoE KLINNS Lab
Department of Computer Science
    and Technology
Xi'an Jiaotong University
    Xi'an, China

Dmitry Ishchenko
ABB US Corp. Research Center
USA

Victor C. M. Leung
Department of Electrical and
    Computer Engineering
The University of British Columbia
Vancouver, Canada

Haoming Li
Department of Electrical and
    Computer Engineering
The University of British Columbia
Vancouver, Canada

Zhao Li
ABB US Corp. Research Center
USA

Bin Liu
Department of Computer Science
    and Technology
Tsinghua University
Beijing, China

Peter X. Liu
Department of Systems and
    Computer Engineering
Carleton University
Ottawa, Canada

Michael J. Neely
University of Southern California
Los Angeles, CA, USA

Marco Polverini
University of Rome Sapienza
Rome, Italy

Yinan Qi
University of Surrey
Surrey, UK

Yantai Shu
Tianjin University
Tianjin, China

Dimitra Simeonidou
University of Essex
Colchester, UK

WenZhan Song
Georgia State University
Atlanta, GA, USA

Rahim Tafazolli
University of Surrey
Surrey, UK

Anna Tzanakaki
Athens Information Technology
    Center
Peania Attikis, Greece

Xiaojun Wang
School of Electronic Engineering
Dublin City University
Dublin, Ireland

Sheng Xiao
University of Massachusetts Amherst
Amherst, MA, USA

Fang Yang
ABB US Corp. Research Center
USA

Yanzhu Ye
Energy Management Department
NEC Laboratories America, Inc.
Cupertino, CA, USA

F. Richard Yu
Department of Systems and
    Computer Engineering
Carleton University
Ottawa, Canada

Li Yu
Zhejiang University of Technology
Hangzhou, China

Beichuan Zhang
Department of Computer Science
Arizona University
Tucson, AZ, USA

Mingui Zhang
Huawei Inc.
China

Xi Zhang
Department of Electrical and
    Computer Engineering
Texas A&M University
College Station, TX, USA

Chang Zhou
China Jiliang University
Hangzhou, China

Zenghua Zhao
Tianjin University
Tianjin, China

Ting Zhu
Binghamton University
Binghamton, NY, USA

Part I

Green Wireless Communications and Networking

Chapter 1

Power-Efficient Last Mile Access Using Fiber-Connected Massively Distributed Antenna (BWA-FMDA) System

Alireza Attar, Haoming Li and Victor C. M. Leung
Department of Electrical and Computer Engineering
The University of British Columbia
{attar, hlih, vleung}@ece.ubc.ca

1.1   Evolution of Cellular Networks and Power Efficiency Considerations

The pervasive presence of wireless and cellular access networks has been made possible by a constant stream of technological innovations. In the case of cellular communications, major technical milestones have been marked by the "x-th generation" labels; two examples include shifting from analog systems of 1st generation cellular networks to digital communication in the 2nd generation, or evolving from 3rd generation cellular systems based on code division multiple access (CDMA) technology to 4th generation based on orthogonal frequency division multiple access (OFDMA). However, to better understand the need for more power efficiency in the next generation of cellular networks, we need to look past technology trends to explore the underlying shifts in three sets of fundamental parameters for any cellular technology, namely power consumption, traffic density and cost of equipment.

1.1.1   Power Consumption

The unprecedented growth of demand for wireless connectivity in the past couple of decades mandates continued expanding and upgrading wireless and cellular networks. It is then not surprising that power needed for operating this growing wireless ecosystem has significantly risen. As an example, Vodafone, one of the largest mobile service providers globally, has reported that over 57% of the company's total 397 GWh energy consumption (gas and electricity) in the United Kingdom (UK) was used to power over 12,000 Base Stations (BS). This is in sharp contrast to only 112 GWh of power consumed by other network equipment and 57 GWh of power used for running 8 major offices and 347 retail spaces during the 2007/2008 operation year [1].

A surge (reduction) in wireless network power consumption can be directly translated into increasing (decreasing) carbon dioxide (CO_2) emission. Another recent report by Vodafone estimates the total (sum of mature and emerging market shares) gross CO_2 emissions of this firm at 961,982 tons by BSs and 346,304 tons by its other network equipment as compared with 157,258 tons by offices and retail stores for 2009/2010 [2]. Based on these reported measurements it is clear that operations of the access network, and more specifically BSs, significantly outweigh other entities in the overall power consumption and CO_2 emissions of any given wireless operator.

Given the considerable impact of BSs in the power consumption pattern of any cellular network, we can further examine how this consumed power is distributed within a typical cell site. The main components of a typical cell site include a power supply unit and a cooling system, and the BS units including base band processing units and radio frequency (RF) units such as power amplifier (PA), low noise amplifier (LNA) and antenna feed. The exact specifications of BSs differ from vendor to vendor. The authors in [3] estimate that of the total power consumption of a typical cell site, 43% is consumed by the cooling system followed by 41% by the BS itself. Within the BS, the shares of the main power sinks are the feeder at 44%, RF conversion and PA at 15% and signal processors at 9%.

Next, we study the changing pattern of data consumption by wireless end users, which is the main driving force behind the expansion of wireless/cellular access networks by the service providers.

1.1.2 Traffic Density

The explosive growth of bandwidth-hungry applications in conjunction with the ever-increasing wireless and mobile coverage around the world has contributed to the exchange of staggering amounts of data traffic. Increasingly a major portion of this data traffic is carried over last-mile access networks. A good measure of the growth of cellular coverage is the penetration rate which varies over different geographical regions. In Canada, approximately 24.04 million subscribers translate to a 71.78% penetration rate [4]. In a country like Sweden where the penetration rate is estimated at 123.53%, the total number of mobile subscriptions exceeds the total population [4].

As mentioned at the beginning of this section, besides increasing penetration rates of wireless/cellular networks, the user traffic load and pattern has also changed over the years. While the main traffic load was attributed to voice in the past, recent reports from operators around the world demonstrate a shifting traffic pattern toward packet data, especially from applications such as social media. For instance the operator Three, which has more than 6.2 million customers in the UK, reported recently a traffic volume of 2,500 terabytes over its mobile broadband network in only one month (June 2010) [5]. This cellular operator identified three social networking and gaming websites among the top five traffic generating websites, during the reported period. To cope with such massive data traffic loads, many wireless operators have abandoned their "unlimited" data access plans and have shifted toward tiered pricing strategies. However, tiered pricing by itself might not be able to contain the growing thirst for wireless traffic by end users.

The average traffic generated by all mobile users is estimated at 114 MB per user per month, without tiered pricing schemes and around 131 MB per user per month with tiered pricing, according to Cisco [6]. Thus it seems tiered pricing has indeed increased the traffic volume, a counter-intuitive result. To understand the cause of this surprising growth of traffic we need to distinguish between two types of end users, namely a small percentage of very high volume data consumers and the majority of users who exchange much lower traffic volumes. The latter type of user, identified in the Cisco report as those who generated at least 10 KB of data traffic per month, increased their total data consumption from 217 MB to 243 MB per month upon introduction of tiered pricing. In effect, tiered pricing encourages lower traffic generating users to consume more bandwidth but only regulates the bandwidth occupancy of a small subset of users, i.e., the highest traffic generating subscribers.

Combining this traffic load share per user with the cellular penetration rate, given at the beginning of this subsection, and taking into account the average population density for dense urban, residential urban and rural areas will result in reliable estimates of expected average traffic over a given locale. However, besides the growing data volumes and the increased power consumption of cellular networks in order to operate a higher number of BSs, a third factor, namely the operator's costs and revenues also affect the evolution of wireless systems.

1.1.3 Equipment Cost

The final parameter considered in our analysis of a typical cellular network is the equipment cost. This latter factor plays a significant role in the adoption of different technologies and solutions. If the final cost of a green communication strategy is substantially higher than a traditional solution, the operators might have less incentive to adopt the green technology. Unlike power consumption and traffic density, there are very few openly reported price quotes for macro BSs. The vendors will negotiate the price with an operator based on the exact BS specifications and the quantity of the order, among other factors. One 2004 report cites prices in the range of $10,000-$30,000 per BS, with possibly much higher prices for the state-of-the-art equipment [7]. A micro BS is expected to cost around one tenth of a macro BS, and similarly a femto BS might cost a tenth of a micro BS.

Costs of BSs are a part of the capital expenditure (CAPEX) for a given operator. Other CAPEX components include the costs of cell site acquisition or rental, spectrum licensing and upgrading the equipment with new technologies emerging at an ever faster pace. The needs to optimize the cell planning as well as running and maintaining the cell sites constitute parts of an operator's operating expenditure (OPEX). In shifting from traditional technologies to green alternatives major changes in the equipment costs seem unlikely to emerge due to the existing market competition forces. On the other hand, the benefits of such a shift can be realized through the power efficiency sustained by green technologies that can potentially reduce the OPEX of an operator.

1.1.4 The Goals and Organization of This Chapter

The objective of this chapter is to introduce a novel solution, named broadband wireless access with fiber-connected massively distributed antennas (BWA-FMDA), to deliver green last-mile access. The advantages of BWA-FMDA architecture are its flexibility of deployment, scalability of coverage from a few meters for indoor access to several kilometers for outdoor communications and superior performance in terms of throughput as well as power efficiency. The focus of this chapter, however, is mainly on power saving capabilities of BWA-FMDA compared with state-of-the-art last-mile access solutions. In particular we compare the power consumption model for several last-mile solutions and demonstrate that through integrating optical fibers with wireless access a more power efficient access solution can be envisioned which also enhances the network throughput.

The rest of this chapter is organized as follows. In Section 1.2 the BWA-FMDA architecture is introduced in detail. Then, Section 1.3 elaborates on the power consumption model that forms the basis of comparison of BWA-FMDA and other last-mile solutions. Extensive simulation results are presented in Section 1.4, before Section 1.5 concludes this chapter.

We now proceed to introduce our proposed solution for next generation cellular networks, which provides a cost effective framework that can address the challenges arising due to higher wireless data volume and correspondingly increasing network power consumption.

1.2 BWA-FMDA Architecture

Our last-mile architecture, referred to as BWA-FMDA is depicted in Fig. 1.1. This system is composed of 3 components, namely the antenna elements (AEs), the optical communication medium and the central processing entity. In the following subsections we elaborate on each component. The interested readers can refer to [8] for in-depth descriptions of the system and its performance results. But first we briefly cover recent advances in radio over fiber (RoF) solutions.

1.2.1 A Brief Background on RoF Solutions

Integration of wireless communications and optical fibers has evolved considerably over the past couple of decades. The majority of initial studies aimed to design hotspot solutions, i.e., to increase the feasible data throughput over a relatively small locale such as areas covered by micro- or picocells [9], [10]. It is then no surprise that communicating wireless local area network (WLAN) signals over fiber has received the most attention from the research community. The possibility of accommodating several WLAN signals using RoF technology was demonstrated in [11] and [12]. Design of indoor and outdoor access network based on IEEE 802.11x is discussed in [13], [14] and [15]. More

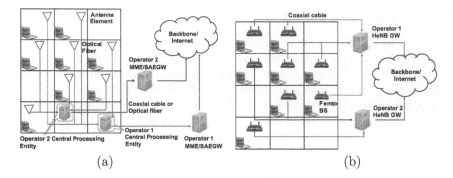

(a)   (b)

Figure 1.1: Comparison of (a) BWA-FMDA architecture and (b) femtocell architecture with two service providers. In this figure, GW stands for gateway, MME/SAE denotes Mobility Management Entity/System Architecture Evolution and HeNB refers to Home enhanced NodeB, from 3GPP's LTE-A terminology.

recently, the impact of integration of optical fibers with IEEE802.11x has been investigated in [16], [17] and [18]. RoF solutions for other medium access control (MAC) techniques such as OFDMA have been studied recently [19]. Other studies have considered the possibility of transmitting wideband CDMA (WCDMA) signals over fiber, e.g., [20].

It is therefore technically possible to develop an access network with any MAC standard over a RoF platform. The potential of RoF technology is not limited to hotspot solutions and can be extended to cellular and other last-mile solutions [8], [21].

1.2.2   Antenna Elements

AEs in BWA-FMDA architecture are responsible to transmit and receive the RF signal over the air. Except for broadband PA and LNA, and optical-to-RF and RF-to-optical conversion capabilities, there is no signaling processing capability embedded in an AE. In particular, channelization of the transmit/receive signals is not performed at the AE. Instead, all signals within a wide spectrum received by the antenna are converted to a broadband optical signal and sent to the central processing entity for processing via the optical communication medium (i.e., optical fiber). Likewise, RF signals already channelized at the central processing entity are sent to the AE for amplification and emission via an optical fiber. Each subscriber can randomly place the AE within his/her home, similar to a wireless router. Therefore no cell planning cost is incurred in the deployment of a BWA-FMDA network. We assume 0-dBi omni-directional antennas and assume that the central processing entity can support up to 8 AEs. Thus if more AEs will be required to cover a building, the operator needs to deploy additional processing entities. Finally due to low-loss optical fiber medium, we can assume the distance between

AEs and central processing entity, using off-the-shelf equipment, can be of the order of many hundreds of meters, which suffices to cover most buildings in practice.

1.2.3 Optical Communication Medium

A major differentiating point of BWA-FMDA with other last-mile solutions is that the radio signals are carried over optical fibers between the central processing entity and the AEs. In contrast to all-wireless last-mile techniques, BWA-FMDA shortens the communication link by utilizing the high-bandwidth low-loss characteristics of optical communications. Utilizing RoF techniques [9], RF signals following any cellular or broadband access standards can be transported over the optical fiber medium. This optical medium can be implemented in practice as a passive optical network (PON) [22] or active optical network. We have adopted the PON model here due to its cost-effectiveness and ease of deployment. Within the PON framework, time division multiplexing (TDM) and wavelength division multiplexing (WDM) techniques can be utilized to separate communications to various AEs.

1.2.4 Central Processing Entity

The proposed BWA-FMDA integrates the capabilities of a distributed network, based on its massively distributed antenna topology and the centralized processing capability at its central processing entity. This centralized processing facilitates an extensive array of sophisticated signal processing techniques which results in an unparalleled throughput delivery performance in a power efficient manner. As an example, while femtocells will operate independently of each other and even of the macro BS of their service providers, BWA-FMDA can coordinate AEs within a building to deliver Coordinated Multi-Point (CoMP)-like performance at the femtocell scale [8].

Deployment of our proposed BWA-FMDA also adds to both OPEX and CAPEX of an operator. The equipment costs are currently higher than micro BS due to the economy of scale. There is also the need for installation of optical fiber in the premises; however, this latter cost can be part of the long-term enhancement of operator's network. For instance many service providers offer bundled services, whereby cable TV, landline phone, digital subscriber line Internet access, and at times cellular subscriptions, are offered in single contracts. Most service providers are planning to upgrade their existing wireline access network by Fiber-to-the-x (FFTx) solutions. BWA-FMDA will become a more cost efficient solution as deployment of FTTx proliferates and the number of optical fiber-served residential and commercial buildings increases.

1.3   Power Consumption and Simulation Models

We argue that shortening the last-mile communication link has a doubly positive effect of enhancing the data rate for end users, on the link level, as well as conserving the power consumption on the network level. Therefore the overall power efficiency of the wireless network will increase. A comparison between deployment of macro and micro BSs in cellular communications has demonstrated the power efficiency improvement feasible through network densification [23]. We will further demonstrate that our proposed BWA-FMDA architecture can outperform micro BS and femtocell approaches in green communications over the last mile. In the following subsections, we will first explain the power consumption model used to benchmark the above mentioned wireless access approaches and then introduce our numerical results.

1.3.1   Power Consumption Model

We will compare three communication link shortening solutions, namely micro BS, femtocell and BWA-FMDA (Fig. 1.2).

Micro BS: We propose a power consumption model for a micro BS, which integrates the micro BS model in [24] with the cooperative BS model in [23]. The total consumed power at a micro BS takes into account two components. First, link budget considerations mandate adjustment of transmitted power, P_{tx}, which has a linear relation with power consumption. Further signal processing adds to the overall power consumption, the level of which depends on the number of AEs participating in Multiple Input Multiple Output (MIMO) transmissions by each micro BS, denoted by N_c. The overall micro BS power consumption, $P_{microBS}$, is given by

$$P_{microBS} = 5.5P_{tx} + P_{sp}, \tag{1.1}$$

where

$$P_{sp} = P_{spB}(0.87 + 0.1N_c + 0.03N_c^2) \tag{1.2}$$

denotes power consumption of signal processing. In (1.2), P_{spB} is set at 32 watts, the linear term of N_c represents energy overhead due to MIMO pilots, and the quadratic term stands for MIMO signal processing overhead.

Femto BS: Generally femto BSs are reduced-scale plug-and-play access points. The energy consumption of one femto BS unit is assumed constant at 18 watts, which is a typical energy consumption value in commercial femtocell access points.

BWA-FMDA: As detailed in Section 1.3, the BWA-FMDA architecture is comprised of two energy-consuming components, namely the RoF transmission system and the central processing entity. For the power consumption model of the RoF segment, we adopt the model presented in [25].

The laser diode needs a +5 dBm output power from its laser drive amplifier (LDA) for successful operation. The required output power of the LDA can be translated into the required input power using the amplifier efficiency. LDAs, however, generally demonstrate a low power efficiency level. Similar to [25] we assume 2.2% power efficiency in this study, which takes into account the required backoff to ensure linearity of the LDA. The laser diode itself will consume around 80 mW power, but it also requires a driver circuit for stability of its bias current, which in turn adds an extra 60 mW power consumptions to our model. At the other end of fiber medium a similar power consumption model is needed for the photodiode. In this case, we can assume 3 mW power consumption by the photodiode and an extra 80 mW power consumed by the corresponding transimpedance amplifier (TIA). Finally the output RF PA feeding the antenna can be assumed to have similar efficiency as the laser diode's LDA. We can summarize the power consumption model for RoF segment of BWA-FMDA, denoted by P_{RoFD}, as follows.

Denote by P_{LDA} and $P_{LDAOutt}$ the input and output power for the LDA, respectively. Further assume γ_{LDA} represents the LDA efficiency. The power consumption of laser diode (LD), photodiode (PD), and RF PA at an AE are denoted by P_{LD}, P_{PD} and P_{RFPA}, respectively. Therefore, the power per RoF link is given by:

$$
\begin{aligned}
P_{RoFD} &= (P_{LDA} + P_{LD} + P_{PD}) + P_{RFPA}, \\
P_{LDA} &= P_{LDAOut}/\gamma_{LDA}, \\
P_{LD} &= 140 \ \text{mW}, \\
P_{PD} &= 83 \ \text{mW}, \\
P_{RFPA} &= P_{tx}/\gamma_{RFPA},
\end{aligned}
\tag{1.3}
$$

where γ_{LDA} and γ_{RFPA} indicate the efficiency of LDA and RF PA, both fixed at 2.2%, and P_{tx} represents the transmission power per AE, which is usually determined by the desired wireless coverage or peak downlink transmission rate.

The power consumption for the signal processing at the central processing entity can be based on micro BS models. A CoMP cluster in BWA-FMDA performs the same MIMO signal processing as a micro BS. We have slightly modified (1.2) to model the BWA-FMDA energy consumption at the central processing entity as

$$
P_{sp} = 0.87 P_{spB} + 32(0.1 N_c + 0.03 N_c^2) \times N_{CoMP},
\tag{1.4}
$$

where N_{CoMP} is the number of CoMP clusters. The total energy consumption is then given by,

$$
P_{BWA} = P_{sp} + 1 \cdot N_{AE}.
\tag{1.5}
$$

We consider several AE placements and frequency reuse patterns and investigate the effect of transmission power on energy efficiency in our simulations.

1.3.2   Signaling Overhead

There are two types of signaling overhead incurred in distributed MIMO communications such as CoMP or BWA-FMDA, where we are mostly interested in their effect on power consumption of the network. First, by transmitting pilot signals and analyzing the corresponding responses at the receivers each node participating in distributed MIMO maintains an updated table of channel state information (CSI) for each receiver. The effect of this type of signaling on the power consumption of micro BSs are captured by the term "$0.1 \times N_c$" in (1.2). A similar term also appears in (3.45), which models our proposed BWA-FMDA.

A second type of signaling overhead is associated with the coordination of resource allocation at multiple nodes and the scheduling mechanism of distributed MIMO communications. The effect of this class of signaling overhead on the power consumption in CoMP is represented in (1.2) by the term "$0.03 \times N_c^2$". A similar term for BWA-FMDA is introduced in (3.45).

Note that the actual overhead (e.g., as measured by the number signaling packets) pertinent to this second type of signaling for BWA-FMDA is significantly lower than CoMP methods due to its centralized processing architecture. However, we used a similar level of power consumption for BWA-FMDA due to the complexity of signal processing in MIMO communications.

1.3.3   Simulation Model

We consider a 4-floor office building with 6 rooms per floor. The rooms are divided into two stripes, divided by a 2-meter-wide corridor. The dimension of each room is $12 \times 6 \times 3$ m^3. We assume two units of User Equipment (UE) with fixed separation $d_{UE}=12$ meters, as depicted in Fig. 1.2, and focus on saturated downlink traffic. To provide wireless access to UEs in the building, we have three options: micro BS, femtocell and BWA-FMDA. The following describes AE or access unit placement within each access strategy and details the frequency reuse pattern in our simulated scenarios.

Micro BS: There is one micro BS equipped with N_c co-located 17dBi directional antenna elements, where $N_c \in \{2,4,6\}$. The maximum of N_c is assumed to be 6, which is in-line with ITU M.2135-1-2009 guidelines [26] for evaluation of radio interface technologies for IMT-Advanced. A universal frequency reuse (UFR) pattern is adopted for the micro BS model, whereby the total available bandwidth is available throughout the building.

Femtocell: Each femto BS forms a closed subscriber group (CSG) and therefore operates independently. We assume that there exists one femto BS, equipped with one AE in each room. The femto BS is placed in the center of the room and UFR is assumed.

BWA-N_c-N_c-1: Denotes a BWA-FMDA network that is equipped with N_c AEs which form a "single" CoMP cluster, where $N_c \in \{2,4,6,8\}$. A UFR regime is assumed and further it is assumed that AEs are placed on the ceiling

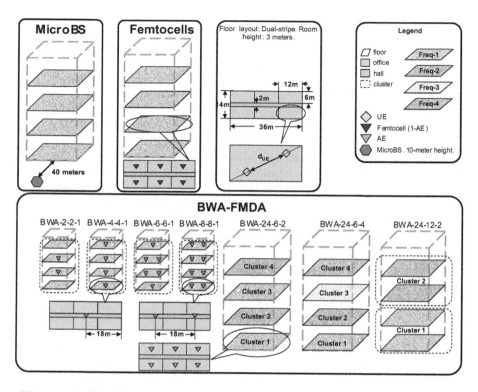

Figure 1.2: Simulation scenarios for microBS, femtocell, and BWA-FMDA.

of the corridors.

BWA–N_r–N_c–N_f: Denotes a BWA-FMDA network that is equipped with N_r ceiling-mounted AEs, every N_c of which forms one CoMP cluster. However, as opposed to BWA–N_c–N_c–1, AEs are placed within the rooms in the building. If the number of clusters, N_{CoMP} in (3.45), is larger than the number of frequency bands, N_f, neighboring clusters use different frequencies; otherwise, each cluster uses a different channel and thus has no inter-CoMP interference. Detailed configurations are illustrated in Fig. 1.2. Note that using the above notation, the femtocell approach can be regarded as a special case of BWA-FMDA denoted as BWA-24-1-1.

Every device is attached to only one channel with 1.4 MHz bandwidth in the 2.5 GHz band. When considering frequency reuse patterns other than UFR, multiple 1.4-MHz channels are used. In such circumstances simulated throughput is divided by the number of channels for fair comparison among different frequency reuse modes. The channel model of each link is composed of static pathloss, spatially correlated static shadowing, and time-varying frequency-selective fading.

Shadowing and Fading: While fading channels in the three studied access technologies are modeled as tapped delay lines with parameters from

ITU-R M.1225 [27], pathloss and shadowing models vary with systems and different AE placements. Links between the micro BS and UEs inside the building follow the non-line-of-sight hexagonal option in the UMi outdoor-to-indoor channel model in ITU-R M.2135-1-2009. In femtocells, link pathloss follows M.1225 with the penetration loss model from COST231 [28], i.e., 18.3 dB loss between two floors and 6.9 dB loss per wall.

In the femtocells, there are two links from two UEs in each room and their corresponding femto BS. We model the shadowing in this case as spatially correlated with an exponential correlation function of the UE separation. The shadowing is assumed to be independent for two UEs in different rooms. Given two links from two femtocells to one UE, we always assume that these two links have independent shadowing due to the large femtocell separation distance.

Further in the BWA-FMDA network, when AEs are placed in corridors, the indoor hotzone model in M.2135-1 is used to describe the pathloss and shadowing. When an AE is placed in each room, as in BWA–N_r–N_c–N_f, the pathloss and shadowing models follow those in femtocells. The noise factor at each UE receiver is assumed to be 7 dB. All other channel propagation parameters follow M.2135-1 unless otherwise noted.

Scheduling: In the femtocell model, all 24 femtocells run an exhaustive search algorithm with Proportional Fairness (PF) [29] to maximize the proportional fairness in each Resource Block (RB), which is represented by the sum of logarithms of averaged rates. In BWA-FMDA, however, a PF exhaustive search is not practical due to the huge number of combinations of UEs to be beamed. We therefore use a proportional fairness scheduler based on a semi-orthogonal user selection (PF-SUS) algorithm [30].

In BWA-24-6-2 mode, as shown in Fig. 1.2, only two frequencies are used while there are four CoMP clusters. Each CoMP cluster therefore suffers interference from another CoMP cluster deployed on the next adjacent floor. However, for simplicity of beamforming, water-filling was conducted as if there is no interference from outside of the CoMP cluster.

In the scheduling techniques considered for different wireless access solutions, we assume that scheduled UEs are independent among all RBs, thereby fully exploiting the frequency diversity. The subcarrier in the middle of each RB is used by the scheduler to predict the throughput of that RB.

1.4   Numerical Results

Based on the detailed simulation model described in the previous section, we now demonstrate the advantages of the proposed BWA-FMDA architecture in both energy efficiency and spectral efficiency in comparison with the micro BS and femtocell approaches. The spectral efficiency is obtained by dividing the system throughput by the "actual" signal transmission bandwidth. More specifically, following the air interface of evolved UMTS Terrestrial Radio Access Network (e-UTRAN), i.e., 3GPP Long Term Evolution (LTE) Release

8 [31], [32], we chose to use 1.4 MHz channel bandwidth with 1.08 MHz allocated for data communications. The throughput of a given UE at a given time slot is the sum of Shannon capacities of all subcarriers used by this UE. Note that all packets are transmitted in a slotted manner, where each slot lasts 1 millisecond and each simulation lasts 5 seconds.

1.4.1 Spectral Efficiency (bps/Hz)

First, we discuss the spectral efficiency of the three investigated last-mile access methods. As shown in Fig. 1.3, the spectral efficiency of BWA-FMDA as a function of transmit power per AE outperforms both femtocell and micro BS approaches. More specifically, independent femtocells provide a shorter communication link than micro BS and consistently deliver a higher throughput than micro BS with 2, 4 or 6 AEs. BWA–N_c–N_c–1 configuration achieves twice the spectral efficiency of corresponding N_c-AE micro BS. The spectral efficiency of BWA–N_c–N_c–1, where a single CoMP cluster exists, linearly increases with N_c when $N_c < 8$. When N_c reaches 12, the increment in spectral efficiency slows down, indicating larger conditional numbers in multi-user MIMO downlink operations.

When compared with 2 and 4 AE BWA-FMDA with UFR, femtocell is more spectrally efficient due to the shorter radio links. However, under UFR regime with a higher number of BWA-FMDA AEs, i.e., 6 and 8 AEs, if the transmit power per AE, P_{tx}, is more than 0.01 mW/MHz, the proposed BWA-FMDA demonstrates a superior spectral efficiency. In fact as P_{tx} increases, both femtocell and the BWA-FMDA configuration labeled BWA-24-6-2 show no sign of spectral efficiency improvement because they are interference-

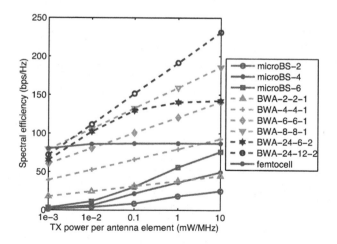

Figure 1.3: Spectral efficiency vs. transmit power per AE.

limited, whereas the spectral efficiencies of micro BS and other BWA-FMDA systems increase logarithmically, which agrees with Shannon's capacity theory. The increment rate is linearly proportional to the number of spatial channels, which is equal to N_c, the number of AEs participating in multi-user MIMO communications. The BWA-24-12-2 system configuration sustains a faster spectral efficiency increase than others because it has 12 spatial channels while, e.g., micro BS with 2 AEs, denoted as microBS-2 in Fig. 1.3, has only two spatial channels.

For the cases of BWA-24-12-2 and BWA-24-6-2, where higher frequency reuse distances are exploited, again BWA-FMDA outperforms the femtocell approach when P_{tx} is sufficiently large.

Finally, we note that while setting up more complex CoMP clustering strategies, such as BWA-24-12-2, results in more efficient spectral utilization, even simpler UFR solutions such as BWA-8-8-1 can sustain a significantly higher throughput than the state-of-the-art last-mile solutions.

1.4.2   Energy Efficiency (bit/Joule)

Delivering a high spectral efficiency by BWA-FMDA is necessary to motivate operators to adopt this solution, but it still remains to verify the green credentials of this technology. The presented energy efficiency results in Fig. 1.4 demonstrate the power conservation capabilities of micro BS, femtocell and BWA-FMDA for the same scenarios as Fig. 1.3.

For almost all the transmit power levels examined in our simulation scenarios, BWA-FMDA manifests a higher energy efficiency than both micro BS and femtocells. In particular, the energy efficiencies of all BWA-N_c-N_c-1 configurations considered are higher than the corresponding microBS-N_c systems. The gain is less than twofold because the BWA-FMDA systems have slightly higher energy consumption due to RoF transmissions.

However, the higher signal processing-related power consumption when forming multiple CoMP clusters in BWA-FMDA, such as in BWA-24-6-2 and BWA-24-12-2, makes these BWA arrangements less power efficient than single-cluster CoMP modes, for instance BWA-8-8-1. This is due to the fact that CoMP clustering sublinearly increases the spectral efficiency, due to frequency reuse, but linearly increases the energy consumption. For relatively high transmit power levels, near or above 1 mW/MHz, even micro BS outperforms the multi-cluster BWA-FMDA scheme, in the absence of inter-micro-cell interference considerations. Femtocells are only power efficient in lower transmit power ranges, i.e., below 0.01 mW/MHz. The lower energy efficiency of femtocells is largely caused by the inter-femtocell interference management and relatively large (component) energy consumption when compared with active optical fibers.

Finally, Fig. 1.5 demonstrates the total power consumption comparison for micro BS, femtocell and BWA-FMDA. Note only micro BS can use a larger transmit power, i.e., $P_{tx} > 10$ mW/MHz, whereas indoor systems such

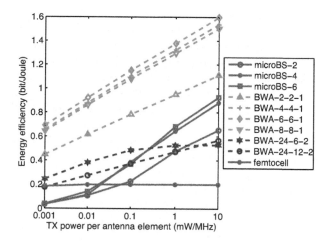
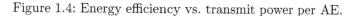

Figure 1.4: Energy efficiency vs. transmit power per AE.

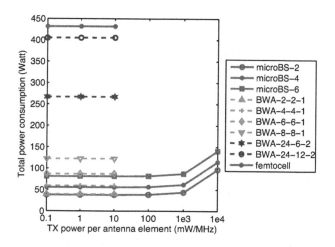

Figure 1.5: Power consumption vs. transmit power per AE.

as femtocells and BWA-FMDA have to limit their transmission power to meet radio spectrum regulations and lower health concerns.

Taking microBS-6 as an example, we can extrapolate its spectral efficiencies at $P_{tx}{=}10^3$ and $P_{tx}{=}10^4$mW/MHz per AE as 116.49 and 136.72 bps/Hz, respectively. The corresponding energy consumption equals 87.54 and 141 watts. Thus, we can calculate the energy efficiency as 1.3307 bit/Joule when $P_{tx}{=}10^3$ mW/MHz and 0.9696 bit/Joule when $P_{tx}{=}10^4$ mW/MHz. Therefore, there exists a sudden drop on the micro BS energy efficiency for P_{tx} larger than certain values in the range $[10^3, 10^4]$ mW/MHz. In practice, as P_{tx} becomes too large, the excessive micro BS transmit power will cause severe interference to neighboring cells.

Overall it can be determined that the proposed BWA-FMDA architecture configured in a single-cluster CoMP mode with a larger number of AEs will yield considerable advantages in both spectral and power efficiency and constitute the "greenest" last-mile access solution among the studied technologies.

1.4.3   Tradeoff between Energy Efficiency and Spectral Efficiency

The tradeoff between energy efficiency and spectral efficiency is shown in Fig. 1.6 and 1.7, where P_{tx} varies within a large range, from 10^{-3} to 10^6 mW/MHz, thus requiring the dynamic RoF link power consumption model presented in Section 1.3.1 for better accuracy.

We first notice that MicroBS-N and BWA–N–N–1 follow the same tradeoff trend because both are non-interference-limited systems and both apply M-antenna MIMO to serve M mobile terminals at a time. BWA–N–N–1 provides higher spectral and energy efficiency than micro BS in any P_{tx} and any antenna configuration.

For each antenna configuration, there is an optimal P_{tx} that maximizes energy efficiency. We can observe the fifth P_{tx} point produces the largest energy efficiency, which corresponds to $P_{tx} = 10$ mW/MHz/AE. In fact, the optimal P_{tx} can be approximated through an analysis for each fixed antenna configuration since throughput can be extrapolated by Shannon capacity formula at high SNR regimes and energy consumption follows a deterministic quadratic model. However, the value $P_{tx} = 10$ mW/MHz/AE is already larger than the maximum transmission power in indoor WLAN and may cause health concerns. Thus, it is safe to state that for indoor non-interference-limited systems, using as much power per antenna element as possible provides the best energy efficiency.

As expected, indoor interference-limited systems such as "femtocell" and "BWA-24-6-2" have the same spectral efficiency and decreasing energy efficiency when P_{tx} increases to be much larger than noise. In fact, the energy efficiency in BWA-24-6-2 plummets when P_{tx} is large enough to dominate over noise. Apparently, BWA-24-N-1 does not have a good energy efficiency as a

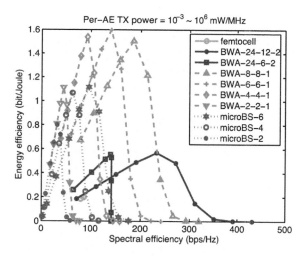

Figure 1.6: Energy efficiency vs. spectral efficiency under different P_{tx}. $P_{sp} \propto N^2$.

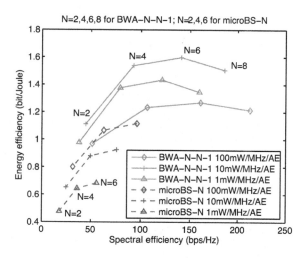

Figure 1.7: Energy efficiency vs. spectral efficiency under different N. $P_{sp} \propto N^2$.

large amount of energy is consumed in MIMO processing.

$$P_{sp} \propto N^2$$

In Fig. 1.7, we vary N while fixing P_{tx} to gain insights on choosing N to improve energy efficiency while satisfying a given spectral efficiency constraint. Choosing $N=6$ provides the best energy efficiency among the configurations we investigated, in either BWA–N–N–1 or micro BS systems. Regardless of P_{tx} being used, when a given spectral efficiency constraint is smaller than the one provided by $N = 6$, reducing N decreases energy efficiency, and when a given spectral efficiency constraint is larger than the one provided by $N = 6$, reducing N increases energy efficiency. Note that femtocells only have one data point in Fig. 1.6 and 1.7 because its power consumption comes from real product data and is fixed in our simulations.

From system design perspective, the above results reveal the following insights:

a) System choice:

Indoor CoMP femtocell systems outperform micro BS in both spectral and energy efficiency.

b) N choice:

Using 4-AE or 6-AE MIMO is a good choice for BWA–N–N–1 system as long as they satisfy the spectral efficiency requirement. Using 4-AE has an additional advantage on system complexity and thus incurs smaller system costs. BWA-24-12-2 system provides the highest spectral efficiency; however, its relatively poor energy efficiency and processing complexity makes it less attractive.

c) P_{tx} choice:

Indoor systems are subject to more stringent wireless transmission power constraint than outdoor systems. Our results show that using the highest possible P_{tx} generates the highest energy efficiency (and certainly the highest spectral efficiency) because the major contribution of energy consumption is from MIMO processing rather than wireless transmission.

A good energy consumption model is urgently needed to guide practical design of multiuser MIMO systems. When the energy consumption model changes, the above results may change dramatically. For example, when energy consumption is proportional to N^3, choosing $N=2$ produces the highest energy efficiency, as shown in Fig. 1.8 and 1.9.

1.5   Conclusions

There is a growing concern over the power consumption of wireless networks and its effect on increasing the greenhouse gas emission levels worldwide. At the same time, end users are demanding higher data rates and seamless connectivity as part of their daily life. To accommodate the conflicting requirements of power conservation and better wireless service, shortening the last-

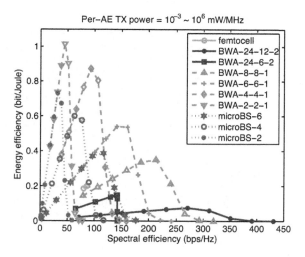

Figure 1.8: Energy efficiency vs. spectral efficiency under different P_{tx}. $P_{sp} \propto N^3$.

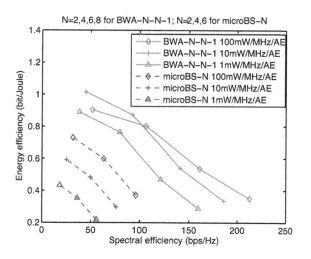

Figure 1.9: Energy efficiency vs. spectral efficiency under different N. $P_{sp} \propto N^3$.

mile communication link seems a viable strategy. However, as demonstrated in this chapter, the most spectral and power efficient network densification solution might not be all-wireless. In fact we have demonstrated through extensive simulations with as much practical assumptions as we could, that a fiber-connected distributed antenna system can outperform state-of-the-art solutions such as micro BS and femtocells. Our proposed architecture named BWA-FMDA can enhance spectral efficiency by forming CoMP clusters in serving the end users. Further, BWA-FMDA also delivers a more power efficient wireless access strategy than existing solutions in the literature, particularly when a single CoMP cluster with a large number of AEs is configured.

Acknowledgments

This work was supported in part by the Canadian Natural Sciences and Engineering Research Council through grant STPGP 396756, and a UBC Postgraduate Scholarship.

Bibliography

[1] Vodafone UK, *Corporate Responsibility Report 2007/2008*, Available Online at http://www.vodafone.com/content/dam/vodafone/uk_cr/previous_reports/report0708.pdf, Accessed March 2011.

[2] Vodafone Group Plc, *2010 Sustainability report*, Available Online at http://www.vodafone.com/content/index/about/sustainability/publications.html, Accessed March 2011.

[3] G. Koutitas and P. Demestichas, "A Review of Energy Efficiency in Telecommunication Networks," *17th Telecommu. Forum TELFOR'09*, 24-26 Nov. 2009, Serbia, Belgrade.

[4] List of mobile network operators, *Wikipedia*, Available Online at http://en.wikipedia.org/wiki/List_of_mobile_network_operators#By_region, accessed Jul. 2011.

[5] Three press release, Social Media Dominates Mobile Broadband Activity, 27 Oct. 2010.

[6] Cisco Visual Networking Index: Global Mobile Data Traffic Forecast Update, 2010–2015, *Cisco White Paper*, 1 Feb. 2011.

[7] Gabriel Brown, "What Price a 3G Base Station?," Available Online at http://www.lightreading.com/document.asp?doc_id=51942, accessed Feb. 2011.

[8] H. Li, J. Hajipour, A. Attar and V. C. M. Leung, "Efficient HetNet Implementation Using Broadband Wireless Access with Fibre-connected Massively Distributed Antennas Architecture," *IEEE Wireless Commun. Mag.*, Vol. 18, pp. 72-78, June 2011.

[9] T.-S. Chu and M. J. Gans, "Fiber optic microcellular radio," *IEEE Trans. Veh. Tech.*, vol. 40, no. 3, pp. 599-606, 1991.

[10] M. Sauer, A. Kobyakov, and J. George, "Radio Over Fiber for Picocellular Network Architectures," *Journal of Lightwave Technology*, vol. 25, no. 11, pp. 3301-3320, 2007.

[11] A. Nkansah, A. Das, C. Lethien, J.-P. Vilcot, N. J. Gomes, I. J. Garcia, J. C. Batchelor, and D. Wake, "Simultaneous dual band transmission over multimode fiber-fed indoor wireless network," *IEEE Microw. Wireless Compon. Lett.*, vol. 16, no. 11, pp. 627–629, 2006.

[12] T. Niiho, M. Nakaso, K. Masuda, H. Sasai, K. Utsumi, and M. Fuse, "Transmission performance of multichannel wireless LAN system based on radio-over-fiber techniques," *IEEE Trans. Microwave Theory Tech.*, vol. 54, no. 2, pp. 980-989, 2006.

[13] A. Das, A. Nkansah, N. J. Gomes, I. J. Garcia, J. C. Batchelor, and D. Wake, "Design of low-cost multimode fiber-fed indoor wireless networks," *IEEE Trans. Microwave Theory Tech.*, vol. 54, no. 8, pp. 3426-3432, 2006.

[14] K. K. Leung, B. McNair, L. J. Cimini Jr., and J. H. Winters, "Outdoor IEEE 802.11 cellular networks: MAC protocol design and performance," in *Proc. IEEE ICC 2002*, vol. 1, 2002.

[15] S. Hwang, H. Kim, B. Kim, S. K. Kim, J. Lee, H. Lee, Y. Kim, G. Lee, S. Kim, and Y. Oh, "RoF Technologies for In-Building Wireless Systems," *IEICE Transactions On Electronics*, IEICE, vol. 90, pp. 345-350, 2007.

[16] M. G. Larrodé, A. M. J. Koonen, and P. F. M. Smulders, "Impact of radio-over-fibre links on the wireless access protocols," in *Proc. NEFER-TITI Workshop*, Brussels, Belgium, Jan. 2005.

[17] A. Das, M. Mjeku, A. Nkansah, and N. J. Gomes, "Effects on IEEE 802.11 MAC Throughput in Wireless LAN Over Fiber Systems," *Journal of Lightwave Technology*, vol. 25, no. 11, pp. 1-8, 2007.

[18] B. L. Dang and I. Niemegeers, "Analysis of IEEE 802.11 in Radio over Fiber Home Networks," in *Proc. IEEE Conference on Local Computer Networks 30th Anniversary (LCN) 2005*, Washington, D.C., USA, 2005, pp. 744-747.

[19] H. Kim, J. H. Cho, S. Kim, K. U. Song, H. Lee, J. Lee, B. Kim, Y. Oh, J. Lee, and S. Hwang, "Radio-Over-Fiber System for TDD-Based OFDMA Wireless Communication Systems," *Journal of Lightwave Technology*, vol. 25, no. 11, pp. 1-9, 2007.

[20] H. Nasoha and S. M. Idrus, "Modeling and performance analysis of WCDMA radio over fiber system," In *Proc. APACE 2007*, 4-6 Dec. 2007, Melaka, Malaysia.

[21] P. K. Tang, L. C. Ong, A. Alphones, B. Luo and M. Fujise, "PER and EVM Measurements of a Radio-Over-Fiber Network for Cellular and WLAN System Applications," *Journal of Lightwave Technology*, vol. 22, no. 11, pp. 2370, 2004.

[22] J. R. Stern, J. W. Ballance, D. W. Faulkner, S. Hornung, D. B. Payne, and K. Oakley, "Passive optical local networks for telephony applications and beyond," *Electronics Letters*, vol. 23, no. 24, pp. 1255-1256, 1987.

[23] A. J. Fehske, F. Richter, and G. P. Fettweis, "Energy efficiency improvements through micro sites in cellular mobile radio networks," *In Proc. 2nd Workshop on Green Commun.*, Hawaii, USA, Dec. 2009.

[24] A. J. Fehske, P. Marsch and G. P. Fettweis, "Bit per Joule Efficiency of Cooperating Base Stations in Cellular Networks," *In Proc. 3rd Workshop on Green Commun.*, Miami, Florida, 6-10 Dec. 2010.

[25] M. Crisp, R. V. Penty, I. H. White, and A. Bell, "Wideband Radio over Fiber Distributed Antenna Systems for Energy Efficient In-Building Wireless Communications," *In Proc. IEEE* VTC'10-Spring, 16-19 May 2010, Taipei, Taiwan.

[26] Available online, http://www.itu.int/pub/R-REP-M.2135/en

[27] Available online, http://www.itu.int/rec/R-REC-M.1225/en

[28] Available online, http://www.lx.it.pt/cost231/final_report.htm

[29] F. P. Kelly, A. K. Maulloo, and D. K. H. Tan, "Rate control for communication networks: shadow prices, proportional fairness and stability," *Journal of Operational Research*, vol. 49, Mar. 1998, pp. 237-252.

[30] T. Yoo and A. Goldsmith, "On the optimality of multiantenna broadcast scheduling using zero-forcing beamforming," *IEEE Journal on Selected Areas in Communications*, vol. 24, no. 3, pp. 528-541, March 2006.

[31] 3GPP TS 36.106 V8.6.0 Evolved Universal Terrestrial Radio Access (E-UTRA); FDD repeater radio transmission and reception, Available Online: http://ftp.3gpp.org/specs/html-info/36106.htm

[32] 3GPP TS 36.101 V8.15.0 Evolved Universal Terrestrial Radio Access (E-UTRA); User Equipment (UE) radio transmission and reception, Available Online: http://ftp.3gpp.org/specs/html-info/36101.htm

Chapter 2

Wireless Networks Resources Trading for QoS Performance Guaranteed Green Communications

*Xi Zhang and +Wenchi Cheng
*Texas A&M University, xizhang@ece.tamu.edu
+Texas A&M University, wccheng@neo.tamu.edu

Recently, the CO_2 emission of information and communication technologies (ICTs) has attracted much research attention [1–3]. For wireless access networks, which cost significant energy in wireless communications, a great deal of work has been done for reducing the energy consumption [4,5]. Authors of [4] investigated the impact of deployment strategies on the power consumption of mobile radio networks. Authors of [5] studied the impact of reducing the cell size on the energy performance of an HSDPA RAN [6], and then proposed to save energy by putting some cells into sleep mode. Authors of [7] proposed the energy efficient spectrum allocations for two tier cellular networks by rationally using subcarriers. Authors of [8] investigated the possibility of reducing the energy consumption of a cellular network by switching off some cells during the periods in which they are under-utilized when the traffic level is low. The above-mentioned works mainly explore new techniques to reduce the energy consumption of wireless networks. However, how to minimize the energy consumption of a wireless network while satisfying the required QoS performance metrics has been neither well understood, nor thoroughly studied.

To remedy the above-mentioned problems, we develop the Demanding-Based Resources Trading (DBRT) model for Green Communications in this chapter. We propose a mechanism to minimize the energy consumption of wireless networks without compromising the quality-of-service (QoS) for users. Applying the Demanding Based Communications model, we develop a novel scheme – Wireless Networks Resources Trading, which characterizes the trading relationships among different wireless resources for a given number of QoS performance metrics. According to wireless networks resources trading relationships, different wireless resources can be consumed to meet the same set of QoS performance metrics. Therefore, to minimize the energy consumption for given QoS performance metrics, we can trade the other type of wireless networks resources for the energy resources while satisfying the demanded QoS performance metrics. Based on our developed wireless networks resources trading relationships, we derive the optimal energy-bandwidth, energy-time, energy-space, and energy-code wireless networks resources trading relationships for green wireless networks. We also conduct two example-case studies to show how to use the available bandwidth or the acceptable delay bound to achieve the minimum energy consumption while guaranteeing the required QoS performance metrics in wireless networks.

2.1 Demanding-Based Communications Model

For wireless networks, available resources can be classified into the following five categories: time, bandwidth (frequency), space, energy, and code. Maximizing QoS performance for wireless users usually implies costing more wireless resources. This is beneficial to users, but harmful to environments and operators. For environments, more energy consumption results in more CO_2

emission. For operators, the more bandwidth, time, energy, space, and complex code, the more the cost.

As a result of the above reasons, it's desirable for operators to support the Demanding-Based service for users, which implies satisfying the QoS requirements of users while minimizing the resources costs. The minimum resources minimize the CO_2 emission and the cost of operators. In wireless networks, the model of Demanding-Based Communications can be further formulated as follows:

$$
\begin{aligned}
&\mathbf{argmin}\ \{Consumed\ Resources\} \\
&\text{S.T.} \\
&Users'\ Obtained\ Service\ Performance \\
&\geq Users'\ Required\ Service\ Performance
\end{aligned}
\tag{2.1}
$$

For green wireless networks, our objective is to minimize energy resource while focusing on the demanded service in terms of throughput and delay. Therefore, Eq. (2.1) can be detailed as follows:

$$
\begin{aligned}
&\mathbf{argmin}\ \{Consumed\ Energy\} \\
&\text{S.T.} \\
&Implemented\ Throughput \geq Users'\ Required\ Throughput \\
&Implemented\ Delay \leq Users'\ Required\ Delay\ Bound
\end{aligned}
\tag{2.2}
$$

2.2 Resources Trading in Wireless Networks

The available resources of wireless networks are always limited. The five categories of wireless resources can be consumed to satisfy the demanded users' performance. Many papers have shown that using these five resources for high throughput [9,10] implies that a certain level of required QoS performance can be achieved by consuming different resources. Therefore, there are trading relationships among the five resources. In this section, we develop the scheme of resource trading. Then, focusing on minimizing the energy resource, we derive the relationships between energy resource and the other resources. Based on the trading relationships between energy resource and the other resources, we develop the optimal bandwidth, delay, space, and code trading strategies for green wireless networks, respectively.

2.2.1 Wireless Resource Trading

There are various tradeoffs to improve the QoS performance in wireless networks.This implies that the wireless networks can consume different amounts of bandwidth, time, space, energy, and code resources, respectively, for guaranteed QoS performance. The guaranteed QoS performance achieved by consuming one type of resource can also be obtained by consuming different types of resources. Fig. 2.1(a) shows the general wireless networks resources trading.

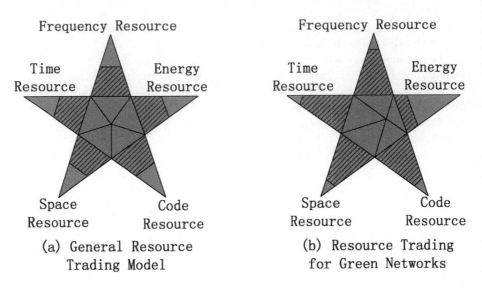

(a) General Resource   (b) Resource Trading
    Trading Model   for Green Networks

Figure 2.1: Wireless resources trading model.

The center pentagon area represents the demanded QoS performance of mobile users. The different areas represent the available volume of five resources, respectively, among which the shaded area represents the amount of used resources for the demanded performance and the un-shaded area represents the amount of saved resources for different types of resources.

For green wireless networks, the goal of resource trading is to consume the minimum energy resource for the required QoS performance. Fig. 2.1(b) shows that the resource trading relationship for green wireless networks under the same demanded performance as Fig. 2.1(a). We can see from Fig. 2.1(b) that the other four resources are traded more than the energy resource for the given same required QoS performance as shown in Fig. 2.1(a).

2.2.2 The Tradeoff between Energy and Bandwidth/ Delay

When considering the bandwidth and the transmission time, with capacity approaching channel codes, such as LDPC codes and turbo codes, the channel capacity of an AWGN channel, denoted by R_{bt}, is given by

$$R_{bt} = W \log_2 \left(1 + \frac{Pg}{W N_0}\right) \tag{2.3}$$

where W is the channel bandwidth, P is the transmitted power, g is the channel gain, and N_0 is the noise spectral density. If the time needed to transmit one bit is t, the corresponding data rate is $R = 1/t$. Thus, using Eq. (2.3), we get the energy consumption for transmitting one bit, denoted by E_{tr}, as follows:

$$E_{tr} = Pt = \frac{\left(2^{\frac{1}{Wt}} - 1\right) W N_0 t}{g} \tag{2.4}$$

From Eq. (2.4), we observe that E_{tr} monotonically decreases when W or t increase. This is not beneficial for resource trading because not only the demanded performance cannot be satisfied, but also infinite bandwidth or time resources are needed if minimizing E_{tr}. However, this is because we only consider transmitting power. For green wireless networks, we must also consider operating power, which will cause the relationship between energy resource and bandwidth resource to be different from that of only considering transmitting power. The relationship between the energy resource and the time resource will also be different from that of only considering transmitting power.

Once considering the operating power of wireless networks, to transmit with the maximum bandwidth or the maximum delay is no longer the optimal case for minimum energy consumption. In this case, the circuit energy consumption increases when the bandwidth or the transmission time (delay) increases. We derive the overall energy consumption per bit with the consideration of the bandwidth and the delay, denoted by E_{bt}, as follows:

$$E_{bt} = E_{tr} + E_{cir} = Pt + P_c t = \frac{\left(2^{\frac{1}{Wt}} - 1\right) W N_0 t}{g} + W P_{cir} t + P_{sb} t \tag{2.5}$$

where E_{tr} and E_{cir} are the energy consumption of transmitting and circuit, respectively, P is the transmit power, P_c is the circuit power, including all system power consumption except for transmit power, P_{cir} is the part of circuit power consumption which is related to bandwidth W, and P_{sb} is the average static part of circuit power for every bit which is not related to W.

Clearly, there are two independent variables in Eq. (2.5): the bandwidth W and the delay t, which affect the energy consumption. Ignoring the Available Resource Limitation (RAL)[1] of the wireless network, the bandwidth and the time resource can trade for energy resource. Fig. 2.2 jointly shows the relationship by trading bandwidth resource and time resource for energy resource. Fig. 2.3(a) shows the partial zooming-in of Fig. 2.2 from $W = 0.3$ and $t = 0.3$ to $W = 0.5$ and $t = 0.5$. Fig. 2.3(b) shows the partial zooming-in of Fig. 2.2 from $W = 0.5$ and $t = 0.5$ to $W = 1$, $t = 1$. From these two figures, we can observe that the relationship between W/t and E is not that the larger W/t, the smaller E. Table 2.1 shows the data for $t = 1$ and $W = 1$ in Fig. 2.2, respectively. In Table 2.1(a), the minimal energy consumption is 2.7725e-06. In Table 2.1(b), the minimal energy consumption is 1.4448e-06. Fig. 2.2, Fig. 2.3,

---

[1]RAL represents the maximum available resource of the system. For example, for a wireless network, the available bandwidth is 20MHz and the acceptable delay of the data is 0.1s, then the RAL of the wireless system is 20MHz for bandwidth resource and 0.1s for time resource.

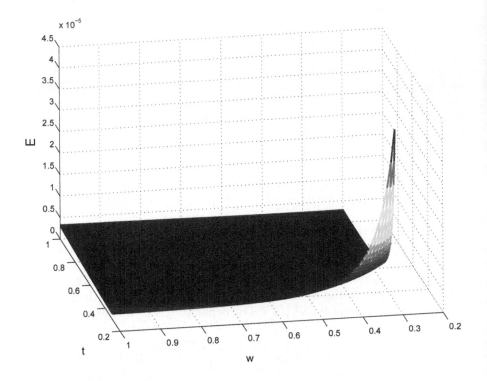

Figure 2.2: Energy per bit versus bandwidth per bit and delay time.

and Table 2.1 imply that there exists the minimum energy consumption by trading other resources. Because we need to derive the trading relationships between energy consumption and bandwidth/delay, respectively, we analyze each trading relationship with the fixed value of another resource.

For fixed t, the relationship between E and W is given by (setting $t = 1$ for simplicity)

$$E_{bt} = \frac{\left(2^{\frac{1}{W}} - 1\right) W N_0}{g} + W P_{cir} + P_{sb} \tag{2.6}$$

For fixed W, the relationship between E and t is expressed as (setting $W = 1$ for simplicity)

$$E_{bt} = \frac{\left(2^{\frac{1}{t}} - 1\right) t N_0}{g} + t P_{cir} + t P_{sb} \tag{2.7}$$

In Fig. 2.4, the three plots marked with TP-d=600m, TP-d=800m, and TP-d=1000m show the relationship between E_{tr} and W when the transmission distance is 600m, 800m, and 1000m, respectively. The three plots marked

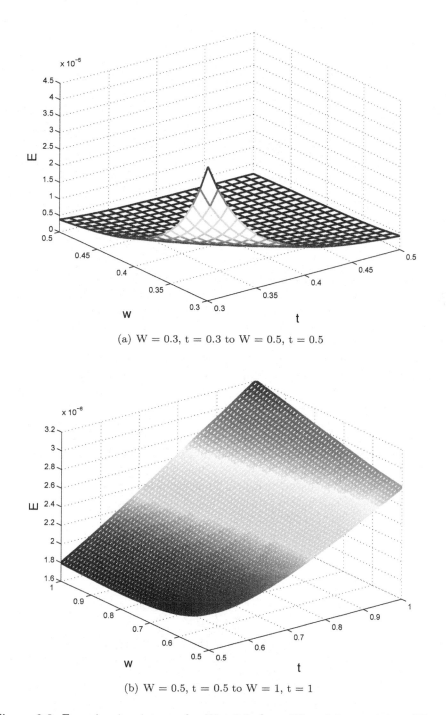

(a) W = 0.3, t = 0.3 to W = 0.5, t = 0.5

(b) W = 0.5, t = 0.5 to W = 1, t = 1

Figure 2.3: Zooming-in pictures for Fig. 2.2. from W = 0.3, t = 0.3 to W = 0.5, t = 0.5 and from W = 0.5, t = 0.5 to W = 1, t = 1, respectively.

Table 2.1: Data for $t = 1$ and $W = 1$ in Fig. 2.2, respectively

(a) Data for $t = 1$ in Fig. 2.2.

Fixed Delay	$t = 1$
$W = 0.1$	2.2560e-05
$W = 0.2$	3.4400e-06
$W = 0.3$	2.8448e-06
$W = 0.4$	2.7725e-06
$W = 0.5$	2.8000e-06
$W = 0.6$	2.8610e-06
$W = 0.7$	2.9369e-06
$W = 0.8$	3.0205e-06
$W = 0.9$	3.0205e-06
$W = 1.0$	3.2000e-06

(b) Data for $W = 1$ in Fig. 2.2.

Fixed Bandwidth	$W = 1$
$t = 0.1$	2.0760e-05
$t = 0.2$	1.8400e-06
$t = 0.3$	1.4448e-06
$t = 0.4$	1.5725e-06
$t = 0.5$	1.8000e-06
$t = 0.6$	2.0610e-06
$t = 0.7$	2.3369e-06
$t = 0.8$	2.6205e-06
$t = 0.9$	2.9088e-06
$t = 1.0$	3.2000e-06

with OP-d=600m, OP-d=800m, and OP-d=1000m show the relationship between E_{bt} and W when the transmission distance is 600m, 800m, and 1000m, respectively. We can observe that when taking operating power into account, for fixed transmission distance d, there exists the minimum energy per bit for the available bandwidth.

In Fig. 2.5, the three curves marked with TP-d=600m, TP-d=800m, and TP-d=1000m show the relationship between E_{tr} and t when the transmission distance is 600m, 800m, and 1000m, respectively. The three curves marked with OP-d=600m, OP-d=800m, and OP-d=1000m show the relationship between E_{bt} and t when the transmission distance is 600m, 800m, and 1000m, respectively. We can find that taking operating power into account, for fixed transmission distance d, there exists the minimum energy per bit for the given available delay.

Taking the derivative over Eq. (2.6) with respect to W and setting the result to zero, we can obtain

$$-\frac{N_0 2^{\frac{1}{W}} \log 2}{Wg} + \frac{\left(2^{\frac{1}{W}} - 1\right) N_0}{g} + P_{cir} = 0 \qquad (2.8)$$

Simplifying Eq. (2.8), we can obtain

$$2^{\frac{1}{W_o}} \left(1 - \frac{\log 2}{W_o}\right) = \frac{N_0 - gP_{cir}}{N_0} \qquad (2.9)$$

where W_o is the optimal bandwidth for the minimum energy per bit.

Taking the derivative over Eq. (2.7) with respect to t and setting the derivative to zero, we can obtain

$$-\frac{N_0 2^{\frac{1}{t}} \log 2}{tg} + \frac{\left(2^{\frac{1}{t}} - 1\right) N_0}{g} + P_{cir} + P_{sb} = 0 \qquad (2.10)$$

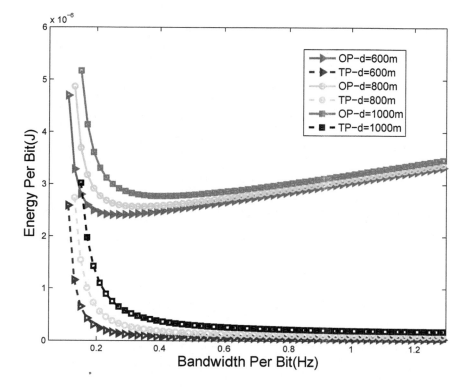

Figure 2.4: Relation between energy per bit and bandwidth per bit.

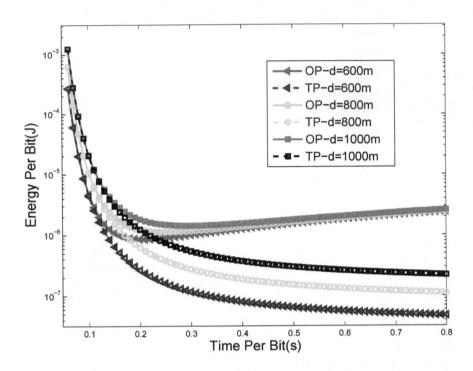

Figure 2.5: Relation between energy per bit and delay time.

Simplifying Eq. (2.10), we can obtain

$$2^{\frac{1}{t_o}}\left(1 - \frac{\log 2}{t_o}\right) = \frac{N_0 - g(P_{cir} + P_{sb})}{N_0} \tag{2.11}$$

where t_o is the optimal t for the minimum energy per bit.

Under the free space propagation model, for fixed G_t, G_r, λ, and L, the channel gain, denoted by $g(d)$, is the function of transmission distance d which can be expressed as

$$g(d) = \frac{G_t G_r \lambda^2}{(4\pi)^2 d^\alpha L} \tag{2.12}$$

where G_t and G_r are the transmit and receive antenna gains, respectively, λ is the wavelength, and L is the system loss unrelated to propagation ($L \geq 1$).

Therefore, for fixed d, there exists the optimal W and t for the minimum energy per bit, which are W_o and t_o. Fig. 2.4 and Fig. 2.5 show that the curves of Eq. (2.6) and Eq. (2.7) are convex. Thus, there exist unique solutions for W_o and t_o, respectively. The optimal strategy for the minimum energy per bit is to choose different W_o and t_o pairs for different transmission distances adaptively. However, in real wireless networks, the available bandwidth can be smaller or larger than W_o, and the acceptable delay can also be smaller or larger than t_o. Therefore, our next problem is how to use the available bandwidth or the acceptable delay trading for minimum energy consumption? For this purpose, we develop Algorithms 1 and 2 to search for optimal W_o and t_o, respectively. Methods in Algorithms 1 and 2 are similar, but for searching W_o and t_o, respectively.

Algorithm 1 Adaptive W_o Strategy.

Input: W_a
Output: W_{user}
 1: **for** i = 1:κ **do**
 2:    Calculate d_i
 3:    Calculate $W_o(d_i)$ by Eq. (2.9)
 4:    **if** $W_a > W_o(d_i)$ **then**
 5:       $W_{user}(d_i) = W_o(d_i)$
 6:    **else**
 7:       $W_{user}(d_i) = W_a$
 8:    **end if**
 9: **end for**

Algorithm 2 Adaptive t_o Strategy.

Input: t_a
Output: t_{user}
 1: **for** i = 1:κ **do**
 2:    Calculate d_i
 3:    Calculate $t_o(d_i)$ by Eq. (2.11)
 4:    **if** $t_a > t_o(d_i)$ **then**
 5:       $t_{user}(d_i) = t_o(d_i)$
 6:    **else**
 7:       $t_{user}(d_i) = t_a$
 8:    **end if**
 9: **end for**

In Algorithms 1 and 2, κ and d_i denote the number of users in the wireless networks and the distance between the BS and user i, W_a and t_a represent the available bandwidth and acceptable delay for each bit, respectively. The parameters $W_{user}(d_i)$ and $t_{user}(d_i)$ represent the bandwidth and the delay which the user i should trade for the minimum energy per bit, respectively, $W_o(d_i)$ and $t_o(d_i)$ represent the optimal bandwidth and delay for user i, respectively.

2.2.3   The Tradeoff between Energy and the Number of Antennas

To derive the tradeoff between energy and the number of antennas, we assume that the energy is averagely distributed among transmit antennas. Our purpose is to derive the relationship between energy and the number of antennas. We do not pursue the optimal power allocation schemes among antennas. With power averagely distributed among transmit antennas, the AWGN channel capacity, denoted by R_{an}, is given by

$$R_{an} = MW \log_2 \left(1 + \frac{Pg}{MWN_0} \right) \tag{2.13}$$

where $M = \min\{N_t, N_r\}$ is the minimal number of transmit antennas and receive antennas. The time to transmit one bit is t. Thus, using Eq. (2.13), we can get the transmitting energy consumption per bit, denoted by E_{ta}, as follows:

$$E_{ta} = Pt = \frac{\left(2^{\frac{1}{MWt}} - 1 \right) MWN_0 t}{g}. \tag{2.14}$$

Thus, the overall energy consumption per bit can be obtained as follows:

$$\begin{aligned} E_a &= Pt + P_{ca}t = \frac{\left(2^{\frac{1}{MWt}} - 1 \right) MWN_0 t}{g} \\ &\quad + (N_t P_t t + P_{ti} t) + (N_r P_r t + P_{ri} t). \end{aligned} \tag{2.15}$$

where P_{ca} is the total circuit power consumption, P_t is the transmit circuit power consumption corresponding to N_t transmit antennas, P_{ti} is the transmit circuit power consumption independent of N_t transmit antennas, P_r is the receive circuit power consumption corresponding to N_r transmit antennas, and P_{ri} is the transmit circuit power consumption independent of N_r receive antennas.

For fixed t and W, we can derive the relationship between E_a and M as follows (setting $t = 1$ and $W = 1$ for simplicity):

$$E_a = \frac{\left(2^{\frac{1}{M}} - 1 \right) MN_0}{g} + (N_t P_t + P_{ti}) + (N_r P_r + P_{ri}). \tag{2.16}$$

To simplify the analytical manipulations, we assume the number of transmit antennas and the number of receive antennas are equal, i.e., $N_t = N_r = M$. Thus, Eq. (2.16) can be simplified as follows:

$$E_a = \frac{\left(2^{\frac{1}{M}} - 1 \right) MN_0}{g} + (MP_t + P_{ti}) + (MP_r + P_{ri}). \tag{2.17}$$

2.2.4 The Tradeoff between Energy and Coding Schemes

Different coding schemes yield different coding gains. To derive the relationship between the energy and coding schemes, we denote the coding gain as

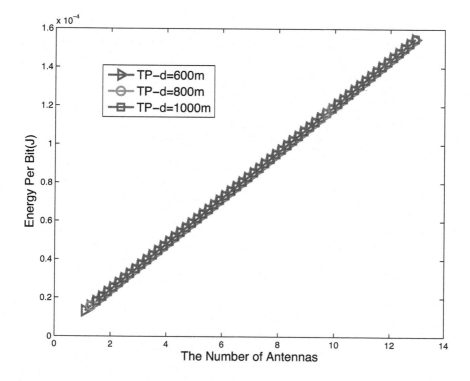

Figure 2.6: Relation between energy per bit and the number of antennas.

G. Thus, we can write the channel capacity, denoted by R_c, as follows:

$$R_c = W \log_2 \left(1 + \frac{GPg}{WN_0} \right) \tag{2.18}$$

The time to transmit one bit is t and the corresponding data rate is $R_c = 1/t$. Thus, using Eq. (2.18), we can get the transmitting energy consumption per bit, denoted by E_{tc}, as follows:

$$E_{tc} = Pt = \frac{\left(2^{\frac{1}{Wt}} - 1 \right) WN_0 t}{Gg}. \tag{2.19}$$

Thus, the overall energy consumption per bit can be obtained as follows:

$$E_{oc} = Pt + P_c t = \frac{\left(2^{\frac{1}{Wt}} - 1 \right) WN_0 t}{Gg} + P_c t. \tag{2.20}$$

where P_c is the circuit power consumption.

For fixed t and W, we can derive the relationship between E and M as follows (setting $t = 1$ and $W = 1$ for simplicity):

$$E_{oc} = \frac{N_0}{Gg} + P_c. \tag{2.21}$$

2.3   Example-Case Study of Using Resource Trading in Cellular Networks

To evaluate our adaptive strategies in cellular networks, we consider a classical hexagonal deployment which is shown in Fig. 2.8. As shown in Fig. 2.8, the given 57 cells are deployed and the radius of each cell is 1000m. For each cell C_i ($1 \le i \le 57$), B_i and U_i denote the base station (BS) and mobile user (MU) in each cell, respectively. The parameter d_i denotes the current distance from B_i to U_i. The data sent by base station B_i is transmitted with power P_i. The system bandwidth is 20MHz and the total number of users is 500. Because different wireless networks have different available bandwidths and acceptable delays, we will show the relationship between the energy consumption and the available bandwidth and the relationship between the energy consumption and the acceptable delay, respectively. The related simulation parameters are listed in Table 2.2.

Fig. 2.9 shows the energy per bit versus available bandwidth under different n-steps. The optimal adaptive strategy uses the continuous value of W_o. This implies that for different d, the wireless network uses different optimal bandwidths obtained by Algorithm 1. In Fig. 2.9, n-step represents that n types of bandwidths, denoted by $W_o(i)$ ($1 \le i \le n$), can be used in the

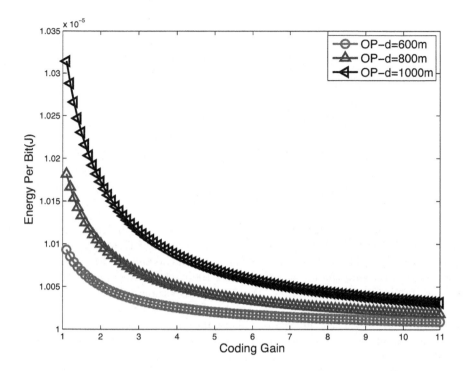

Figure 2.7: Relation between energy per bit and coding gain.

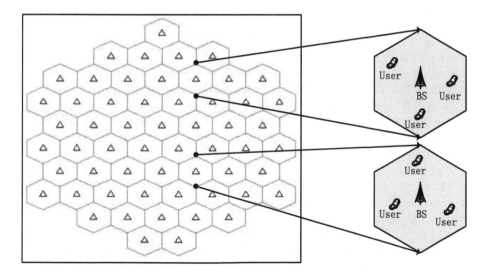

Figure 2.8: Hexagonal network deployment.

Table 2.2: Simulation Parameters

Parameter	Value
Carrier Frequency f_c	2.4GHz
Cell Radius R	1000m
Transmit Antenna Gain G_t	1
Receiver Antenna Gain G_r	1
Circuit Power P_{cir}	$1 \times 10^{-6} W/Hz$
Static Power P_{sb}	$2 \times 10^{-6} W$
System Loss L	2.5
PSD of the Local Noise N_0	8×10^{-21}
Path-Loss Exponent α	3

wireless networks. The parameter $W_o(i)$ $(1 \leq i \leq n)$ can be calculated using Algorithm 1. When n is fixed, the users located in the i-th step use the same bandwidth $W_o(i)$ $(1 \leq i \leq n)$. For instance, $n = 5$, the bandwidth W_{user} for the user located outside the circle with radius $1000(i-1)/5$ and in the circle with radius $1000i/5$ is $W_o(i)$ $(1 \leq i \leq 5)$. Fig. 2.9 also shows that when available bandwidth increases, the energy per bit decreases in the case of small available bandwidth. However, the energy per bit finally decreases to a constant, which is the minimum energy per bit for the used strategy. This is because when the available bandwidth is small, most users cannot use the optimal bandwidth for them. When the available bandwidth becomes larger, more users can use their optimal bandwidths. When all users' optimal bandwidths are smaller than the available bandwidth, the energy per bit decreases to the constant. The larger n, the smaller energy per bit. This is because the larger n, the smaller difference between $W_o(i)$ $(1 \leq i \leq n)$ and W_o. Hence more users can make use of their optimal bandwidths, decreasing the energy per bit. Fig. 2.9 also shows that the energy per bit does not decrease when available bandwidth is large to some extent. For example, with 2-step, when available bandwidth is larger than 0.3Hz, the energy per bit remains as 2.4e-6J.

Fig. 2.10 shows the curves of the energy per bit versus acceptable delay under different n-steps. For different d, the system uses the optimal delays t_o obtained by Algorithm 2. In Fig. 2.10, n-step represents that n types of delays, denoted by $t_o(i)$ $(1 \leq i \leq n)$, can be used in the wireless networks. The parameter $t_o(i)$ $(1 \leq i \leq n)$ can be calculated using Algorithm 2. When n is fixed, the users located in the i-th step use the same delay $t_o(i)$ $(1 \leq i \leq n)$. For instance, $n = 5$, the used delay t_{user} for the user located outside the circle with radius $1000(i-1)/5$ and in the circle with radius $1000i/5$ m is $t_o(i)$ $(1 \leq i \leq 5)$. Fig. 2.10 also shows that when the acceptable delay increases, the energy per bit decreases to a constant, which is the minimum energy per bit. This is because when the acceptable delay is small, most users cannot wait for the optimal delay. When the acceptable delay becomes larger, more users can use their optimal delay. When all users' optimal delays are smaller

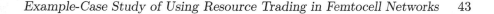

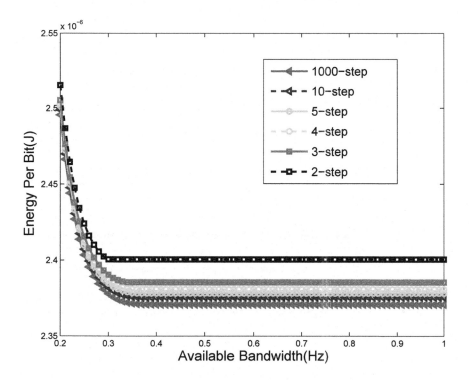

Figure 2.9: Energy per bit versus available bandwidth (n-step in the legend indicates that n can be 2, 3, 4, 5, 10, and 1000).

than the acceptable delay time, the energy per bit decreases to the constant. The larger n, the smaller the energy per bit. This is because the larger n, the smaller difference between $t_o(i)$ ($1 \leq i \leq n$) and t_o. Thus, more users can make use of their optimal delay, decreasing the energy per bit. The energy per bit does not decrease when the acceptable delay is large to some extent. For example, with 2-step, when acceptable delay is larger than 0.22s, the energy per bit remains as 8.3e-7J. The minimum energies per bit in Fig. 2.9 and in Fig. 2.10 are different. This is because we employ the fixed bandwidth for optimal delay and the fixed delay time for optimal bandwidth, respectively.

2.4   Example-Case Study of Using Resource Trading in Femtocell Networks

Resource trading can be efficiently used in macro-femto cellular networks. In this section, we propose to use the bandwidth resource trading for energy saving in macro-femto cellular networks. We define a new metric, called green factor, to evaluate the power efficiency of macro-femto networks.

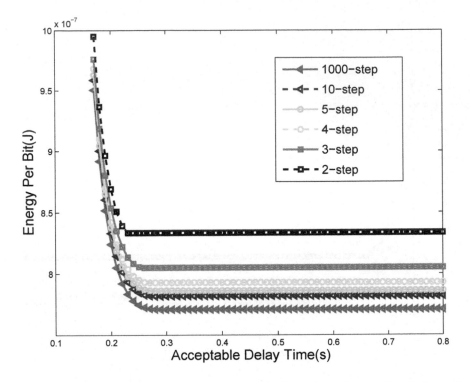

Figure 2.10: Energy per bit versus acceptable delay time (n-step in the legend indicates that n can be 2, 3, 4, 5, 10, and 1000).

2.4.1 System Model

In two-tier macro-femto cellular networks, the first tier and second tier are macrocells and femtocells, respectively. The central macrocell is a hexagonal region $A_m = 3\sqrt{3}R_c^2/2$ with a central BS, whose height is H. Two rings of interfering macrocells are around the central macrocell. Every macrocell has its theoretic spatial coverage $S = A_m H/3 = \sqrt{3}HR_c^2/2$. The central macrocell is overlaid with femtocell access points (FAPs) of radius R_f and height h_f, which are randomly distributed according to a homogeneous Spatial Poisson Point Process (SPPP) Ω_f with intensity λ_f. The mean number of femtocells per cell site is $N_f = \lambda_f S$. Users are assumed to be uniformly distributed inside each cell site. Fig. 2.11 shows the two-tier macro-femto cell (MU: macrocell user; FU: femtocell user).

When deploying femtocell networks, the access method, which refers to the rights of the users when making use of the femtocells, needs to be determined. Two main different strategies for femtocell access have been proposed so far [11]:

1). *Public Access*: where all the users can access all the femtocells of a

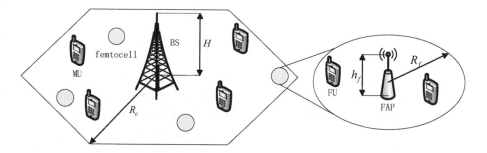

Figure 2.11: The two-tier macro-femto cell.

given operator;

2). *Private Access*: where only the subscriber of the femtocell and a list of invited users can access a given femtocell.

It has been shown that the public access will provide a better network performance than the private access. However, those who pay for a femtocell are generally not willing to share their resources with any other users. Operators are inclined to use femtocells with private access. We assume that femtocells operate with private access. Let $U = U_m + N_f U_f$ denote the average number of users per cell with U_m and U_f referring to the number of outdoor users per macrocell and indoor users per femtocell, respectively. The downlink channel between each BS and its users (each FAP and its users) is composed of a fixed distance dependent path loss, a slowly varying component modeled by lognormal shadowing, and Rayleigh fast fading with unit average power. The thermal noise is ignored for simplicity.

Available spectrum consists of F subchannels each with bandwidth W Hz. All subchannels are assigned equal transmit power. Each user is assumed to track their Signal-to-Interference Ratio (SIR) in each subchannel. Each BS assigns rate adaptively based on the received SIR per user, i.e., assigns i bps/Hz when SIR lies in $[\tau_i, \tau_{i+1}), 1 \le i \le I$ in each subchannel, where I is the maximum available transmission rate. Thus, the throughput per subchannel is given by

$$T = \sum_{i=1}^{I-1} i \Pr\{\tau_i \le SIR < \tau_{i+1}\} + I \Pr\{SIR \ge \tau_I\} \qquad (2.22)$$

Some existing spectrum efficiency metrics for wireless networks such as $b/s/Hz$ spectral efficiency, $b/s/Hz/m^2$ area spectral efficiency (ASE) [12], $b/s/Hz/antenna$, etc., have been used for many years. For cellular access networks, previous research was prone to use $b/s/Hz/m^2$ ASE to evaluate the spectrum efficiency. However, the ASE has its limitation in measuring femtocell networks. In the vertical direction of femtocell networks, there may exist multiple femtocells. For example, one house is a femtocell with another femtocell on the top of the house. Thus, the spatial spectral efficiency is urgently needed.

There are also already some energy efficiency metrics for wireless networks, such as b/J energy efficiency, $b/TNEU$ power efficiency, where $TNEU$ refers to the amount of signal energy identical to the variance of the complex-valued AWGN samples recorded at the receiver [13]. To evaluate the spectrum efficiency and energy efficiency of macro-femto cellular networks, we propose two metrics:

(1). *Spatial Spectral Efficiency:* the data rate of each user per bandwidth unit per space unit supported by a cell;

(2). *Green Factor:* the b/J power efficiency per subchannel.

2.4.2 Model for Energy Consumption in Downlink Channels

Because of the significant energy supply for BSs, we often ignore the energy saving on downlink channels and pay more attention to uplink channels. However, with the emerging Green Radio concept, we need to pay more attention to downlink energy saving because BSs cost the most energy in cellular access networks. Moreover, from operators' perspective, reducing the downlink energy consumption not only minimizes the environmental impact of the industry, but also benefits for economical reasons. For example, reduced energy consumption directly translates to lower operating expenditure (OPEX) [14]. For reducing the CO_2 emission and improving the interests of operators (lower OPEX), we consider cellular downlink energy consumption. Compared with BSs and FAPs, which cost the most energy in the cell, mobile users and femtocell users cost little energy to receive the signal from BSs and FAPs. Thus, ignoring the energy consumption of mobile users and femtocell users is reasonable.

Until now, there is no exact equation which can characterize the relationship between the radiated power of BS P_{tx} and the average consumed power P_M. It is difficult to derive the exact relationship between P_{tx} and P_M because the power consumption of BS depends on a large number of factors, such as amplifier and feeder losses, cooling, signal processing, battery backup, etc. The power consumption of amplifier and feeder losses as well as cooling is proportional to P_{tx}. The power consumption of signal processing and battery backup is constant with P_{tx}. Letting P_{lm} and P_{cm} be the linear part and the constant part, respectively, we propose the general model for power consumption as follows:

$$P_M = P_{lm} + P_{cm} = aP_{tx} + P_{cm} \qquad (2.23)$$

where a is a constant. The typical values for a and P_{cm} can be found in [10]. Similar to BSs, FAPs can also employ the model:

$$P_F = P_{lf} + P_{cf} = bP_{txf} + P_{cf} \qquad (2.24)$$

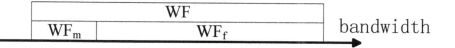

Figure 2.12: The spectrum allocation in a cell.

where P_{lf} and P_{cf} represent the linear and constant parts of the power consumption of FAPs, respectively, b is a constant, P_{txf} is the transmitting power of FAPs. However, the components of P_{lf} and P_{cf} are different from P_{lm} and P_{cm}, respectively. For example, there is no cooling device for FAPs.

According to the above analyses of BSs' and FAPs' power consumptions, we can obtain the entire downlink power consumption P_{system} as follows:

$$P_{\text{system}} = P_M + N_f P_F \tag{2.25}$$

where P_M and P_F are given by Eq. (2.23) and Eq. (2.24), respectively.

2.4.3   Spectrum Allocation for Green Radio

With the participation of FAPs in the cell, the spectrum should be repartitioned for energy saving. Existing spectrum allocation strategies in femtocell networks should be divided into two different ones. In a splitting spectrum network, femtocells use different frequency bands from those employed by macrocells. This avoids interference between macrocells and femtocells, but needs additional frequency bands, decreasing the spectrum efficiency of cellular networks. In a shared spectrum network, femtocells use the same frequency band as macrocells. This is beneficial for increasing the spectrum efficiency, but causes serious interference between macrocells and femtocells. All of these strategies aim at improving spectrum efficiency, but do not consider the energy saving. For energy efficient networks, we propose a novel split spectrum allocation strategy, which maximizes the green factor of the cell under the given spectrum efficiency constraint.

For a cell with frequency bandwidth WF, once being participated by FAPs, some MUs, which originally belong to the BS are handed off to FAPs, i.e., becoming FUs. Due to this change, excessive subchannels will be allocated for MUs now because the required spectrum efficiency of FUs can be provided by FAPs. Thus, it is desirable to allocate some subchannels to femtocells, which originally belong to macrocell. Fig. 2.12 shows the spectrum partition in a cell with WF_m and WF_f belonging to macrocells and femtocells, respectively. According to recent surveys, over 50 percent of calls and 70 percent of data services will take place indoors using femtocell in the future [15]. Thus, more subchannels should be allocated to femtocells for getting required spectrum efficiency as shown in Fig. 2.13.

However, due to interference among femtocells, it is not desirable to use the entire bandwidth WF_f in each femtocell. Thus, we consider the part

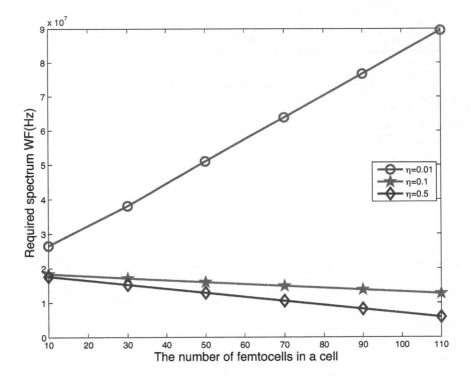

Figure 2.13: Required spectrum WF meeting a target average data rate of 0.1Mbps for MUs.

frequency access of femtocells. The frequency access ratio is $r_f = F_a/F_f$, where F_a subchannels are active for each FAP.

Define T_m and T_f as the throughput (in b/s/Hz) in each subchannel provided by BSs and FAPs, respectively. Thus, we have

$$
\begin{cases}
T_m = \displaystyle\sum_{i=1}^{I-1} i[Z_m(\tau_{i+1}) - Z_m(\tau_i)] + I[1 - Z_m(\tau_I)] \\[2mm]
T_f = \displaystyle\sum_{i=1}^{I-1} i[Z_f(\tau_{i+1}) - Z_f(\tau_i)] + I[1 - Z_f(\tau_I)]
\end{cases}
\tag{2.26}
$$

where $Z_m(\tau_i) = \Pr\{\mathrm{SIR}_m \leq \tau_i\}$ and $Z_f(\tau_i) = \Pr\{\mathrm{SIR}_f \leq \tau_i\}$ denote spatially cellular SIR distribution for macrocell and femtocell users, respectively. There are some strategies to calculate T_m and T_f [16].

Macrocell BS transmits over WF_m and FAPs transmit over WF_f, where $WF_m = rWF$, $WF_f = (1-r)WF$, and r is the bandwidth ratio between femtocell and macro-femto cell. Thus, the two-tier throughput (b/s) per subchannel can be calculated as follows:

$$T_c = W[rT_m + N_f r_f (1 - r)T_f(r_f, N_f)] \tag{2.27}$$

Then, the cellular throughput can be obtained by multiplying both sides of Eq. (2.27) by F. We consider to achieve the maximum value of green factor under the required spectrum efficiency. Thus, we can formulate the problem as follows:

$$\max_r \{\text{green factor}\} = \max_r \left\{ \frac{W[rT_m + (1 - r)N_f r_f T_f(r_f, N_f)]}{P_{\text{system}}} \right\} \tag{2.28}$$

subject to the throughput constraint:

$$\frac{T_m r F}{U_m}(1 - \eta) = \frac{T_f(r_f, N_f)(1 - r)F r_f}{U_f}\eta \tag{2.29}$$

where η is the ratio of throughput per user in the first tier to throughput per user in the second tier. To calculate the green factor, we firstly obtain the equation for r as follows:

$$r = \frac{U_c T_f(r_f, N_f)\eta}{U_c T_f(r_f, N_f)\eta + U_f T_m(1 - \eta)} \tag{2.30}$$

Then, using Eq. (2.28) and Eq. (2.30), we can obtain the optimal spectrum allocation which maximizes the green factor. The maximum green factor corresponds to r_f and N_f. In the next section, we will present the simulation results.

2.4.4   Simulations Results and Performance Analyses

We consider macro-femto cellular networks, which consists of a central macro-cell BS with two rings of 18 interfering macrocells and several FAPs. FAPs are assumed to be scattered according to SPPP. Users are placed randomly in the cell area following a uniform distribution. Table 2.3 shows the simulation parameters.

Fig. 2.13 shows the minimum required spectrum WF, satisfying a target data rate 0.1Mbps for MUs. The corresponding data rates for FUs are 10Mbs, 1Mbps, and 0.1Mbps for $\eta = 0.01$, $\eta = 0.1$, and $\eta = 0.5$, respectively. Thus, the required spatial spectrum efficiencies are $0.1Mbps/WF$ for MUs, $10Mbps/WF$, $1Mbps/WF$, and $0.1Mbps/WF$ for FUs, respectively. As illustrated in Fig. 2.13, the minimum required spectrum WF for $\eta = 0.01$ increases when the number of femtocells in a cell increases. In this case, femtocells are fully utilized. The minimum required spectrum WF for $\eta = 0.1$ and $\eta = 0.5$ decreases when the number of femtocells in a cell increases. In this case, femtocells are under-utilized.

Figs. 2.14–2.20 plot the relationship between green factor and the frequency access ratio γ_f. From Figs. 2.14–2.20, we can make the following observations:

Table 2.3: Simulation Parameters

Parameters	Values
Macrocell/Femtocell radius	288m,40m
Total users in a cell	300
Users per femtocell	2
BS transmitting power	20W
FAP transmitting power	0.1W
Wall partition loss	2dB
Energy model coefficient a, b	21.54, 7.84
Energy model constant P_{cm}, P_{cf}	354.44, 71.50
The ratio of throughput	$\eta = 0.01; 0.05; 0.1; 0.2; 0.3; 0.4; 0.5$
Femtocell numbers in a cell	$N_f = 10; 30; 50; 70; 90; 110$

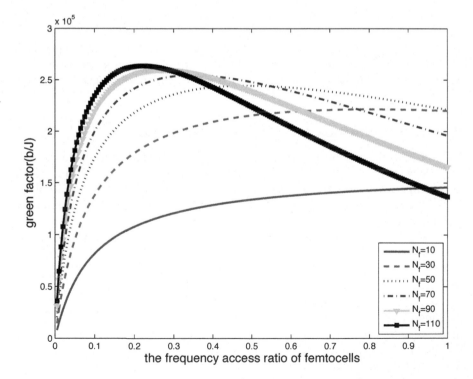

Figure 2.14: Green factor versus frequency access ratio of femtocells ($\eta = 0.01$).

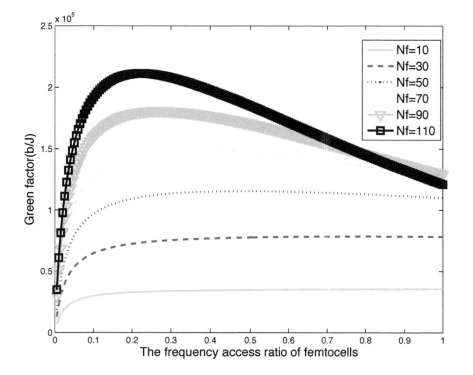

Figure 2.15: Green factor versus frequency access ratio of femtocells ($\eta = 0.05$).

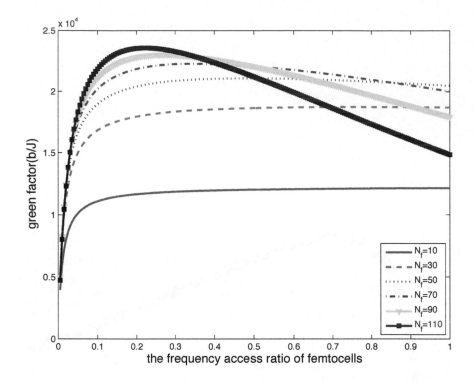

Figure 2.16: Green factor versus frequency access ratio of femtocells ($\eta = 0.1$).

1. When N_f is less than 30, the green factor increases when r_f increases. When N_f is larger than 50, the green factor first increases, and then decreases when r_f increases. The more femtocells, the larger spectrum efficiency can be obtained. But, the more femtocells, the larger interference among femtocells. Thus, it is desirable to fully utilize femtocells when N_f is small and under-utilize femtocells to achieve the maximum green factor when N_f is large.

2. The maximum green factor increases when N_f increases. However, for different N_f, the maximum green factor corresponds to different r_f. Thus, adjusting r_f is needed to achieve the maximum green factor.

3. When η is small ($\eta = 0.01$ or 0.05), the maximum green factor for different N_f will be obtained at different r_f. For example, in Fig. 2.14 for $N_f = 10$ and 110, the green factor has the maximum value when r_f is 1 and 0.2, respectively. While η is relatively large ($\eta = 0.5$), the maximum green factor can be obtained when we take $r_f \approx 0.2$ for different N_f. This is because when the throughput required by the FUs is large ($\eta = 0.01$ or 0.05), more femtocells are needed to reduce the energy consumption of the wireless networks. When the throughput required by the FUs

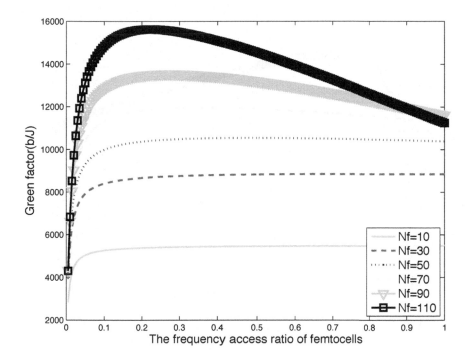

Figure 2.17: Green factor versus frequency access ratio of femtocells ($\eta = 0.2$).

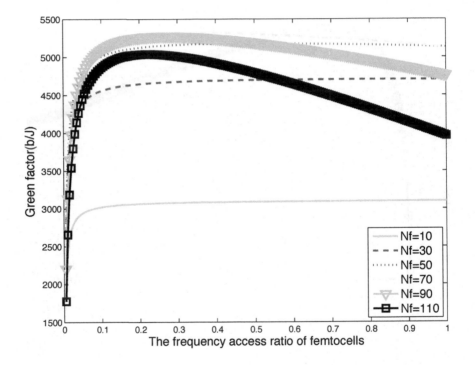

Figure 2.18: Green factor versus frequency access ratio of femtocells ($\eta = 0.3$).

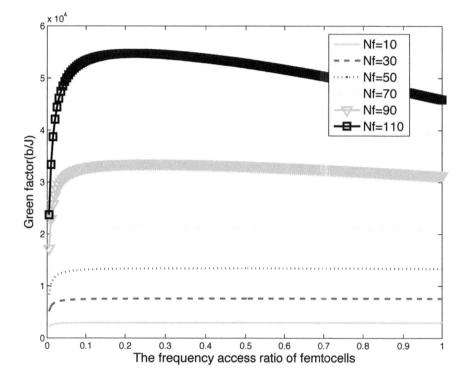

Figure 2.19: Green factor versus frequency access ratio of femtocells ($\eta = 0.4$).

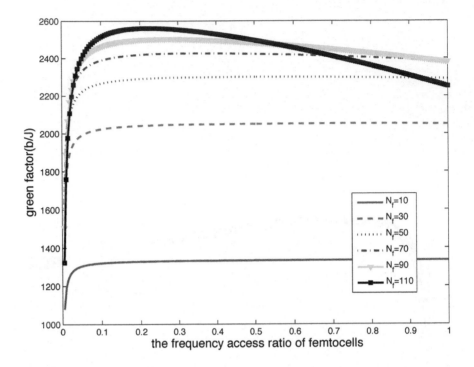

Figure 2.20: Green factor versus frequency access ratio of femtocells ($\eta = 0.5$).

is small ($\eta = 0.5$), the interference among femtocells determines the maximum green factor.

2.5   Conclusions

We proposed the wireless networks resource trading strategy for QoS performance guaranteed green communications. We developed the demanding-based resources trading model. For the given QoS performance requirements demanded, any wireless resources can be saved by consuming the other types of wireless resources instead. To reduce the energy consumption of the wireless networks, we derived the optimal energy-bandwidth, energy-time, energy-space, and energy-code wireless networks resources trading relationships for green wireless networks. We show that there exists the minimum energy per bit for the available bandwidth and the given delay constraint, respectively, when taking operating power into account. We also demonstrate that the minimum energy per bit increases when the number of antennas increases and the minimum energy per bit decreases with the coding gain increasing. The example-case studies of using our proposed resource trading strategy in one-tier and two-tier cellular networks indicate that our proposed resource trading strategies are highly efficient in implementing green communications over wireless networks. Our simulations evaluations show that our proposed modeling techniques and schemes can minimize the energy consumption for given QoS performance metrics by trading the other types of wireless networks resources for the energy resources while satisfying the required QoS performance metrics.

Bibliography

[1] Andy Nolan, "Global action plan, an inefficient truth," *http://www.globalactionplan.org.uk/green-it*, 2007.

[2] Fletcher S., "Green radio @ sustainable wireless networks," *http://kn.theiet.org/magazine/rateit/communications/green-radio-article.cfm*, Jun. 2009.

[3] G. P. Fettweis and E. Zimmermann, "Ict energy consumption - trends and challenges," in *Proceedings of the 11th International Symposium on Wireless Personal Multimedia Communications*, Lapland, Finland, Sep. 2008.

[4] A. J. Fehske, F. Richter, and G. P. Fettweis, "Energy efficiency improvements through micro sites in cellular mobile radio networks," in *GLOBECOM Workshops, 2009 IEEE*, Hawaii, USA, Nov. 2009, pp. 1–5.

[5] B. Badic, T. O'Farrell, P. Loskot, and J. He, "Energy efficient radio access architectures for green radio: Large versus small cell size deployment," in *Vehicular Technology Conference Fall (VTC 2009-Fall), 2009 IEEE 70th*, Anchorage, Alaska, USA, Sep. 2009, pp. 1–5.

[6] H. Holma and A. Toskala, Eds., *HSDPA/HSUPA for UMTS: High Speed Radio Access for Mobile Communications*, John Wiley & Sons, 2006.

[7] W. Cheng, H. Zhang, L. Zhao, and Y. Li, "Energy efficient spectrum allocation for green radio in two-tier cellular networks," in *GLOBECOM, 2010 IEEE*, Miami, USA, Dec. 2010.

[8] M. A. Marsan, L. Chiaraviglio, D. Ciullo, and M. Meo, "Optimal energy savings in cellular access networks," in *Communications Workshops, 2009. ICC Workshops 2009. IEEE International Conference on*, Jun. 2009, pp. 1–5.

[9] G. J. Foschini, and M. J. Gans, "On limits of wireless communications in a fading environment when using multiple antennas," *Wireless Personal Communications*, vol. 6, No. 3, pp. 311-335, Mar. 1998.

[10] V. Tarokh, H. Jafarkhani, and A. R. Calderbank, "Space-time block coding for wireless communications: performance results," *IEEE Journal on Selected Areas in Communications*, vol. 17, no. 3, pp. 451-460, Mar. 1999.

[11] D. L. Perez, A. Valcarce, G. D. L. Roche, E. Liu, and J. Zhang, "Access methods to WiMAX femtocells: a downlink system-level case study," in *11th IEEE ICCS*, Guangzhou, China, Nov. 2008, pp 1657-1662.

[12] M. S. Alouini, and A. J. Goldsmith, "Area spectral efficiency of cellular mobile radio systems," *IEEE Trans. Veh. Technol.*, vol. 48, no. 4, pp. 1047-1066, July 1999.

[13] J. Akhtman, and L. Hanzo, "Power versus bandwidth efficiency in wireless communications: the economic perspective," *IEEE VTC Fall*, Anchorage, Alaska, USA, 2009.

[14] S. Fletcher, "Green Radio@ Sustainable Wireless Networks," VCE Core5 Programme. Available: *http://kn.theiet.org/magazine/rateit/ communications/green-radio-article.cfm*

[15] G. Mansfield, "Femtocells in the US Market-Business drivers and consumer propositions," *FemtoCells Europe*, ATT, London, UK, June 2008.

[16] V. Chandrasekhar, and J. G. Andrews, "Spectrum allocation in tiered cellular networks," *IEEE Trans. Commun.*, vol. 57, no. 10, Oct. 2009, pp. 3059-3068.

Chapter 3

Green Relay Techniques in Cellular Systems

Yinan Qi, Fabien Héliot, Muhammad Ali Imran and Rahim Tafazolli
University of Surrey, yinan.qi@surrey.ac.uk

There has been escalated demand for rapid, low-latency and high-rate communication of information at homes and business premises in the past decade. As the need for high speed access by end-users evolved, particularly fuelled by the widespread adoption of the wireless internet (smart phone), some standards such as High-Speed Downlink Packet Access (HSDPA) [1], 3rd Generation Partnership Project Long Term Evolution (3 GPP LTE) as well as its major enhancement LTE-Advanced for mobile phone networks [2,3], and IEEE standard for mobile broadband wireless access, also known as "mobile Worldwide Interoperability for Microwave Access (WiMAX)" [4], have been proposed to provide high speed data transmission.

One of the main objectives of future wireless systems is to provide high data rate with a uniform coverage by adapting itself to multipath fading, path-loss and shadowing conditions. In the pursuit of finding schemes that will provide a solution to minimize these effects, various granular and distributed network architectures based on relaying techniques are emerging. However, due to the increasing concern about the CO_2 contribution from the ICT (Information and Communications Technology) industry [5,6], the future wireless systems should also be energy efficient to meet the increasing demand for flexible use of emerging green technologies, where relay is no doubt one of the strongest candidates.

This chapter analyzes the relaying technique at link and system levels from both spectrum efficiency (Se) and energy efficiency (EE) perspectives. A thorough investigation will be provided for a variety of approaches at the relay node (RN) to forward information including amplify-and-forward (AF), decode-and-forward (DF) and compress-and-forward (CF). Advanced relaying schemes, where the conventional relaying schemes are combined in a hybrid manner to adapt to the variations of the channel states, are introduced and investigated. Furthermore, the relaying techniques are combined with retransmission protocols for packet oriented communication systems and a study from spectrum and energy efficiency perspectives is presented. Finally, this chapter also addresses the challenge of designing and positioning RNs in a state-of-the-art wireless cellular system, namely LTE system, coupled with practical power consumption models.

The rest of the chapter is organized as follows: Section 3.1 introduces the broad topic of relaying. In the next section, the analysis of a relay-assisted system is presented from the spectrum and energy efficiency perspectives. Relay is further combined with retransmission protocols and its performance is studied in Section 3.3. The energy efficiency of relaying techniques in cellular networks is investigated in Section 3.4 and the final section summarizes the findings of the chapter and identifies the potential future works.

3.1   Introduction

In recent years, many projects focusing on cooperative communications have been launched by the 7th Framework Programme of the European Commission, such as FIREWORKS (FlexIble RElay Wireless OFDM-based networks) [7] and ROCKET (Reconfigurable OFDMA-based Cooperative NetworKs Enabled by Agile SpecTrum Use) [8]. Recently, the "green" potential of cooperative communication for improving the energy efficiency (EE) has drawn considerable attention toward this topic from both academia and in-

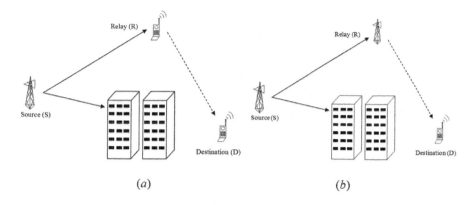

Figure 3.1: A relay system in the urban environment. (a) Mobile relay (b) fixed relay.

dustry. For instance, theoretical analysis of relaying technology from an EE point of view has been recently conducted in [9,12]. Moreover, major research projects focusing on green communications, such as EU FP7 project EARTH (Energy Aware Radio and network tecHnologies) [13], have considered relay as one of their main research tracks.

The basic idea behind relaying is to use some radio nodes, called relays, in order to enable more spectrum and energy efficient communications. RNs can be specifically devoted network nodes or can be other user devices in the vicinity, as is depicted in Fig. 3.1. For instance in an urban environment, the direct source-destination link can be impaired because of buildings and with the help of a fixed or mobile RN, another independent link, i.e., source-RN-destination, can be established for improving the communication between the source and destination nodes. This link does not suffer from shadowing effect because it is obstacle free. There are two basic working modes for a relay system: full-duplex and half-duplex. In full-duplex mode, the relay is assumed to be able to transmit and receive simultaneously. Due to the large dynamic range between the incoming and outgoing signals through the same antenna elements, the full-duplex mode is regarded as difficult to implement in reality. In contrast, in half-duplex mode, the relay is assumed to work in a time-division manner, where it either receives or transmits at a given time and band instance. Compared with the full-duplex mode, the half-duplex mode is more practical. Therefore, we will focus here on the time-division half-duplex relay channel. The entire frame is separated into two phases: the relay-receive phase (phase 1) and the relay-transmit phase (phase 2). During the first phase, the source broadcasts to the relay and the destination. In phase 2, the source and the relay transmits to the destination simultaneously, as is shown in Fig. 3.2. The duplexing ratio in percentage, denoted by α, represents the ratio of the duration of the first phase to the duration of the total transmission.

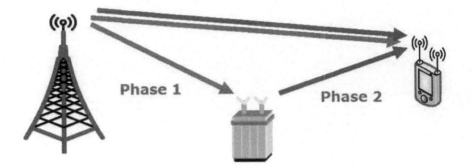

Figure 3.2: Half-duplex relaying.

The relaying model was first introduced by van der Meulen [14], where a communication system with three nodes denoted as source (S), relay node (RN) and destination (D), respectively, was investigated. There are three links between these nodes and each link has an input/output pair. Cover and El Gamal have substantially developed the relaying model and provided three basic relaying principles [15]:

1. Cooperation: If the relay receives a better signal than the destination, it is able to cooperate with the source by sending a signal that contains a perfect source signal to the destination.

2. Facilitation: If the relay sees a corrupted version of what the destination sees, the relay transmits constant signal which is known to the source and the destination to open the channel between the source and the destination.

3. Observation: Alternative information can be sent by the relay. This information does not comprise a perfect source signal, thus precluding pure cooperation, and is not constant, thus precluding simple facilitation. Instead, the relay forwards what it has observed to the destination.

While the optimal relaying strategy in wireless networks has not yet been fully understood, several relaying schemes developed based on the principles (especially for cooperation and observation) have been suggested in the literature. Among them, the simplest scheme is amplify-and-forward (AF) [16]- [19], in which the relay just forwards what it has received with a proper amplification gain. Another well developed scheme is decode-and-forward (DF) [20]- [27], where the relay decodes, re-encodes and forwards the received signal to the destination. DF has shown improvement in terms of the achievable rate and outage behavior when the relay is close to the source, [20] and [27]- [28]. Both schemes share a common weakness: their performance is constrained by the quality of the source-relay link. If this link endures deep fading, for AF

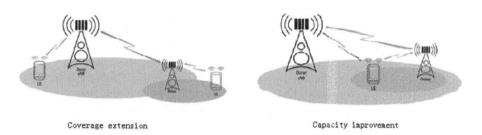

Coverage extension   Capacity improvement

Figure 3.3: Relay deployment approaches.

the relay will forward nothing but its own noise, and for DF the relay will not be able to successfully decode and forwarding of erroneous messages will lead to error propagation at the destination. To cope with this problem, an effective solution for the case of a weak source-relay link based on the observation principle in Cover's paper [15] is to forward the relay's observation to the destination [29]-[31].Instead of decoding, the relay quantizes the received signal, compresses the quantized version and forwards the compressed version to the destination. Studies on CF reveal that it outperforms DF when the relay is close to the destination, or generally speaking, the relay-destination link is very strong. The conventional forwarding mechanisms can be combined in a hybrid fashion to avoid the shortcomings of the original ones. In [32] and [33], the hybrid AF/DF forwarding scheme is able to improve the considered system's frame error rate (FER). The hybrid approach has been further extended by taking CF into consideration in [29] and [34]. The proposed hybrid scheme avoids the shortcomings of AF and DF and exhibits remarkable improvement in outage performance. In wireless broadband networks, relay is being studied as a technology that offers the possibility to extend coverage and increase capacity as shown in Fig. 3.3, allowing more flexible and cost-effective deployment options [35]. The link between the donor base station (BS) and the RN is called backhaul link and the link between the RN and the user equipment (UE) is called access link as shown in Fig. 3.3.

Many standards have included relay as part of their study items as follows

1. IEEE 802.16j (16j Relay) [36],

2. IEEE 802.16m (Relay and Femtocell) [37],

3. 3GPP LTE-A (Rel-10 LTE-A Relay) [38].

Based on its functionalities, a number of different classifications exist for RNs. Relay is usually classified as inband relay when the same carrier frequency is used for the backhaul and access links and outband relay when different carrier frequencies are used for these links. Relay can also be categorized with respect to protocol layer functionalities. A Layer 1 RN, also known as repeater, simply picks up the donor BS signal, amplifies and forwards it into its own coverage area. In contrast with Layer 1 RN, a Layer 2 RN has medium access control (MAC) layer functionality; it can decode received signals and re-encode transmitted signals in order to achieve higher quality in the relay cell area. A Layer 3 RN would include functionalities like mobility management, session set-up and handover and as such acts as a full service BS. This type of relay adds more complexity to the implementation and further increases the delay. In 3GPP standardization [35], RNs are distinguished between Type 1 and Type 2. Type 1 relays are RNs operating on Layer 3, i.e., protocol layer up to Layer 3 for user data packets is available at the RN. Such Layer 3 RNs have all the functions that a BS has, and they effectively create their own identity number (Cell-ID) and own synchronization and reference signals. Type 1 RNs are considered visible to the UEs, thus being called non-transparent RNs. In contrast, Type 2 RNs will not have their own Cell-ID. Consequently the UEs will not be able to distinguish between transmitted signals from the BSs and the RNs. Type 2 RNs operate in Layer 3 or Layer 2, depending on the particular solution/implementation and are transparent to UEs.

An important aspect of the design of spectrum and energy efficient relaying schemes requires an understanding of the measurable spectrum and energy metrics. The spectrum efficiency (SE) can be easily evaluated in terms of achievable rate with the metric bit/second/Hz (bps/Hz). It is important for us to adopt a well-defined and relevant performance metrics to quantify and assess the energy efficiency performance. To capture the energy consumption perspective in the analysis, we employ the energy consumption index bits to energy consumption ratio, defined as the total number of bits that were correctly delivered divided by the energy consumption during the observation period and measured in [bits/J]. The bits to energy consumption ratio metric focuses on the amount of bits spent per Joule and is hence an indicator of bit delivery energy efficiency. The inverse of the bits to energy consumption ratio, measured by Joule per bit, can also be used for evaluation purposes indicating how much energy is needed to convey one bit [39]. In this chapter, we use both to help the reader understand the energy efficiency problem from different perspectives.

3.2   Spectrum and Energy Efficiency Analysis of Relay-Assisted Systems

In wireless networks, relaying techniques have been traditionally used to extend the coverage of communication systems. However, in recent years, other relaying schemes to assist in the communications between the source and destination via some cooperation protocols have emerged. By controlling medium access between the source and the relay coupled with appropriate modulation or coding strategies in such cooperative schemes, it has been found that the diversity gain of the system can be improved. In particular, three basic relaying schemes have been proposed including amplify-and-forward (AF), decode-and-forward (DF) and compress-and-forward (CF).

Among various cooperative protocols, AF is one of the basic modes and has attracted lots of attention in recent years due to its simplicity and low expenses. AF is especially suitable for systems with strict energy consumption constraints on RNs because no baseband processing is needed; thus the RF links can be saved and the energy consumption can be further reduced in addition to the saved baseband processing energy. Being a non-regenerative scheme, AF only allows RN to amplify and forward the received signals to the destination without any coding or decoding process. Laneman has, in his landmark paper [27], pointed out that AF is able to achieve second-order diversity in very high SNR regions. Another work indicated that the ergodic capacity of AF can be higher than that of DF with certain SNR settings [16]. In particular, when the source-relay link is statistically worse than the other two links, AF is more spectrum efficient than DF.

Two other basic relaying schemes have been introduced in the classic work of Cover and El Gamal [15]. One is DF, where the RN decodes the received message completely, re-encodes the message and fully or only partially forwards the decoded message to the destination. It has been proved that DF is able to achieve the capacity of the degraded relay channel. The information-theoretical analysis of the outage behavior in Laneman's work [25] indicates that fixed decode-and-forward (FDF), where the RN always forwards the decoded message even when the decoding is unsuccessful, does not achieve diversity. In contrast, selective DF in which RN keeps silent when it is not able to decode the message achieves full cooperative diversity in high SNR regions. Another basic relaying scheme is CF [29]. In CF, the operation of RN is quite different. Instead of decoding, RN compresses the received signal and transmits the compressed version to the destination. As a result, at the destination, the signal received from the source during phase 1 will serve as side information to reconstruct RN's observation of the signal from the source. Then the destination tries to decode the message by joint processing of the received signal and the reconstructed observation of the RN. To help the readers better understand this point, we can just assume an extreme case where the RN and the destination can fully share their observations without any distortion. In such a case, the RN and destination can enjoy full receive diversity. In contrast to DF, which normally exhibits superior performance when the RN is close to the source, CF is desired under the condition of a weak source to RN link and a strong RN to destination link.

However, the wireless channels suffer from fading occasionally. Even when the RN is close to the source or destination, the instantaneous SNRs of the links are not always apposite for one single relaying scheme. For instance, the RN cannot successfully decode every block no matter how close it is to the source. Thus DF cannot be guaranteed to be successful all the time. We need to develop a flexible relay system which is able to adapt itself to the dynamics of the channel. Inspired by this idea, advanced relaying schemes were proposed in [29] and [32–34], where the conventional forwarding mechanisms can be combined in a hybrid fashion to avoid the shortcomings of the original ones. There is another hybrid structure introduced in [40], where the multilevel coding concept is incorporated. This structure exploits the fact that bit-error-rates (BERs) of different coding levels exhibit considerable difference and the authors propose to use DF in the levels where decoding is more likely to be successful, and in the meantime CF in the levels where the error probability of decoding increases significantly. However, this scheme is just a preliminary one and therefore not included in this chapter.

3.2.1 System Model

Consider a relay-assisted system where the source transmits a W bit message $w \in \{1, ..., 2^W\}$ to the destination. During the first phase, the source encodes the message into αn symbols $x^1(w)[1], ..., x^1(w)[\alpha n]$ and broadcasts to the RN and the destination. The received signals at the RN and the destination are given by

$$y_r[i] = \frac{c_1}{\sqrt{K_t} d_1^{\zeta/2}} x^1(w)[i] + z_r[i], \tag{3.1}$$

$$y_d^1[i] = \frac{c_0}{\sqrt{K_t} d_0^{\zeta/2}} x^1(w)[i] + z_d[i],$$

respectively, where c_0 and c_1 are the channel gains of the S-D and S-RN links, modelled by circularly symmetric complex Gaussian distribution with zeros means and unit variances, d_0 and d_1 are distances of the S-D and S-RN links, respectively, ζ is the path-loss exponent, K_t is a constant indicating the physical characteristics of the link and the power amplifier as in [9] and [41], and z_r and z_d are the additive noise modelled by circularly symmetric complex Gaussian distribution with zeros means. The superscript indicates phase when necessary. The encoded message should be subject to the power constraint

$$\frac{1}{\alpha n} \sum_{i=1}^{\alpha n} |x^1[i]|^2 \leq P_s, \tag{3.2}$$

During phase 2, the RN transmits $(1 - \alpha)n$ encoded symbols $x_r[1], ..., x_r[(1 - \alpha)n]$; while at the same time the source transmits symbols $x^2(w)[1], ..., x^2(w)[(1 - \alpha)n]$. The signal received by the destination is

$$y_d^2[i] = \frac{c_0}{\sqrt{K_t} d_0^{\zeta/2}} x^2(w)[i] + \frac{c_2}{\sqrt{K_t} d_2^{\zeta/2}} x_r[i] + z_d[i], \tag{3.3}$$

where d_2 and c_2 are the distance and channel gain of the RN-D link, respectively, and c_2 follows the same distribution as c_0 and c_1. It should be noted that we did not specify the operation of the RN because it depends on the employed relaying schemes.

3.2.2 Spectrum Efficiency Analysis

With a well defined signal model, we can study the spectrum efficiency of the relay-assisted systems in terms of achievable rate. We also analyze their outage probability which indicates the unsuccessful reception at the destination and will be used in the energy efficiency analysis. Starting from non-hybrid relaying schemes, the analysis will be further developed to hybrid ones.

AF: In AF, the duplexing ratio is fixed at 0.5 and we assume that the source transmits exactly the same symbols during two phases. The signal transmitted by the RN is

$$x_r[i] = \beta y_r[i] \tag{3.4}$$

$$= \beta \left(\frac{c_1}{\sqrt{K_t} d_1^{\zeta/2}} x^1(w)[i] + z_r[i] \right),$$

where β is the amplification gain and subject to the power constraint

$$\beta \leq \sqrt{\frac{P_r}{\frac{|c_1|^2 P_s}{K_t d_1^\zeta} + N_0 B}}. \tag{3.5}$$

Here N_0 is the noise power spectral density and B is the bandwidth. The received signal at the destination can be given by

$$y_d^2[i] = \frac{c_0}{\sqrt{K_t} d_0^{\zeta/2}} x^2(w)[i] + \frac{c_2}{\sqrt{K_t} d_2^{\zeta/2}} \beta y_r[i] + z_d[i] \tag{3.6}$$

$$= \frac{c_0}{\sqrt{K_t} d_0^{\zeta/2}} x^2(w)[i] + \frac{c_2}{\sqrt{K_t} d_2^{\zeta/2}} \beta \left(\frac{c_1}{\sqrt{K_t} d_1^{\zeta/2}} x^1(w)[i] + z_r[i] \right)$$

$$+ z_d[i] \tag{3.7}$$

$$= \left(\frac{c_0}{\sqrt{K_t} d_0^{\zeta/2}} + \beta \frac{c_1 c_2}{K_t (d_1 d_2)^{\zeta/2}} \right) x(w)[i] + \left(\beta \frac{c_2}{\sqrt{K_t} d_2^{\zeta/2}} z_r[i] + z_d[i] \right),$$

where $x(w)[i] = x^1(w)[i] = x^2(w)[i]$ for $1 \leq i \leq n/2$. The achievable rate R_{AF} can be derived as in [10]. The closed-form expression for R_{AF} is difficult to calculate since carrier phase synchronization is not assumed here but the trivial upper bound has been derived in [42] as

$$\bar{R}_{AF} = B log_2 \left(1 + \frac{|c_0|^2 P_s}{N_0 B K_t d_0^\zeta} + \frac{P_s \left(\frac{\beta^2 |c_1 c_2|^2}{K_t^2 (d_1 d_2)^\zeta} + \frac{|c_0|^2}{K_t d_0^\zeta} \right)}{N_0 B \left(1 + \frac{\beta^2 |c_2|^2}{K_t d_2^\zeta} \right)} \right). \tag{3.8}$$

$\bar{R}_{AF}$ is a function of amplification gain β and can be rewritten as

$$\bar{R}_{AF} = B log_2 \left(1 + \frac{|c_0|^2 P_s}{N_0 B K_t d_0^\zeta} + \frac{|c_1|^2 P_s}{N_0 B K_t d_1^\zeta} + \frac{|c_1|^2 P_s \left(\frac{|c_0|^2 d_1^\zeta}{|c_1|^2 d_0^\zeta} - 1 \right)}{N_0 B K_t d_1^\zeta \left(1 + \frac{\beta^2 |c_2|^2}{K_t d_2^\zeta} \right)} \right). \tag{3.9}$$

Apparently, in order to maximize $\bar{R}_{AF}$ and taking the constraint (3.5) into account, β should be chosen as

$$\beta = \begin{cases} \sqrt{\dfrac{P_r}{\frac{|c_1|^2 P_s}{K_t d_1^\zeta} + N_0 B}} & \text{if } |c_0|^2 d_1^\zeta \le |c_1|^2 d_0^\zeta \\ 0 & \text{otherwise.} \end{cases} \tag{3.10}$$

It reveals the fact that the RN need not always amplify and forward its reception. It only forwards when the S-RN link is of higher quality than a certain threshold. This makes sense because once the S-RN link is weak, a large part of the received signal of the RN is comprised of noise and nothing but noise is forwarded to the destination if the RN insists on transmitting. In such a case, the final decoding is not benefiting from the RN's forwarding but rather being affected adversely. Thus we can draw the conclusion that if the RN cannot help, it should just stop doing so.

DF: As we mentioned before, DF is more complicated than AF in the sense that the RN needs to decode, re-encode and forward the received message. The received message at destination during phase 2 is

$$y_d^2[i] = \frac{c_0}{\sqrt{K_t} d_0^{\zeta/2}} x^2(w)[i] + \frac{c_2}{\sqrt{K_t} d_2^{\zeta/2}} x_r(w)[i] + z_d[i], \tag{3.11}$$

where $1 \le i \le (1-\alpha)n$. The encoding at the RN and the destination can be carefully designed to form a MISO code to enjoy the transmit diversity. The achievable rate of DF can be given as [28]

$$R_{DF} = min\{R^1, R^2\}, \tag{3.12}$$

where

$$R^1 = \alpha B log_2 \left(1 + \frac{|c_1|^2 P_s}{N_0 B K_t d_1^\zeta}\right), \tag{3.13}$$

$$R^2 = \alpha B log_2 \left(1 + \frac{|c_0|^2 P_s}{N_0 B K_t d_0^\zeta}\right)$$

$$+ (1-\alpha) B log_2 \left(1 + \frac{|c_0|^2 P_s}{N_0 B K_t d_0^\zeta} + \frac{|c_2|^2 P_r}{N_0 B K_t d_2^\zeta}\right).$$

The first term R^1 implies that the RN has to successfully decode the message before forwarding it.

CF: CF is even more complicated than DF. Instead of amplifying or decoding, the RN tries to help the destination by forwarding its own observation y_r, which requires a source coding scheme for the continuous y_r. During the first phase, the signals received by the RN and the destination both originate from the same source and contain a common term $x^1(w)$. Thus these two signals are correlated and this fact provides the possibility of transmitting the observation of the RN at a reduced rate, i.e., y_r can be compressed.

Wyner-Ziv coding is an efficient method for compressing correlated continuous sources which are separately located [43–44]. It is an extension of the work of Slepian and Wolf in [45], where discrete distributed source coding is investigated. Although the details of distributed coding might help the readers better understand the compression mechanism in CF, it is out of the scope of this book and interested readers can refer to [43–44] for in-depth knowledge. It has been pointed out that nested multi-dimensional lattice code can be used for Wyner-Ziv coding [46–47]. However, nested multi-dimensional lattice code requires high complexity and is not feasible for practical implementation. A more feasible two step structure consists of a quantizer and a Slepian Wolf encoder, i.e., compressor, is proposed in [48]. Firstly, the quantizer converts continuous reception into discrete bin index. Secondly, the Slepian Wolf encoder compresses the bin index.

Coming back to our system, at the end of the first phase, the RN quantizes the received signal y_r into some intermediate bin index u, which is then compressed into index v by the Slepian Wolf encoder. The index v is encoded into symbols $x_r(v)[1],...,x_r(v)[(1-\alpha)n]$ and forwarded during the second phase. Meanwhile, the source transmits $x^2(w)$ independent with $x_r(v)$. In this regard, a multiple access channel is formed and the received signal at the destination is

$$y_d^2[i] = \frac{c_0}{\sqrt{K_t d_0^{\zeta/2}}} x^2(w)[i] + \frac{c_2}{\sqrt{K_t d_2^{\zeta/2}}} x_r(v)[i] + z_d[i], \qquad (3.14)$$

The destination starts from decoding the Slepian Wolf coded bin index v by treating $x^2(w)$ as noise. The destination then, with the help of the side information y_d^2, decompresses v and uses the decompressed u to reconstruct the observation of the RN, denoted as $\hat{y}_r$. The destination then subtracts $x_r(v)$ from y_d^2 to get

$$\bar{y}_d^2[i] = y_d^2[i] - \frac{c_2}{\sqrt{K_t d_2^{\zeta/2}}} x_r(v)[i]. \qquad (3.15)$$

Then the destination performs final decoding by joint processing of $y_d^2, \bar{y}_d^2$ and the estimated $\hat{y}_r$. The achievable rate of CF is derived in [42] as

$$R_{CF} = \alpha B log_2 \left(1 + \frac{|c_0|^2 P_s}{N_0 B K_t d_0^\zeta} + \frac{|c_1|^2 P_s}{(N_0 B + \sigma_U^2) K_t d_1^\zeta} \right)$$
$$+ (1-\alpha) B log_2 \left(1 + \frac{|c_0|^2 P_s}{N_0 B K_t d_0^\zeta} \right), \qquad (3.16)$$

where σ_U^2 is the compression noise variance. In order to maximize the achievable rate, the compression noise should be minimized as follows. During phase 2, the destination can successfully decode v if

$$R_0 \leq I(X_r; Y_d^2) = log_2 \left(1 + \frac{\frac{|c_2|^2 P_r}{K_t d_2^\zeta}}{N_0 B + \frac{|c_0|^2 P_s}{K_t d_0^\zeta}} \right), \qquad (3.17)$$

where R_0 is transmission rate of the compressed information. On the other side, the minimum transmission rate for the compressed signal, such that it can be recovered with the smallest distortion, is given in [28] as,

$$H(Y_r+U|Y_d^1)-H(Y_r+U|Y_r) = log_2 \left(\frac{\frac{|c_1|^2 P_s N_0 B}{K_t d_0^\zeta}}{N_0 B + \frac{|c_0|^2 P_s}{K_t d_0^\zeta}} + N_0 B + \sigma_U^2 \right) - log_2(\sigma_U^2)$$

(3.18)

where U is an auxiliary random variable following circularly symmetric complex Gaussian distribution with zero mean and variance σ_U^2. In order to satisfy both constraints, we have

$$(1 - \alpha)R_0 \leq \alpha \left(H(Y_r + U|Y_d^1) - H(Y_r + U|Y_r) \right),$$

(3.19)

leading to

$$\sigma_U^2 \geq \frac{\frac{|c_1|^2 P_s N_0 B}{K_t d_0^\zeta} + N_0 B \left(N_0 B + \frac{|c_0|^2 P_s}{K_t d_0^\zeta} \right)}{\left(2^{\frac{(1-\alpha)R_0}{\alpha B}} - 1 \right) \left(N_0 B + \frac{|c_1|^2 P_s}{K_t d_1^\zeta} \right)},$$

(3.20)

where σ_U^2 can be minimized when equal holds.

Hybrid Schemes: In a fading environment, the instantaneous SNRs always change in a wide range. Thus no single relaying scheme is able to outperform others all the time. Other than AF, DF and CF, some advanced hybrid relaying schemes, where the conventional forwarding mechanisms are combined in a hybrid fashion, will be addressed. These hybrid schemes are able to adapt themselves to the variations of the time-varying channel and achieve either transmit or receive diversity depending on the current channel state. This sub-section will investigate these hybrid schemes. We focus on DF/CF hybrid relaying and all the derivations can be easily extended to other hybrid schemes because they are sharing the same mechanism.

The operation of hybrid DF/CF relaying in the first phase is the same as any non-hybrid one. However, during phase 2, the RN starts by trying to decode the received message and re-encodes and forwards it once the decoding is successful; otherwise, it compresses and forwards its observation instead. In other words, RN's transmission depends on its decoding results and the destination receives

$$y_d^2[i] = \begin{cases} \frac{c_0}{\sqrt{K_t d_0^{\zeta/2}}} x^2(w)[i] + \frac{c_2}{\sqrt{K_t d_2^{\zeta/2}}} x_r(w)[i] + z_d[i], & \text{if } \alpha BI(X^1; Y_r) \geq \frac{W}{T} \\ \frac{c_0}{\sqrt{K_t d_0^{\zeta/2}}} x^2(w)[i] + \frac{c_2}{\sqrt{K_t d_2^{\zeta/2}}} x_r(v)[i] + z_d[i], & \text{otherwise.} \end{cases}$$

(3.21)

where $I(X^1; Y_r)$ is the achievable rate of the S-RN link and is given as

$$I(X^1; Y_r) = log_2 \left(1 + \frac{|c_1|^2 P_s}{N_0 B K_t d_1^\zeta} \right),$$

(3.22)

and T is the duration of the whole frame. It leads to the rate

$$R_{CDF} = \begin{cases} R_{DF}, & \text{if } \alpha BI(X^1; Y_r) \geq \frac{W}{T} \\ R_{CF}, & \text{otherwise.} \end{cases} \tag{3.23}$$

3.2.3   Energy Efficiency Analysis

The total energy consumption of a relay system is composed of both transmission power and circuitry energy consumption of all involved nodes. It should be noted that the optimization of a system's energy efficiency must be subject to certain quality of service (QoS) constraint. In this chapter, the QoS constraint to be satisfied by a communication system is defined in terms of the maximum tolerable probability of unsuccessful reception of a message, i.e., target outage probability.

Let us define the average transmission energy consumed by a two node system for transmission of one bit as E_b, the overall energy efficiency optimization should take into account circuitry power consumption. This optimization problem can be described as following: given the QoS constraint in terms of target outage probability and the packet duration constraint, find the optimal transmission energy E_b and packet duration T. It can be generally formulated as

$$\text{minimize } E(E_b, T) \tag{3.24}$$
$$\text{subject to } p_{out} \leq p_t, \, T \leq T_{max},$$

where p_{out} is the outage probability, p_t is the target outage probability and T_{max} is the maximum transmission time of each frame. However, in a three node relay-assisted system, not only the transmission energy E_b of the source should be optimized, but RN's E_b should also be optimized as well. The problem amounts to

$$\text{minimize } E(E_{b,s}, E_{b,r}, T) \tag{3.25}$$
$$\text{subject to } p_{out} \leq p_t, \, T \leq T_{max},$$

where $E_{b,s}$ and $E_{b,r}$ are the transmission energy per bit at the source and RN, respectively. In the following parts, we will study the energy efficiency of various relaying schemes.

AF: Define the transmission energy per bit at the source during phase 1 as $E_{b,s}^1$. In the first phase, the source broadcasts and the RN and the destination receive. In order to transmit message w with W bits, the source spends energy $W E_{b,s}^1$. In addition, the transmitting electronic circuitry of the source also contributes energy $P_{ct}T/2$ to total energy expenditure. Here P_{ct} is the power (in watts) consumed by the transmitting electronic circuitry. The total energy expenditure needs to take both the transmission and reception sides into account. In particular, the electronic receiving circuitries of the RN and the destination also consume energy $P_{cr}T/2$ per node, where P_{cr} measures power (in watts) consumed by receiving circuitries. Here we made two assumptions:

1. The transmitting circuitries of the source and the RN consume the same amount of power;

2. The receiving circuitries of the RN and the destination consume the same amount of power.

These are simplified assumptions only for the purpose of theoretical analysis. More accurate power consumption models will be introduced later in the next section. The total energy consumption during phase 1 can be expressed as

$$E^1 = \left(WE_{b,s}^1 + \frac{P_{cr}T}{2}\right) + P_{cr}T, \tag{3.26}$$

where the first term in the parentheses is the total energy consumption on the transmitting side (source) and the second term stands for the energy consumption on the receiving side (RN and destination) during phase 1. During phase 2, the RN is actively transmitting as well as the source and the destination is the only receiving node. Suppose that the source spends $E_{b,s}^2$ for each bit, the total energy consumption is given by

$$E^2 = \left(WE_{b,s}^2 + \frac{P_{cr}T}{2}\right) + \frac{(P_r + P_{cr})T}{2} + \frac{P_{cr}T}{2}, \tag{3.27}$$

where P_r is the RN transmission power. The first and second terms are the source and RN energy consumption, respectively, and the last term stands for the energy consumed by the destination's receiving circuitry. Although the source might have different transmission energy per bit during two phases, it is more applicable to assume that its transmission power is constant during the whole frame, i.e.,

$$P_s = P_s^1 = P_s^2, \tag{3.28}$$

where P_s^1 and P_s^2 are transmission power during phase 1 and 2, respectively, and

$$P_s^1 = \frac{WE_{b,s}^1}{\alpha T}, P_s^2 = \frac{WE_{b,s}^2}{(1-\alpha)T}. \tag{3.29}$$

Since $\alpha = 1/2$ in AF, we have $E_{b,s} = E_{b,s}^1 = E_{b,s}^2$, the overall energy efficiency can be expressed in terms of Joule/bit as

$$E(E_{b,s}, E_{b,r}, T) = \frac{2WE_{b,s} + WE_{b,r} + \frac{3}{2}(P_{cr} + P_{ct})T}{W} \tag{3.30}$$

$$= 2E_{b,s} + E_{b,r} + \frac{3(P_{cr} + P_{ct})T}{2W},$$

where $E_{b,r} = (P_rT/2)/W$. The outage probability is given by

$$p^{out} = \Pr\left\{\bar{R}_{AF}(\gamma_0, \gamma_1, \gamma_2) < \frac{W}{T}\right\}, \tag{3.31}$$

where γ_i are random variables and $\gamma_i = |c_i|^2$, $0 \le i \le 2$. Considering the complexity of the expression for $\bar{R}_{AF}(\gamma_0, \gamma_1, \gamma_2)$, it is difficult to calculate the closed-form expression for outage probability. We can resort to the characteristic function of $\bar{R}_{AF}(\gamma_0, \gamma_1, \gamma_2)$

$$\phi(s) = \mathbf{E}\left\{exp\left(-s\bar{R}_{AF}(\gamma_0, \gamma_1, \gamma_2)\right)\right\}, \tag{3.32}$$

where $\mathbf{E}\{\}$ is the expectation function and the expectation is with respect to γ_i. The outage probability can be expressed by the Laplace inversion formula

$$p^{out} = \Pr\left\{\bar{R}_{AF}(\gamma_0, \gamma_1, \gamma_2) < \frac{W}{T}\right\} \tag{3.33}$$

$$= \frac{1}{2\pi j}\int_{d_1-j\infty}^{d_1+j\infty}\frac{\phi(s)e^{sW/T}}{s}ds,$$

where d_1 is a proper constant and $j = \sqrt{-1}$. This integration can be approximated as [49]

$$p^{out} = \frac{1}{2\pi j}\int_{d_1-j\infty}^{d_1+j\infty}\frac{\phi(s)e^{sW/T}}{s}ds \tag{3.34}$$

$$= \sum_{i=1}^{M}K_i\frac{\phi(z_iT/W)}{z_i},$$

where z_i are the poles of the Padé rational function, K_i are the corresponding residues and M is an arbitrary integer that determines the approximation accuracy. Now we rewrite the trivial upper bound $\bar{R}_{AF}$ as a function of packet duration T, $E_{b,s}$ and $E_{b,r}$

$$\bar{R}_{AF} = Blog_2\left(1 + \frac{2\gamma_0 W E_{b,s}}{TN_0 BK_t d_0^\zeta} + \frac{2W E_{b,s}\left(\frac{\beta^2\gamma_1\gamma_2}{K_t^2(d_1 d_2)^\zeta} + \frac{\gamma_0}{K_t d_0^\zeta}\right)}{TN_0 B\left(1 + \frac{\beta^2\gamma_2}{K_t d_2^\zeta}\right)}\right), \tag{3.35}$$

where $E_{b,s} = (P_s T/2)/W$. It is easy to see that the outage probability is also a function of T, $E_{b,s}$ and $E_{b,r}$. With this dependence, the energy efficiency of AF can be explicitly optimized as

$$\text{minimize } E(E_{b,s}, E_{b,r}, T) \tag{3.36}$$

$$\text{subject to } p_{out}(E_{b,s}, E_{b,r}, T) \le p_t, \ T \le T_{max}.$$

DF: In DF, the duration of phase 1 is αT. The energy expenditure including circuitry energy consumption can be given by

$$E^1 = \left(WE_{b,s}^1 + \alpha P_{ct}T\right) + 2\alpha P_{cr}T. \tag{3.37}$$

where the first two terms in the parentheses account for energy consumed for transmission of the message, mainly by the power amplifier, and the last term in the summation is the energy consumption of the receiving circuitries at the RN and the destination. During phase 2, the source and the RN transmit and the destination receives with total consumed energy

$$E^2 = \left(WE_{b,s}^2 + (1-\alpha)P_{ct}T\right) + (1-\alpha)(P_r + P_{ct})T + (1-\alpha)P_{cr}T. \quad (3.38)$$

Let $P_s = P_s^1 = P_s^2$, the overall energy efficiency is expressed as

$$
\begin{aligned}
E(E_{b,s}, E_{b,r}, T) &= \frac{\frac{WE_{b,s}}{\alpha} + WE_{b,r} + (2-\alpha)P_{ct}T + (1+\alpha)P_{cr}T}{W} \\
&= \frac{E_{b,s}}{\alpha} + E_{b,r} + \frac{(2-\alpha)P_{ct}T + (1+\alpha)P_{cr}T}{W}, \quad (3.39)
\end{aligned}
$$

where $E_{b,s} = E_{b,s}^1$.

The outage events of DF can be classified into two categories: RN outage and destination outage. RN outage refers to the case where the RN detects an erroneous message. Forwarding this erroneous message to the destination will corrupt rather than help the final joint decoding, leading to unsuccessful reception. The probability of this kind of outage event is given by

$$\Pr(\text{relay outage}) = \Pr\left(R^1 < \frac{W}{T}\right) \quad (3.40)$$

$$= 1 - exp\left(-\frac{N_0 B K_t d_1^\varsigma}{P_s}\left(2^{\frac{W}{\alpha TB}} - 1\right)\right).$$

Destination outage accounts for the events that the RN decodes and forwards correct information but the destination detects errors. The probability is

$$\Pr(\text{destination outage}) = \Pr\left(R^1 \geq \frac{W}{T}, R^2 < \frac{W}{T}\right) \quad (3.41)$$

$$= \Pr\left(R^1 \geq \frac{W}{T}\right)\Pr\left(R^2 < \frac{W}{T}\right),$$

where R^1 is a function of γ_1, and R^2 is a function of γ_0 and γ_2, therefore independent with R^1. These two categories of outage events are mutually exclusive and the overall outage probability is given as

$$p^{out} = \Pr(\text{relay outage}) + \Pr(\text{destination outage}). \quad (3.42)$$

In order to calculate $\Pr(R^2 < W/T)$, we can also use the Laplace inversion of characteristic function method. The optimization takes the same format as AF in equation.

CF: Following the same line of argument, the overall energy efficiency of CF is

$$E(E_{b,s}, E_{b,r}, T) = \frac{\frac{WE_{b,s}}{\alpha} + (1-\alpha)P_t T + (2-\alpha)P_{ct}T + (1+\alpha)P_{cr}T}{W}.$$

$$(3.43)$$

In contrast to AF or DF, rather than forwarding a W bit message during phase 2, the RN forwards its compressed observation to be reconstructed by the destination. The compression rate determines the number of bits required for reconstruction, denoted by S_c. Equation (3.18) gives the constraint of the maximum bits that can be successfully transmitted in the RN to destination link

$$S_c = (1 - \alpha)BTR_0. \tag{3.44}$$

The above equation implies that the required number of bits S_c is a function of the instantaneous qualities of the RN to destination and source to destination links. The overall energy efficiency can be given as

$$E(E_{b,s}, E_{b,r}, T) = \frac{E_{b,s}}{\alpha} + \frac{S_c E_{b,r}}{W} + \frac{(2 - \alpha)P_{ct}T + (1 + \alpha)P_{cr}T}{W}, \tag{3.45}$$

where $E_{b,r} = (1-\alpha)P_r T / S_c$. Note that with fixed transmission energy per bit, the transmission power is fixed at the source but adaptively adjusted because Sc is based on current channel condition at the RN.

The outage probability can be easily obtained as

$$p^{out} = \Pr\left(R_{CF} < \frac{W}{T}\right), \tag{3.46}$$

which can also be approximated by the same method discussed previously.

Hybrid Schemes: Depending on the decoding status at the RN, DF or CF is employed in hybrid relaying. When the RN detects error, it compresses and forwards its reception and the energy efficiency takes the form of equation (3.45). When RN's decoding is successful, DF is envisaged and the energy efficiency is given by (3.39). The average energy efficiency is the weighted summation of (3.39) and (3.45)

$$E(E_{b,s}, E_{b,r}, T) = \Pr\left(R^1 < \frac{W}{T}\right)\left(\frac{E_{b,s}}{\alpha} + \frac{S_c E_{b,r}}{W} + \frac{(2-\alpha)P_{ct}T + (1+\alpha)P_{cr}T}{W}\right)$$

$$+\Pr\left(R^1 \geq \frac{W}{T}\right)\left(\frac{E_{b,s}}{\alpha} + E_{b,r} + \frac{(2-\alpha)P_{ct}T + (1+\alpha)P_{cr}T}{W}\right)$$

$$= \frac{E_{b,s}}{\alpha} + \frac{(2-\alpha)P_{ct}T + (1+\alpha)P_{cr}T}{W} + \left(\Pr\left(R^1 < \frac{W}{T}\right)\left(\frac{S_c}{W} + 1\right)\right)E_{b,r}. \tag{3.47}$$

The outage probability in the CF mode and DF mode are given, respectively, as

$$p_{CF}^{out} = \Pr\left(R^1 \leq \frac{W}{T}, R_{CF} < \frac{W}{T}\right), \tag{3.48}$$

$$p_{DF}^{out} = \Pr\left(R^1 \geq \frac{W}{T}, R^2 < \frac{W}{T}\right).$$

Considering two modes are mutually exclusive, the overall outage probability is

$$p^{out} = p_{CF}^{out} + p_{DF}^{out}. \tag{3.49}$$

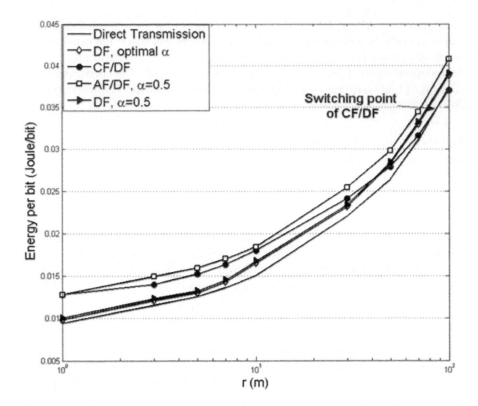

Figure 3.4: Energy per bit in shot range.

3.2.4   Insights and Discussions

We optimize the energy per bit for both the source and relay and compare the energy efficiency performance of various strategies. We assume that the source, RN and the destination are placed in a straight line. The distance between the source and the destination is r and the RN is in the middle with distance $0.9r$ from the source and $0.1r$ from the destination. We use the typical power model settings in [41]: $W=10$, $P_{ct}=98$mW, $P_{cr}=112.4$mW, $K_t=6.05\times10^9, \zeta=3$, $N_0=-171$dBm/Hz, $B=1$ and $T_{max}=10$.

Fig. 3.4 shows the situation where the S-D distance is short ($r \leq 100$m). The direct transmission is more efficient than any of the cooperative strategies until the S-D distance r is around 90m. The reason is that in small distance the circuitry energy consumption is the major part of the overall energy consumption. If the relay is activated, although the transmission energy per bit can be reduced, the overall circuitry consumption is almost doubled. Thus, to optimize the energy consumption, the relay should be de-activated to save the energy. When the S-D distance is increased ($r >100$m) in Fig. 3.5, circuitry energy consumption becomes minor and the relay is able to help to reduce

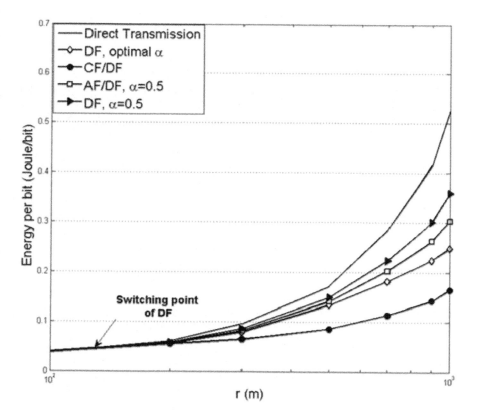

Figure 3.5: Energy per bit in long range.

the overall energy per bit more efficiently. Among all the relaying strategies, the CF/DF based strategy has the best energy efficiency performance. When is fixed, the AF/DF based strategy shows better energy efficiency than the DF based one. We can name, for each cooperative strategy, the point where it shows better energy efficiency performance than direct transmission as the switching point. As shown in Fig. 3.5, the switching point of the hybrid CF/DF strategy is much smaller than other strategies.

Based on these results, we can conclude that the hybrid relay system that enables a pair of terminals (relay and destination) to exploit spatial diversity shows significant improvement in energy efficiency performance in terms of consumed energy per bit. However, compared with direct transmission, the cooperative strategy only shows improved energy efficiency when the destination is not very close to the source. Furthermore, it should be emphasized that this conclusion is highly dependent on the energy consumption model adopted. These issues will be considered in the next sections.

3.3 H-ARQ Relaying and H^2-ARQ Relaying

In this section, relaying schemes will be investigated in a packet-oriented data communication system, where retransmission protocols are needed. We analyze the Hybrid Automatic Repeat re-Quest (HARQ) protocols used in conjunction with non-hybrid and hybrid relaying schemes from the energy efficiency perspective. If the relaying schemes are non-hybrid, this combined strategy is named H-ARQ relaying; otherwise, it is called H^2-ARQ relaying since both the retransmission protocols and relaying schemes are hybrid.

Despite the exploitation of an efficient cooperative relaying strategy the transmitted packet might be lost due to the instantaneous channel condition and noise realization. The packet loss could be even more severe when the system is operating under static (block) fading condition and the transmitter is not able to properly tune its transmission rate due to the lack of sufficient level of channel knowledge. Retransmission techniques based on automatic repeat request, i.e., ARQ [50–52] and its advanced hybrid types that combine forward error correction (FEC) with ARQ by keeping previously received packets for detection, commonly known as hybrid ARQ (HARQ), will be the natural choices to circumvent this problem and guarantee correct data packet delivery to the final destination. Common encoding techniques for HARQ are repetition coding (RC) which chase combing and unconstrained coding (UC) with incremental redundancy (INR), respectively [53–55]. The emphasis of this section is on HARQ, specifically INR as it is capable of offering higher throughput [56]. As the repetition coding based HARQ performs weaker than INR and the extension of the presented analysis to repetition coding is a straightforward practice, it will not be considered in this section.

Being a technology that spans both the MAC and physical (PHY) layers, HARQ protocols are of great interest. Some studies of HARQ protocols regarding throughput analysis, error rate and average delay in two-node communications can be found in [56–57]. The combination of cooperative communications and HARQ has been investigated in [58–61]. In [58], the diversity-multiplexing-delay tradeoff was analyzed and two kinds of diversity including space and ARQ diversity were exploited. Lin Dai studied the application of adaptive cooperation with ARQ in [59], where the relay will not be involved in the cooperation if errors are detected. The work has been further extended to HARQ in [62–63] and [9]. In [9], the energy consumption of a DF based HARQ system is analyzed. However, in these previous works, HARQ is deployed with a non-hybrid forwarding scheme, usually DF, and hence referred to as H-ARQ-Relaying. As shown by previous results, we know that the outage behavior can be greatly improved if the relay adaptively switches between different forwarding schemes. Further improvement can be expected by combining a hybrid relay system with HARQ. This combined strategy is called H^2-ARQ-Relaying since both the retransmission protocols and the relaying schemes are hybrid. In [62], the HARQ strategy is deployed in a relay system where the relay is allowed to switch between AF and DF. Compared with

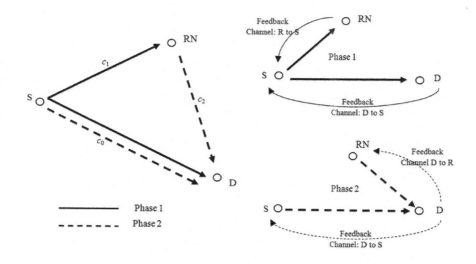

Figure 3.6: Feedback channel.

conventional HARQ strategies based on DF only, the new relay system is able to enjoy a certain level of flexibility and exhibits significant improvement in FER.

In the next section, we focus on the hybrid CF/DF based H²-ARQ-Relaying strategy, although the AF/DF one is also addressed briefly. Only incremental redundancy (INR) based HARQ is considered and the proposed strategies and their spectrum and energy efficiency analysis are the main topics.

3.3.1 H²-ARQ Relaying Strategy

When retransmission protocols are considered, the same message is transmitted until the destination successfully decodes it or the maximum retransmission limit is reached. In the l-th retransmission the signals received by the RN and the destination during the source broadcasting phase are

$$y_{r,l}[i] = \frac{c_{1,l}}{\sqrt{K_t}d_1^{\zeta/2}}x_l^1(w)[i] + z_{r,l}[i], \qquad (3.50)$$

$$y_{d,l}^1[i] = \frac{c_{0,l}}{\sqrt{K_t}d_0^{\zeta/2}}x_l^1(w)[i] + z_{d,l}[i],$$

respectively, where $1 \leq l \leq L$ and L is the maximum retransmission limit. As depicted in the right part of Fig. 3.6, the feedback channels are used to convey the decoding status at the relay and the destination.

If the message is successfully decoded by the destination, where the decoding is based on all previously buffered signals, an acknowledgment (ACK) message is sent back to the source and the RN and the transmission of message w is finished. If the destination detects errors, it broadcasts a not acknowledgment (NAK) message. Once the relay receives the NAK message, it tries to decode. Different actions are envisaged based on its decoding status:

Relay Decoding Success: An ACK is broadcast by the RN to indicate that it will employ the DF mode. The source and the RN then transmit $x_l^2(w)$ and $x_{r,l}(w)$, respectively, and the destination receives

$$y_{d,l}^2[i] = \frac{c_{0,l}}{\sqrt{K_t}d_0^{\zeta/2}}x_l^2(w)[i] + \frac{c_{2,l}}{\sqrt{K_t}d_2^{\zeta/2}}x_{r,l}[i] + z_{d,l}[i]. \qquad (3.51)$$

Afterwards, the source keeps silent during phase 1 of the following frames to save energy. It only transmits with the RN in a more efficient cooperative manner during phase 2 until the source receives ACK or the retransmission limit N is reached.

Relay Decoding Failure: The relay switches to the CF mode and broadcasts a NAK message. At the end of phase 2, the destination attempts to decode by joint processing of current and previous receptions. Upon successful decoding, an ACK is sent back; otherwise, the destination issues a NAK message to start a new frame if $l < L$. If $l = L$, HARQ failure is announced. In this H²-ARQ-Relaying strategy, we need to consider five possible scenarios as follows:

1. Case 1: the information message w is successfully decoded by the destination at the end of phase 1 in the l-th frame. An implication of this scenario is that in previous l-1 transmission, the RN cannot decode and CF is conducted.

2. Case 2: the RN cannot decode but successful decoding occurs at the destination after l CF operations.

3. Case 3: the DF mode is activated in the i-th frame and the destination correctly decodes w in the l-th frame.

4. Case 4: after maximum retransmission, neither the RN nor the destination can decode.

5. Case 5: the DF mode is activated in the i-th frame but the destination cannot correctly decode w after L retransmission.

State diagram composed of a finite number of states can be used to describe the behavior of a system and analyze and represent the events of the system. The H²-ARQ-Relaying can also be represented by a state diagram presented in Fig. 3.7 to demonstrate the process. B_l stands for the state where the source is ready to broadcast w in the l-th frame. The state D_l indicates that the DF mode is activated at the end of phase 1 of the l-th frame. $DF_{m,l}$ is the state where the system has entered the DF mode in the l-th frame and cooperative transmission in phase 2 has been repeated m times but the destination still

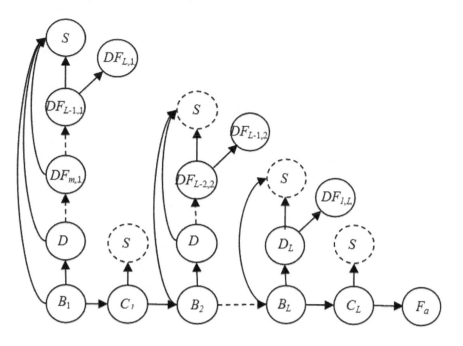

Figure 3.7: State diagram.

detects errors. C_l defines the state in which the relay is ready to conduct CF during phase 2 of the l-th frame. S and F_a are the successful decoding state and HARQ failure state, respectively.

3.3.2 Performance Analysis

Each state in Fig. 3.7 represents a different, transmission rate achieved and amount of energy expenditure consumed. In order to investigate the average behavior in terms of spectrum and energy efficiency for the H²-ARQ-Relaying strategies, we first need to derive the state and transition probabilities of the state diagram. The outage probability introduced and derived in the previous section will help us to find the state and transition probabilities.

Caire has pointed out that the INR based HARQ sends additional coded symbols (redundancy) until successful decoding is achieved [56]. After l transmission, the achievable rate at the destination is expressed as

$$R_d(l) = \sum_{i=1}^{l} R_{d,l}, \qquad (3.52)$$

where $R_{d,l}$ stands for the achievable rate in the l-th frame and can either take the form of Equation (3.12) or (3.16) depending on whether DF or CF operation is conducted. Similarly, the achievable rate at the RN is

$$R_r(k) = \alpha B \sum_{i=1}^{k} log_2 \left(1 + \frac{|c_{1,i}|^2 P_s}{N_0 B K_t d_1^\varsigma} \right). \tag{3.53}$$

We define the outage as the event $A_{l,k} = R_d(l) < W/T, R_r(k) < W/T$. The outage probability $\Pr(A_{l,k})$ can be calculated by resorting to the 2-dimensional characteristic function of $R_d(l)$ and $R_r(k)$,

$$\Psi(\mathbf{s}, l, k) = \mathbf{E} \left\{ exp\left(-s_1 R_d(l) - s_2 R_r(k) \right) \right\} \tag{3.54}$$

$$= \mathbf{E} \left\{ \prod_{i=1}^{k} exp\left(-s_1 R_{d,i}(l) - s_2 R_{r,i}(k) \right) \prod_{i=k+1}^{l} exp\left(-s_1 R_{d,i}(l) \right) \right\},$$

where $\mathbf{s}$ is (s_1, s_2). The outage probabilities can be expressed by the Laplace inversion formula of $\Psi(\mathbf{s}, l, k)$ and approximated as [49],

$$\Pr(A_{l,k}) = \frac{1}{(2\pi j)^2} \int_{d_2-j\infty}^{d_2+j\infty} \int_{d_1-j\infty}^{d_1+j\infty} \frac{\Psi(\mathbf{s}, l, k) e^{\frac{s_1 W}{T} + \frac{s_2 W}{T}}}{s_1 s_2} ds_1 ds_2 \tag{3.55}$$

$$\approx \sum_{i=1}^{M} \sum_{j=1}^{M} K_i K_j \frac{\Psi(\mathbf{z}, l, k)}{z_i z_j}, \tag{3.56}$$

where $\mathbf{z}$ is $(z_i T/W, z_j T/W)$. This approximation can be easily extended to the AF/DF-based strategy, where CF is replaced by AF.

For two events sets e_1 and e_2, if $(e_1 \subseteq e_2)$, $\Pr(e_1, e_2)$ is equal to $\Pr(e_1)$. In the state diagram, the transition probability to state j from any of its adjacent incoming state i is

$$\Pr(state\ i \to state\ j) = \Pr(state\ j | state\ i) = \frac{\Pr(state\ j)}{\Pr(state\ i)}. \tag{3.57}$$

In the state diagram, all the states have this property except the state S. Therefore we can first calculate the state probabilities and then obtain the transition probabilities based on (3.57). For instance, $\Pr(B_l)$ can be easily obtained as $\Pr(A_{l-1,l-1})$. Other state probabilities can be obtained by manipulating $\Pr(A_{l,k})$ as well. With state probabilities, transition probabilities can be further derived. One useful trick to reduce the complexity is to use the fact that the probabilities of all the state transitions emanating from a single state must add up to 1. For instance, we have

$$\Pr(B_l \to S) = 1 - \Pr(B_l \to D_l) - \Pr(B_l \to C_l). \tag{3.58}$$

The spectrum efficiency is evaluated by the expected throughput in terms of bit/second. Unlike previous sections, here we use total amount of bits delivered per unit energy expenditure to evaluate the energy efficiency to give a fresh view.

Using the state and transition probabilities, we can apply the renewal-reward theorem [56] to evaluate the throughput by investigating the random reward Φ and average airtime T_{air}. In case 1, CF is used in previous l-1 transmission, the total airtime is clearly $(l-1)T$ and the total energy consumption is $(l-1)WE(E_{b,s}, E_{b,r}, T)$, where $E(E_{b,s}, E_{b,r}, T)$ takes the form of (3.45). The l-th frame consists of a source broadcasting phase only. The airtime and energy consumption are αT and E^1, respectively. The overall average time and energy expenditure associated with case 1 is given by

$$T_1 = T \sum_{l=1}^{L} P_{BS}(l)\left((l-1)+\alpha\right), \tag{3.59}$$

$$E_1 = \sum_{l=1}^{L} \left((l-1)E_{CF}W + E^1\right),$$

respectively, where E_{CF} and E^1 are given in (3.45) and (3.37), respectively, and $P_{BS}(l) = \Pr(B_l)\Pr(B_l \to S)$. Following the same line of argument, the average airtime and energy expenditure can also be derived for the other four cases. For the purpose of conciseness, their expressions are not shown in this book but readers can refer to [12] for details.

The average airtime is the weighted summation

$$\mathbf{E}\{T_{air}\} = \sum_{i=1}^{5} T_i. \tag{3.60}$$

The average reward is

$$\mathbf{E}\{\Phi\} = (1 - p^{out})W, \tag{3.61}$$

where p^{out} is the probability that the decoding is still unsuccessful after maximum retransmission and given by

$$p^{out} = \Pr(F_a) + \sum_{N=1}^{L} \Pr(DF_{L-N+1,N}). \tag{3.62}$$

The average throughput is

$$\eta = \frac{\mathbf{E}\{\Phi\}}{\mathbf{E}\{T_{air}\}}. \tag{3.63}$$

The total amount of bits delivered is W and bits to energy expenditure ratio reads

$$C_J = \frac{W}{\sum\limits_{i=1}^{5} E_i}. \tag{3.64}$$

Apparently, C_J is still a function of $T, E_{b,s}$, and $E_{b,r}$ and it can be optimized subject to the outage probability and transmission time constraints as we did in the previous section. However, due to the complexity of the problem, this optimization problem has to be solved numerically.

3.3.3   Insights and Discussions

We provide some results for the proposed strategies and compare their performance with some benchmark ones. The maximum retransmission limit L is set to 4. The benchmark strategies and their abbreviations are listed as follows:

1. H-ARQ with direct transmission (DT): The relay keeps silent throughout the transmission.

2. H-ARQ-Relaying with the co-located relay and destination (CRD): We assume that the relay and the destination are connected by a wire such that full receive diversity is achieved.

3. H-ARQ-Relaying with conventional DF (DF): When the relay detects errors, it keeps silent during phase 2; otherwise, the relay and the source transmit simultaneously.

4. H²-ARQ-Relaying with hybrid AF/DF (HAD): The relay performs AF when detecting errors and DF when successfully decoding. Note that in this case, α is also fixed at 0.5 and repetition coding is used when AF is performed.

The CF/DF-based strategy is denoted as HCD. The source and the destination are assumed to be placed in the foci of the ellipse $(d/r)2/(e/2)2 + (y/r)2/(b/r)2 = 1$, for $1 < e < +\infty$. The relay is moving along the ellipse. With this assumption, we are able to investigate the relay system's performance when the relay is not on the line segment between the source and the destination. The same power model is used here.

Comparison of Spectrum Efficiency

Fig. 3.8 gives the throughput of strategies with respect to d/r. Basically, the throughputs are upper bounded by CRD because it is able to enjoy full receive diversity. The CF/DF-based strategy is the closest one to CRD. It can be interpreted as follows: when the relay is moving toward the destination, the S-RN link is getting less reliable and the successful decoding probability at the relay decreases. In such a scenario, if the system uses DF, it is more likely that the relay's decoding fails and the system operates in direct transmission mode during phase 2, thus no diversity can be achieved. In contrast, when the hybrid CF/DF scheme is used, the system will be able to achieve receive diversity through CF. As far as HAD is concerned, it enjoys a certain level of flexibility and its throughput is improved in comparison with DF when α is fixed at 0.5. However, when the S-RN link is of low quality, the relay forwards nothing but mostly its own noise. In addition, HAD suffers from the bandwidth loss due to the use of repetition coding, and its throughput is therefore lower than DF with optimal α. Another common trend of strategies is that when the relay

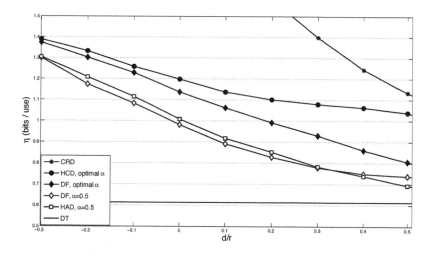

Figure 3.8: Spectrum efficiency.

moves away from the source, their throughputs decrease. For *DF*, this can be explained by the smaller decoding probability of the relay, which causes fewer opportunities for the relay to cooperate. For *HCD*, although CF can be used, the achievable rate depends on the quality of the S-D and S-RN links. The average SNR of the S-RN link is decreasing when the relay moves toward the destination. Hence, the throughput reduces. The AF/DF-based strategies can be explained following the similar argument.

Comparison of Energy Efficiency

We optimize the energy per bit for both the source and the relay and compare the energy efficiency performance of various strategies when the relay is close to the destination (the S-D link and RN-D link are vertical to each other). Fig. 3.9 shows the situation where the S-D distance is from short ($r <100$m) to long (100m$< r <1000$m). The direct transmission is more efficient than any of the cooperative strategies in terms of the energy per bit in short range. The reason has been explained before. Thus, to optimize energy consumption, the relay should be de-activated to save energy. The DF-based strategy consumes less energy than others because in this strategy, the relay is silent when detecting errors and the circuitry energy consumption is saved, while in the hybrid strategies, the relay still transmits, leading to extra circuitry energy consumption. When the S-D distance is increased ($r >100$m) in Fig. 3.10, circuitry energy consumption becomes minor and the relay is able to help to reduce the overall energy per bit. The CF/DF-based strategy has the best energy efficiency performance.

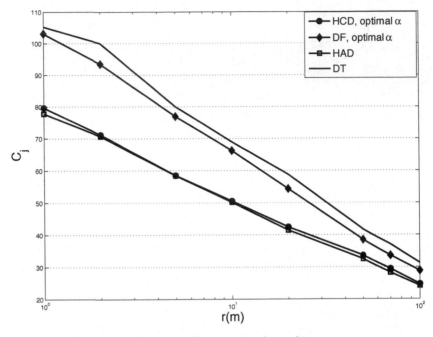

Figure 3.9: Energy efficiency in short distance.

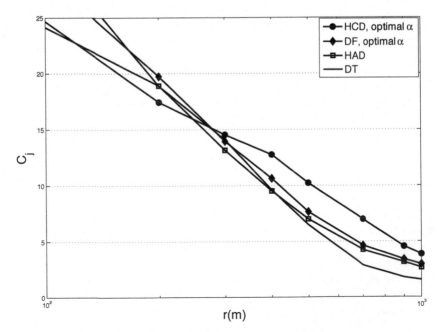

Figure 3.10: Energy efficiency in long distance.

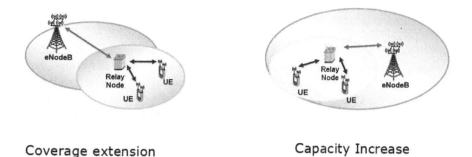

Coverage extension   Capacity Increase

Figure 3.11: RNs in cellular networks.

In this section, we propose novel Hybrid Automatic Repeat re-Quest (HARQ) strategies used in conjunction with hybrid relaying schemes, named H^2-ARQ-Relaying. The strategies allow the relay to dynamically switch between amplify-and-forward/compress-and-forward and decode-and-forward schemes according to its decoding status. The spectrum efficiency of the proposed strategies, in terms of the maximum throughput, is significantly improved compared with their non-hybrid counterparts under the same constraints. The consumed energy per bit is optimized by manipulating the node activation time, the transmission energy and the power allocation between the source and the relay. Numerical results lead to the same conclusion that cooperative HARQ is energy efficient in long distance transmission only.

3.4 Energy Efficient RNs in Cellular Networks

With the previous discussion as basis, this section will study the energy saving potential of relaying schemes in cellular systems. RNs are added for incremental capacity growth, richer user experience and in-building coverage. In particular, RNs are deployed as complementary sites to Macro BSs to improve the cell edge performance, which is one of the main challenges faced by the developing standards. They are normally offering flexible site acquisition with low power consumptions and, using over-the-air link as backhaul connection to the base stations, providing coverage extension and capacity enhancement with little to no incremental backhaul expense as shown in Fig. 3.11. They are especially suitable for the scenarios where terrestrial condition is too harsh for wired backhaul connections.

The first commercial standard incorporating relaying technology is IEEE.801.16j [36]. The relay group developed new BS and RN capabilities to enable relay networks to be realized and provide support for access by legacy devices. Two different operating modes are provided: transparent mode and non-transparent mode. LTE-A is also considering using relaying technology for cost-efficient throughput enhancement and coverage extension and more sophisticated relaying strategies are being incorporated [38].

RNs cover much smaller areas than macro BSs. They may have additional transmission power compared to terminals and yet be much lower compared to a base station because of their limited functionality and lower transmission power. Thus a more energy efficient power consumption model is expected for RNs and as a consequence they can be a promising solution for "green" cellular networks. However, as we discovered in previous results, relay is not always able to help and to what level it can help is highly dependent on the power model, RN's location and number of RNs deployed. The objective of optimizing the usage of RNs needs to answer at least two questions:

1. How many RNs are needed to minimize the energy usage?

2. What are the optimum locations for these relay nodes in order to reduce the energy consumptions?

The rest of the section will address these two questions first and extend our study to other interesting topics such as the comparison of indoor and outdoor relay applications.

3.4.1   Cellular System and Power Model

Before we can investigate the relay-assisted cellular system, we need to introduce the basic concept of a cellular network as well as introducing a well defined power model. A cellular network is defined as a radio network distributed over cells which are joined together to provide radio coverage over a wide geographic area. In this chapter, each cell consists of three sectors and each sector is defined as a hexagon with radius R as shown in Fig. 3.11. The BSs are located in the center of each cell and consist of three directional antennas, each serving a different sector of the cell. The antenna pattern is given as [64]

$$G(\Theta) = G_{max} - min\left\{ 12\left(\frac{\Theta}{\Theta_{3dB}}\right)^2, G_{f2b}\right\} \tag{3.65}$$

$$G_{max} = 14dBi, \Theta_{3dB} = 65°, G_{f2b} = 20dB,$$

where G_{max} is the boresight antenna gain, Θ is the angle between the sector and the mobile (BS-UE) line of sight and the sector boresight, Θ_{3dB} is the 3 dB angle, also defined as beam-width in this work, and G_{f2b} is the antenna front to back ratio.

The RNs are placed in some specific positions as shown in Fig. 3.12. There could be multiple RNs in each sector. In such a case, some interference coordination mechanism between RNs might need to be considered. We assume that each RN has an omni-directional antenna.

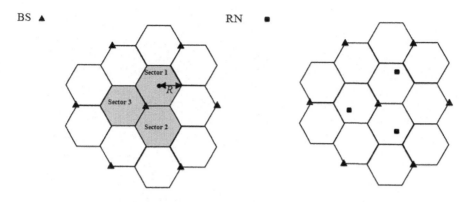

Figure 3.12: Cellular system.

Normally, the received signal from a BS suffers from path-loss, slow fading and fast fading. A simple model taking path-loss into consideration is given as [64]

$$PL(dB) = 15.3 + 37.6 log_{10}(d), \qquad (3.66)$$

where d is the BS-UE distance in meters. The shadowing $SF(dB)$ follows a log-normal law with mean 0 and shadowing standard deviation SD. Based on this propagation model, the long-term power received by a UE from a BS (BS_i) can be expressed in dB as

$$P(BS_i \rightarrow UE) = P_{Tx} + G_{UE} + G - PL(d) - SF(BS_i \rightarrow UE), \qquad (3.67)$$

where P_{Tx} is the transmission power of the BS, G is the antenna gain, d is the BS-UE distance and $SF(BS_i \rightarrow UE)$ is the (correlated) shadowing in dB between the BS and the UE. We can introduce coverage defined as the fraction of cell area where the received power is above a certain threshold. However, in this work, we use another commonly used metric of interest G_{factor} to evaluate coverage

$$G_{factor} = \frac{P(BS_i \rightarrow UE)}{\sum\limits_{j \neq i} P(BS_j \rightarrow UE) + P_{therm}}, \qquad (3.68)$$

where $P(BS_i \rightarrow UE)$ is given in mWatts and P_{therm} is the thermal noise power given in mWatts. The long-term coverage is defined as

$$Cov = \frac{1}{S_a} \int\limits_{S_a} \Pr\left\{G_{factor}(x,y) \geq G_{factor,min}\right\} dx dy, \qquad (3.69)$$

where S_a is the area of a sector.

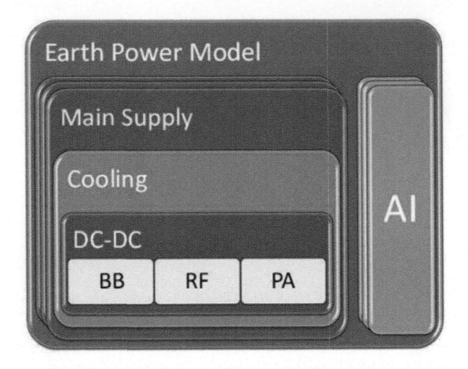

Figure 3.13: Power model.

An accurate power model is essential to evaluate the energy efficiency of a system. Since the energy consumption estimation mainly involves BSs and RNs, we set up two different power models for each type of node. A general power model has been given in [65], where a high-level block diagram with the main radio sub-systems is defined and the power consumption of each sub-system is calculated individually.

As shown in Fig. 3.13, the power consumption is stemming from multiple sub-systems including a lossy Antenna Interface (AI), a Power Amplifier (PA), a Radio Frequency (RF) small-signal transceiver section and a baseband interface (BB), a DC-DC power supply regulation, an active cooling system and finally a main AC-DC power supply for connection to the electrical power grid. The overall power consumption can be broken down in sub-system level to illustrate each sub-system's influence more precisely.

A state-of-the-art energy consumption estimation for a typical commercial BS is given in Table 3.1 as well as the power consumption breakdown Fig. 3.14.

It is interesting to note that in Macro BSs it is mainly the PA that dominates the total power consumption, owing to the high antenna interface losses.

The RN power consumption model is established in [66] and given in Table 3.2. The power breakdown Fig. 3.15 shows that the main source for power

Table 3.1: Macro BS Power Model

Macro			
PA	Max Transmit rms Power	[dBm]	46.0
	Max Transmit rms Power	[W]	39.8
	PAPR	[dB]	8.0
	Peak Output Power	[dBm]	54.0
	Pdc	[W]	128.2
	Power-Added Efficiency	[%]	31.1
TRX	Max Transmit rms Power	[dBm]	-8.0
	TX Pdc	[W]	6.8
	RX Pdc	[W]	6.1
	Total Pdc	[W]	13.0
BB	Radio[inner rx/tx]	[W]	10.8
	LTE turbo [outer rx/tx]	[W]	8.8
	Processors	[W]	10.0
	Total Pdc	[W]	29.5
DC-DC	loss	[%]	8.0
	Pdc	[W]	13.7
Cooling	loss	[%]	12.0
	Pdc	[W]	22.1
Main Supply	loss	[%]	9.0
	Pdc	[W]	18.6
Total 1 Radio		[W]	225.0
Number of Sectors		#	3.0
Number of PAs		#	2.0
Total N Radio		[W]	1350.0

consumption is still the PA. Generically speaking, RNs have significantly lower transmission power, and thus consume much less energy compared to macro BSs. The well-timed power model will form the basis for the energy efficiency study afterwards.

3.4.2 Optimization of RN Deployment

The link level results have revealed that the locations of RNs play an important role in the overall energy efficiency. In cellular systems, the number of RNs deployed in one sector is also an interesting topic worthy of being studied. Most of the analysis in this chapter focuses on downlink communication. Still communication takes place in two orthogonal phases. In the first phase, the BS transmits while the RN and the UEs receive, and in the second phase the BS and the RN transmit while the UEs receive. We assume that the phases are

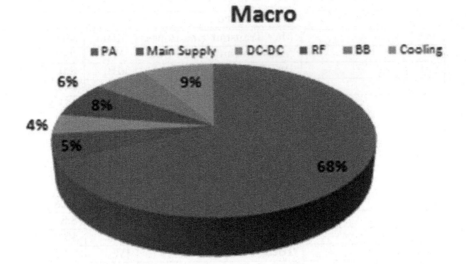

Macro

Figure 3.14: Macro power model breakdown.

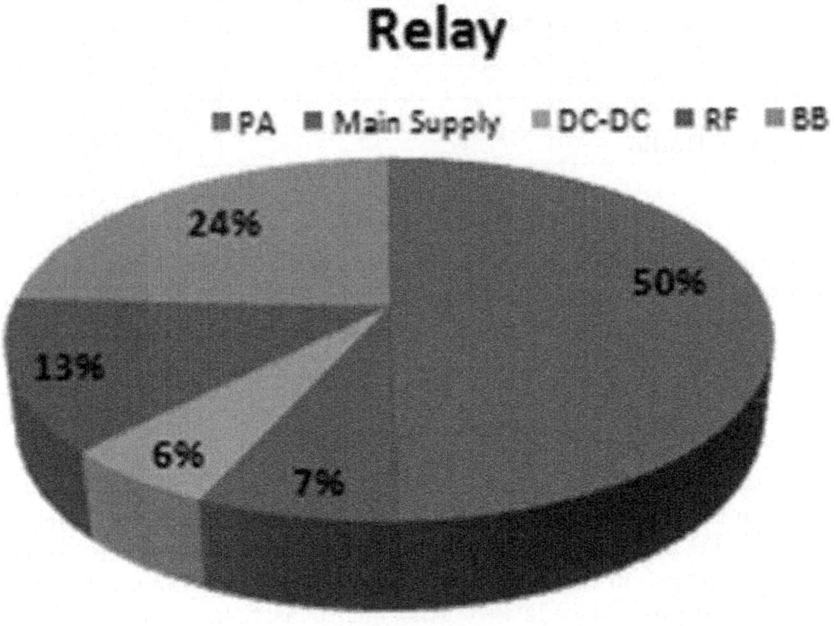

Relay

Figure 3.15: RN power model breakdown.

Table 3.2: RN Power Model

RN				Rural	Urban
PA	Max Transmit rms Power	[dBm]		37.0	30.0
	Max Transmit rms Power	[W]		5.0	1.0
	PAPR	[dB]		8.0	12.0
	Peak Output Power	[dBm]		45.0	42.0
	Pdc	[W]		14.1	2.8
	Power-Added Efficiency	[%]		35.6	35.6
TRX	Max Transmit rms Power	[dBm]		-14.0	-21.0
	TX Pdc	[W]		1.6	0.6
	RX Pdc	[W]		2.2	0.9
	Total Pdc	[W]		3.8	1.5
BB	Radio[inner rx/tx]	[W]		2.3	2.3
	LTE turbo [outer rx/tx]	[W]		2.0	2.0
	Processors	[W]		2.5	2.5
	Total Pdc	[W]		6.8	6.8
DC-DC	loss	[%]		6.4	6.4
	Pdc	[W]		1.6	0.7
Cooling	loss	[%]		0.0	0.0
	Pdc	[W]		0.0	0.0
Main Supply	loss	[%]		7.7	7.7
	Pdc	[W]		2.0	0.9
Total 1 Radio		[W]		28.3	12.8
Number of PAs		#		2.0	2.0
Total N Radio		[W]		56.5	25.5

synchronized so that the first phase and second phase occur simultaneously in all cells.

Assuming that all BSs transmit at the same time and frequency with maximum power, and that the cellular architecture is such that each cell sees the same interference, i.e., neglecting network edge effects, we can focus on a single sector of a single cell. We start from the scenario that there is only one RN associated with each sector and extend out discussion to multiple RNs afterwards.

We start from single RN case. During the first phase, the RN in sector j receives (time indices of the symbols are removed for ease of notation)

$$y_{r,j} = h_j x_{s,j}^1 + \sum_{k \in \Omega, k \neq j} h_k x_{s,k}^1 + z_r, \qquad (3.70)$$

where h_j is the sector j to RN channel (the path-loss, shadowing, antenna gain, etc., are absorbed for ease of notation), $x_{s,j}$ is the transmitted signal from sector j, z_r is the additive Gaussian thermal noise at RN, Ω is the set of all interfering sectors, and superscript stands for phase. Here the second term actually represents the summation of two parts: the intra-cell interference from other sectors within the same BS and inter-cell interferences from the sectors of other BSs. They can be treated in the same way. It should be noted that we do not take RNs in other sectors as interfering nodes because their transmission power is very low compared to the BSs and the interference generated can be ignored. We further assume that only one UE is served by RN in each sector; the received signal is

$$y^1_{UE,j} = h_{UE,j}x^1_{s,j} + \sum_{k \in \Omega, k \neq j} h_{UE,k}x^1_{s,k} + z_{UE}, \qquad (3.71)$$

where $h_{UE,j}$ is the sector j to UE channel and z_{UE} is the additive Gaussian thermal noise at UE. During phase 2, the BS and the RN transmit and the UE receives

$$y^2_{UE,j} = h_{UE,j}x^2_{s,j} + h_{UE,r}x_r + \sum_{k \in \Omega, k \neq j} h_{UE,k}x^2_{s,k} + z_{UE}, \qquad (3.72)$$

where $h_{UE,r}$ is the RN to UE channel. The UE then jointly processes $y^1_{UE,j}$ and $y^2_{UE,j}$ to decode the message, depending on the forwarding scheme employed by the RN. With a similar approach to that followed for the achievable rate derivation, and considering the power consumption model, in each point of the sector the Shannon Capacity and the energy consumption per bit of different relaying schemes can be evaluated. We define the average energy efficiency as

$$\mathrm{E}_{ave} = \frac{\text{total energy consumption}}{\text{average capacity within a sector}}. \qquad (3.73)$$

Without considering fast fading, the equivalent G_{factor} is defined as a function of Shannon Capacity

$$G_{fac,eq} = 2^{C/B} - 1, \qquad (3.74)$$

where C is the Shannon Capacity at one point of the sector. We assume the inter-cell distance is 2000 meters. The equivalent $G_{fac,eq}$ and energy consumption distributions are reported in Fig. 3.16 and 3.17, where DF is applied. If we set the coverage threshold as 0dB, with the RN, the coverage can be improved 8%. However, energy consumption is not reduced but rather increased by using relay. With a RN, the system's capacity can be improved and an extra amount of bits can be delivered. However, adding a RN to the network comes with a cost, i.e., extra energy consumption required for the RN. If the influence of the extra required energy outweighs the benefit, i.e., the improvement in capacity, the energy efficiency of the network is actually degraded. This is exactly what happened in our case.

However, in Fig. 3.17, the RN is just randomly deployed in the sector and only DF is considered. The link level results have indicated that the distance,

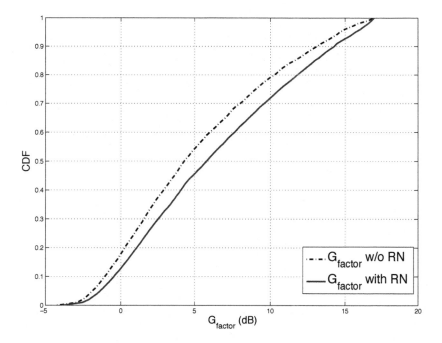

Figure 3.16: CDF of $G_{fac,eq}$.

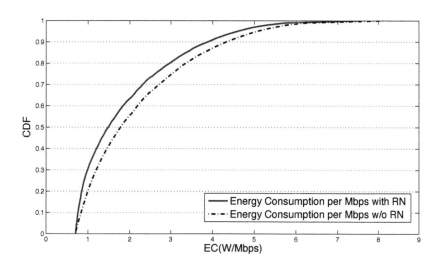

Figure 3.17: CDF of energy consumption per Mbps.

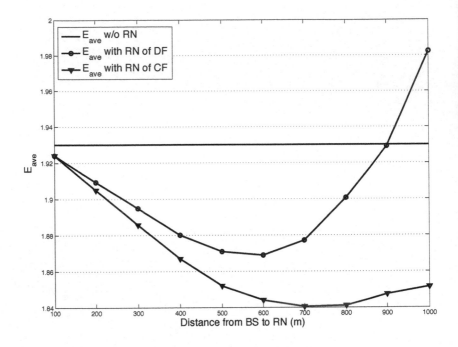

Figure 3.18: E_{ave} with single RN.

designated as d, between the source, i.e., BS, and the RN, plays a key role
in the overall system energy efficiency. Now we optimize the location of the
RN and investigate how the energy efficiency is changed with location and
forwarding schemes. Note that in a two dimensional cellular system, we need
to consider the angle between the BS and the RN line of sight and the sector
boresight as well, denoted as θ_r.

Fig. 3.18 shows E_{ave} when the RN is moving away from the BS toward
the sector edge along the direction with θ_r equal to 60 degrees. As we can see,
the location of the RN changes the system's capacity as well as the energy
efficiency. For a RN using DF, cell edge might not be a good option because
it would be difficult for the RN to decode in such a long distance and the
improvement of capacity is small. With extra energy required, the energy
efficiency is degraded compared to no RN case. The optimal location is in
the middle of the BS and the cell edge and, compared to no RN case, the
energy efficiency is improved. With hybrid relaying, the energy efficiency can
be further improved because the RN has the flexibility of choosing CF when
decoding is difficult. The UE is benefiting from receive diversity when CF is
used. The optimal location of hybrid relaying is close to the cell edge.

Now we move to multiple RN scenario. If phase synchronization is not achieved by the network, multiple RNs within the same sector can have different duplexing schedules which cause interference to each other when some RNs are in transmitting phase but others are in receiving phase. Even with phase synchronization, inter-RN interference can still exist when the same sub-carriers are used by different RNs. In this chapter, we assume that BSs uses some resource allocation schemes which guarantee the orthogonality of multiple RNs within one sector. Since we only consider one UE scenario for simplicity, the UE owns all available resources and the assumption does not make any significant difference. However, in multiple UE case, a central or distributed resource allocation strategy needs to be employed and each RN can only have limited resources.

Supposing N_r RNs within one sector, during phase 1, the BS multicasts to RNs and UE, the received signal at RN i is

$$y_{r,j \to i} = h_{j \to i} x^1_{s,j} + \sum_{k \in \Omega, k \neq j} h_{k \to j} x^1_{s,k} + z_{r,i}, \qquad (3.75)$$

where $h_{j \to i}$ is the sector j to RN i channel, $z_{r,i}$ is the additive Gaussian thermal noise of RN i for $1 \leq i \leq N_r$. At the same time, the UE listens to the BS and it reads (3.71). The second phase has multiple sources and a single destination. We consider the relay selection scheme (RSS). In RSS, the UE is associated with only one RN, denoted as active RN, which has the strongest RN to UE link and other RNs are in idle state. The active RN and BS transmit and the UE receives

$$y^2_{UE,j} = h_{UE,j} x^2_{s,j} + h_{UE,r,i} x_{r,i} + \sum_{k \in \Omega, k \neq j} h_{UE,k} x^2_{s,k} + z_{UE}, \qquad (3.76)$$

where $h_{UE,r,i}$ is the RN i to UE channel. Since the other RNs are not transmitting, they do not generate interference. The UE just acts in the same manner as it does in the single RN scenario. The selected RN can employ any forwarding scheme including a hybrid scheme. The capacity of a sector with two RNs is shown in Fig. 3.19.

As can be seen in Fig. 3.20, adding RNs can improve the capacity in the vicinity of the RN (here only path-loss is considered to give a clear picture). With multiple RNs, it is natural to spread the RNs evenly in a sector to balance the possibility of a UE served by a RN in the whole area. In this regard, we choose θ_r as

$$\theta_{r,i} = \frac{180°}{N_r + 1} i, \text{ for } 1 \leq i \leq N_r. \qquad (3.77)$$

We move the RNs from the BS toward the cell edge and show the average energy efficiency E_{ave} for N_r=2, 4 and 6.

Using N_r RNs does not necessarily mean that the required energy for RNs is increased for N_r times. As we mentioned, RN selection is applied in this multiple RN scenario. Only the UE associated RN is actively receiving and

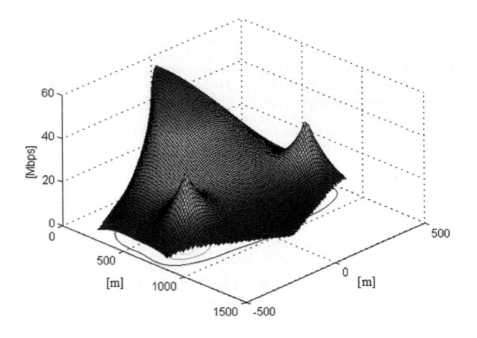

Figure 3.19: Capacity with two RNs.

transmitting while the other RNs are in the idle state. However, those RNs cannot be entirely switched off because a UE with mobility might need access at any time and thus some sub-systems are still working and consuming a certain amount of energy. The idle RN energy consumption can be obtained based on the RN power model and it will determine the optimal number of RNs to be deployed. As shown in Fig. 3.19, an increasing number of RNs is able to improve E_{ave} but the improvement is getting less. For instance, from 1 RN to 2 RNs E_{ave} is improved by about 5% but from 4 RNs to 6 RNs E_{ave} improvement is almost negligible. There are two consequences of deploying more RNs. One is that the system's capacity is improved and the other is that more energy is needed for those idle RNs. If the capacity improvement is more significant, E_{ave} can benefit; otherwise, it will suffer. Our results show that deploying more than 6 RNs in one sector might not be efficient from the energy and deployment cost points of view. The capacity improvement is very limited but the energy consumption is getting more and more significant.

3.4.3 Outdoor-to-Indoor Relaying

On one hand, the use of relay would tend to increase the overall network energy consumption; on the other hand, avoiding wasting a large amount of power for transmitting through walls could save a lot more energy. We investigate

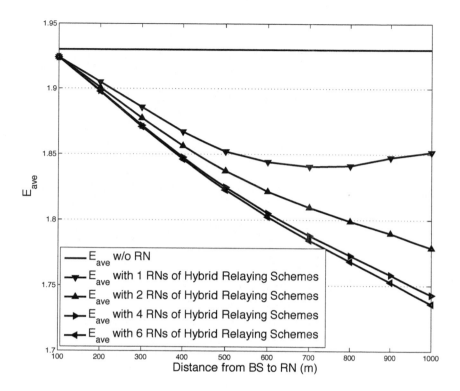

Figure 3.20: E_{ave} with multiple RNs.

here the EE of in-building relaying scenario where a non-regenerative relay is used for relaying the signal of an outdoor BS to an indoor UE and, hence, the access link (RN-UE link) is more reliable than the donor link (BS to RN link) since the RN is meant to be close to the UE. Assuming this scenario, we can derive a simple close-form approximation of the channel capacity by following a similar approach based on random matrix theory as in [67], utilize it for defining the EE of this AF system and, then, comparing its EE against traditional point-to-point (P2P) communication.

In this study, we consider a cooperative MIMO AF system that is composed of three nodes, i.e., a BS with n antennas, a nonregenerative RN with q antennas and a DN with r antennas. For the simplicity of the introduction, we assume a half-duplex relaying scenario with two phases of equal duration. In the first phase, the BS broadcasts its signal to the RN and UE; in the second phase, the RN, which acts as a repeater, transmits an amplified version of the BS signal to the UE. We also assume that the access link is far more reliable than the other two links. In our model, the SNRs of the direct

(BS-UE), donor and access links are defined as γ_0, γ_1 and γ_2, respectively, and $\sigma = \gamma_0/\gamma_1$ stands for the SNR offset between the direct and donor links. According to this system model, it has been indicated in [68] that the maximum achievable SE of the MIMO AF can be approximated as

$$C \approx \frac{1}{2\ln(2)}\left[n\ln\left(\frac{\gamma_0}{d_0}\right) + q\left(\ln(1 + \sigma d_0) + \frac{1}{1 + \sigma d_0} - 1\right)\right.$$
$$\left. + r\left(\ln(1 + d_0) + \frac{1}{1 + d_0} - 1\right)\right] \tag{3.78}$$

when the access link quality is 20 dB higher than the donor link, i.e. $\gamma_2 \gg \gamma_1$, and for large values of n, q and r. In addition, d_0 is the unique nonnegative root of the following polynomial

$$P(d) = d^3\lambda_1^2 + d^2[\lambda_1(\lambda_0\lambda_1(r + q - n) + \lambda_0 + \lambda_1)]$$
$$+ d[\lambda_0\lambda_1(1 + \lambda_0(r - n) + \lambda_1(q - n))] - n\lambda_0^2\lambda_1, \tag{3.79}$$

where $\lambda_i = \gamma_i/n$, $i = \{0, 1\}$. The main purpose of (3.78) is the evaluation and comparison of the capacity of in-building MIMO AF systems in a faster way than time consuming Monte-Carlo simulations, and with a sufficient accuracy such that it can be used in network simulation and optimization. In addition, it can provide upper bounds on the achievable rate of generic cooperative MIMO AF systems. As far as the total power consumption of this MIMO AF system is concerned, it can be characterized as

$$P_{\Sigma,AF} = P_{BS,Tr} + P_{RN,Tr} + P_{RN,Re} + 2P_{UT,Re}. \tag{3.80}$$

according to the two-phase transmission model, where $P_{BS,Tr}$, $P_{RN,Tr}$, $P_{RN,Re}$ and $P_{UT,Re}$ are the consumed power related to BS transmission, RN transmission, RN reception and UE reception, respectively. In [65], the total consumed power of several types of BS for both transmitting and receiving signals has been abstracted from real measurement (see Table 3.1 for macro BS) as

$$P_{BS} = t(\Delta_{P,BS}P_1 + P_{Ov,BS}), \tag{3.81}$$

where t is the number of transmit antennas at the BS, Δ_P accounts for the power amplifier (PA) inefficiency and P_0 is the overhead power, i.e., signal processing overhead, cooling and power supply (PS) losses as well as current conversion losses (see Fig. 3.13). In addition P_1 is the transmit power per PA, i.e., per antenna, at the BS and it varies from 0 to $P_{\max}$. In the updated version of [65], it has been shown that this linear abstraction can also be used for either rural or urban RNs based on the measurement of Table 3.2 such that $P_{RN} = t(\Delta_{P,RN}P_2 + P_{Ov,RN})$, where P_2 is the transmit power per PA at the RN. Moreover, the same linear type of power model has been used in [69] for characterizing the power consumption of a UE. Assuming the linear power model of (3.81) for each node, the total consumed power of the system in (3.80) can be re-expressed as

$$P_{\Sigma,AF} = n(\Delta_{P,BS}P_1 + P_{Ov,BS}) + q(\Delta_{P,RN}P_2 + P_{Ov,RN})$$
$$+ q\varsigma P_{Ov,RN} + 2r\varsigma P_{Ov,UE}, \qquad (3.82)$$

where $\varsigma \in [0,1]$ characterizes the ratio between transmission and reception overhead power. Intuitively, less overhead power will be necessary for receiving than transmitting signals. The EE in bits/J/Hz of this system can be simply defined as the ratio of its SE in (3.78) to its total consumed power in (3.82) such that

$$C_J = \frac{C}{P_{\Sigma}}. \qquad (3.83)$$

Using the values for P_{BS} and P_{RN}, i.e., 1350 W and 25.5 W (urban RN), in Tables 3.1 and 3.2, respectively, and the method in [65] for linearizing the total consumed power of various types of BS, we have obtained the following parameters regarding the abstracted power model of (3.81): $\Delta_{P,BS} = 7.5$, $P_{Ov,BS} = 375$ W as well as $P_{max,BS} = 40$ W for the macro BS and $\Delta_{P,RN} = 6.3$, $P_{Ov,RN} = 6.45$ as well as $P_{max,RN} = 1$ W such that $P_{BS} = 1350$ W and $P_{RN} = 25.5$ W for $t = 2$ as in Tables 3.1 and 3.2, respectively. In addition, we have set $P_{Ov,UE} = 100$ mW according to [69] and the ratio between transmission and reception overhead power as $\varsigma = 0.5$. Utilizing these parameter values, we have then compared the EE of MIMO AF against MIMO P2P in the in-building scenario in Fig. 3.21 and 3.22. Notice that equations (3.78) and (3.82) revert to

$$C \approx \frac{1}{\ln(2)}\left[n\ln\left(\frac{\gamma_0}{d_0}\right) + r\left(\ln(1+d_0) + \frac{1}{1+d_0} - 1\right)\right] \qquad (3.84)$$

and

$$P_{\Sigma,P2P} = n(\Delta_{P,BS}P_1 + P_{Ov,BS}) + r\varsigma P_{Ov,UE}, \qquad (3.85)$$

respectively, for MIMO P2P communication. In Fig. 3.21, we plot the EE of MIMO P2P and MIMO AF as a function of the SNR offset between the direct and donor links, σ, for $n = q = 4$, $P_2 = 1$ W, $P_1 = 40$ W (P2P) and $P_1 = 37.3, 19.3$ and 1.3 W (AF) which have been obtained such that $P_{\Sigma,AF} = P_{\Sigma,P2P}$, $P_{\Sigma,AF} = 0.8P_{\Sigma,P2P}$ and $P_{\Sigma,AF} = 0.6P_{\Sigma,P2P}$, respectively. Assuming the same level of noise at the UE for both the P2P and AF cases, it implies that $\gamma_0 \in [0, 20]$ / $\gamma_1 = 15$ dB for P2P and $\gamma_0 \in [-0.3, 19.7]$ / $\gamma_1 = 14.7$ dB, $\gamma_0 \in [-3.1, 16.9]$ / $\gamma_1 = 11.9$ dB as well as $\gamma_0 \in [-14.8, 5.2]$ / $\gamma_1 = 0.2$ dB that corresponds to $P_1 = 37.3, 19.3$ and 1.3 W, respectively, for AF. Results first indicate that MIMO AF is far more energy efficient than P2P MIMO for the case of $r = 1$. Moreover, it can be remarked by comparing the crossed with the dotted curves that MIMO AF can help to reduce the total consumed power while at the same time slightly increasing the EE, which is most desirable. Indeed, EE being the ratio between SE and

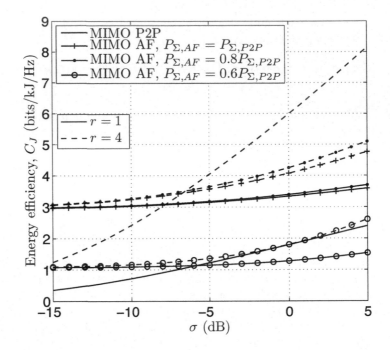

Figure 3.21: EE of MIMO P2P and MIMO AF as a function the SNR offset between the donor and direct links for $n = q = 4$.

the total consumed power, it can be achieved either by increasing the SE while keeping the same P_Σ (or even if P_Σ increases) or reducing the total consumed power while keeping the same C (or even if C decreases). In the context of energy saving, the former approach for being energy efficient is the right one and these results indicate that MIMO AF can do just that. When reducing further $P_{\Sigma,AF}$ to $P_{\Sigma,AF} = 0.6P_{\Sigma,P2P}$, MIMO AF can still be more energy efficient than MIMO P2P when the quality of the donor link is at least 6 dB higher than the quality of the direct link. In the case that $r = 4$, MIMO AF can help to reduce the total consumed power by 20% while being more energy efficient than MIMO P2P as long as $\sigma < -6$ dB.

In Fig. 3.22, we depict the EE, SE and total consumed power of MIMO P2P as well as MIMO AF against the BS transmit power per antenna for various number of antennas and σ values. In addition, we consider that $n = q = 4$, $P_2 = 1$, γ_0 varies from -15 to 15 dB. In the left-hand-side plot (EE plot), results first clearly show the existence of a maximum for the EE, which is not necessarily obtained for the maximum transmit power. Moreover, results confirm that MIMO AF EE gain over MIMO P2P increases as the number of receive antennas at the UE, r, decreases (when comparing the dashed gray with the plain black curves) and as the quality of the donor link increases in comparison with the quality of the direct link (when comparing the plain

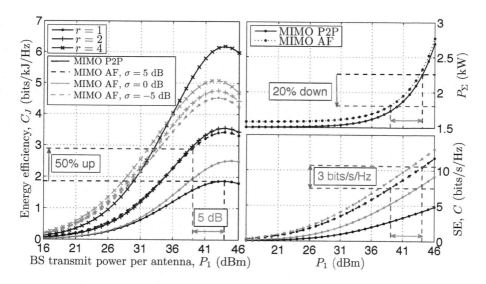

Figure 3.22: EE, SE and total consumed power of MIMO P2P as well as MIMO AF vs. the BS transmit power per antenna in dBm for various number of antennas and σ values.

gray, dashed black and dashed gray dotted curves with the plain black dotted curve for $r = 1$). Looking now at the upper right-hand-side graph (P_Σ plot), it shows that the total consumed power of the MIMO AF as a function of P_1 is always greater than that of MIMO P2P, which is obvious due to the additional RN power consumption in MIMO AF. Thus, it implies that the better performances of MIMO AF against MIMO P2P in terms of EE (EE plot) are not due to power saving but SE improvement, as it is confirmed by the lower right-hand-side graph (SE plot) where the SE gain of MIMO AF over MIMO P2P can be seen for different values of σ and $r = 1$. However, one can also look at these graphs in a different way. Let us assume that we want to use MIMO AF to reduce the total consumed power by 20% in comparison with MIMO P2P when $\sigma = 0$ dB and $r = 1$. It corresponds in the x-axis of the P_Σ plot to a reduction of 5 dBm in the BS transmit power per antenna, which in turn corresponds to an increase of 50% for the EE (see EE plot). As a reminder of the existence of a tradeoff between EE and SE [70], this power saving will come at the cost of a 3 bits/s/Hz reduction (see SE plot).

Overall, the results reveal that relaying can increase the EE in comparison with P2P communication by either power saving or SE improvement in an in-building scenario, especially when the number of receive antennas at the UE is lower than the number of antennas at the BS and RN as well as when the quality of the donor link is higher than the quality of the direct link. Results also indicate that even though transmitting with maximum power implies maximum SE, it does not necessarily imply maximum EE.

3.5 Conclusion and Future Works

In this chapter, a comprehensive investigation of cooperative communication systems has been presented and fundamental understanding from both the theoretical and practical points of view has been achieved. We considered typical relaying schemes including AF, DF CF, and hybrid schemes and studied their spectral and energy efficiency performances at both link and system levels.

Flexible approaches have been taken to combine different forwarding strategies, enabling the system to efficiently and dynamically adapt itself to the variations of the channel. The CF/DF based strategy shows significant improvement in terms of throughput and energy consumption. However, it should be emphasized that compared with direct transmission, the cooperative strategy only shows improved energy efficiency when the distance between the destination and the source is large. The main advantage of the proposed hybrid CF/DF or AF/DF based strategy is that it can achieve either transmit or receive diversity. Another issue we need to emphasize is the power model employed. For a meaningful analysis we need to strike the right balance between realistic modelling and mathematical tractability and we believe that the very important aspect of realistic energy consumption is well captured by our current power model.

Furthermore, relay has been studied in cellular networks. We have used more complex and realistic models for the BS and RN. Analysis has shown that RNs should be deployed at the cell edge to improve the energy efficiency. The number of RNs deployed in a sector should be optimized in terms of energy consumption. Deploying more RNs in a sector might be able to improve the average capacity of the system, but it also causes more energy consumption. When the disadvantage it brings outweighs the benefit, using more RN provides little spectrum improvement but incurs huge energy consumption, leading to low EE. Generally speaking, we can reach the conclusion that green communication is able to benefit from the usage of relaying technology with delicate design.

Although our work has significantly contributed to the improvement of cooperative communication systems both in terms of SE and EE, there are still numerous opportunities for further improvements. In multiple RN case, using coordinated multi-point Tx/Rx of RNs might be more efficient than relay selection and, thus, it is an interesting topic to investigate. The current relay architectures, which are defined in IEEE 802.16m and 3GPP LTE-Advanced, are only optimized for fixed relay, i.e., the RN is attached to a designated BS and becomes a part of the fixed access network. A mobile relay architecture, where relay can change their BS attachment according to operation demand, will promise more resilient and flexible relay deployment. Furthermore, network coding for the two-way relay channel is also regarded as one of the key enabling techniques for green communication.

Bibliography

[1] Ericsson Review, "HSDPA Performance and Evolution," no. 03, Mar. 2006.

[2] Ericsson White Paper, "Long Term Evolution (LTE): An Introduction," Oct. 2007.

[3] Sesia, Toufik, Baker, *LTE - The UMTS Long Term Evolution: From Theory to Practice*, Wiley, 2009.

[4] WiMax Forum at http://www.wimaxforum.org/.

[5] Press release, *EU Commissioner Calls on ICT Industry to Reduce Its Carbon Footprint by 20% as Early as 2015*, MEMO/09/140, Mar. 2009.

[6] *Vodafone Corporate Responsibility Report*, 2008.

[7] http://fireworks.intranet.gr/

[8] http://www.ict-rocket.eu/

[9] I. Stanojev, O. Simeone, Y. Bar-Ness, and D. Kim, "Energy efficiency of non-collaborative and collaborative hybrid-ARQ protocols," *IEEE Trans. Wireless Commun.*, vol. 8, no. 1, pp. 326-335, Jan. 2009.

[10] Yinan Qi , Reza Hoshyar, Imran Ali Muhammad, and Rahim Tafazolli, "Energy Efficiency Analysis of Hybrid-ARQ in Hybrid Relaying Systems," in *Proc. of VTC 2011*.

[11] R. Madan, N. Mehta, A. Molisch, and Jin Zhang, "Energy-Efficient Cooperative Relaying over Fading Channels with Simple Relay Selection," *IEEE Trans. Wireless Commun.*, vol. 7, no. 8, pp. 3013-3025, Aug. 2008.

[12] Yinan Qi, Reza Hoshyar, Muhammad Ali Imran, and Rahim Tafazolli, "H^2-ARQ-Relaying: Spectrum and Energy Efficiency Perspectives," *IEEE Journal on Selected Areas in Communications*, vol. 29, no. 8, pp. 1547-1558, Sep. 2011.

[13] http://www.ict-earth.eu/

[14] E.C. van der Meulen, "Three-terminal Communication Channel," *Adv. Appl. Prob.*, vol. 3, pp. 120-154, 1971.

[15] T. Cover and A.E. Gamal, "Capacity Theorems for the Relay Channel," *IEEE Trans. Inform. Theory*, vol. 25, pp. 572-584, Sep. 1979.

[16] R. U. Nabar, H. Bolcskei, and F.W. Kneubuhler, "Fading relay channels: performance limits and space-time signal design," *IEEE J. Sel. Area. in Comm.*, Aug 2004.

[17] L. Zhao and Z. Liao, "Power Allocation for Amplify-and-Forward Cooperative Transmission over Rayleigh-Fading Channels," *Journal of Communications*, pp. 33-42, vol. 3, No. 3, July 2008.

[18] M. Yu and J. Li, "Is Amplify-and-forward Practically Better Than Decode-and-forward or Vice Versa?" in *Proc. IEEE Int.Conf. Acoustics, Speech, and Signal Processing (ICASSP)*, vol. 3, March 2005, pp. 365-368.

[19] M. R. Souryal and B. R. Vojcic, "Performence of Amplify-and-forward and Decode-and-forward Relaying in Rayleigh Fading with Turbo Codes," *Int. Conf. Acoustics, Speech, and Signal Processing (ICASSP)*, vol. 4, May 2006, pp. 681-684.

[20] M. Janani, A. Hedayat, T.E. Hunter, and A. Nosratinia, "Coded cooperation in wireless communications: space-time transmission and iterative decoding," *IEEE Trans. Signal Processing*, vol. 52, pp 362-371, Feb. 2004.

[21] T.E. Hunter, and A. Nosratinia, "Diversity through coded cooperation," *IEEE Trans. Wireless Commun.*, vol. 5, pp. 283-289, Feb, 2006.

[22] Zhihan Yi and Min Kim, "Diversity order analysis of the decode-and-forward cooperative networks with relay selection," *IEEE Trans. Wireless Commun.*, vol. 7, no. 5, pp. 1792-1799, May 2008.

[23] T. Q. Duong and V. N. Q. Bao, "Performance analysis of selection decode-and-forward relay networks," *IEE Electronics Letters*, vol. 44, pp. 1206-1207, Sep. 2008.

[24] Christos K. Datsikas, Nikos C. Sagias, Fotis I. Lazarakis, and George S. Tombras, "Outage analysis of decode-and-forward relaying over nakagami-m fading channels," *IEEE Signal Processing Letters*, vol. 15, pp. 41-44, 2008.

[25] J. N. Laneman, *Cooperative diversity in wireless networks: Algorithms and architectures*, PhD thesis, Massachusetts Institute of Technology, 2002.

[26] Peyman Razaghi and Wei Yu, "Bilayer Low-Density Parity-Check Codes for Decode-and-Forward in Relay Channels," *IEEE Trans. Inform. Theory*, vol. 53, no. 10, pp. 3723-3739, Oct. 2007.

[27] J.N. Laneman, D.N.C. Tse, and G.W. Wonell, "Cooperative diversity in wireless networks: Efficient protocols and outage behavior," *IEEE Trans. Inform. Theory*, vol. 50, no. 12, pp. 3062-3080, Dec. 2004.

[28] A. Host-Madsen and J. Zhang, "Capacity bounds and power allocation for wireless relay channels," *IEEE Trans. Inform. Theory*, vol. 51, pp. 2020-2040, Jun. 2005.

[29] Michael Kats and Shlomo Shamai, "Relay Protocols for Two Colocated Users," *IEEE Trans. Inform. Theory*, vol. 52, No. 6, pp. 2329-2344, Jun. 2006.

[30] Zhixin Liu, Vladimir and Zixiang Xiong, "Wyner-Ziv Coding for the Half-duplex Relay Channel," *IEEE Conference on Acoustics, Speech and Signal Processing*, vol. 5, pp. 1113-1116, Mar. 2005.

[31] Ruiyuan Hu and Jing Li, "Practical Compress-Forward in User Cooperation: Wyner-Ziv Cooperation," *IEEE International Symposium on Information Theory*, Jul. 2006.

[32] Yonghui Li, B. Vucetic, Zhuo Chen, and Jinhong Yuan, "An improved relay selection scheme with hybrid relaying protocols," in *Proc. IEEE Global Telecommunications Conf. (GLOBECOM'07)*, pp. 3704-3708, New Orleans, USA, Nov. 2007.

[33] Yonghui Li, B. Vucetic, and Jinhong Yuan, "Distributed turbo coding with hybrid relaying protocols," in *Proc. IEEE 19th Int. Symp. on Personal, Indoor and Mobile Radio Communications (PIMRC'08)*, pp. 1-6, Cannes, France, Sep. 2008.

[34] S. Serbetli and A. Yener, "Power allocation and hybrid relaying strategies for F/TDMA Ad Hoc networks," in *Proc. IEEE Int. Conf. on Commun.(ICC'06)*, vol. 4, pp. 1562-1567, Istanbul, Turkey, Jun. 2006.

[35] 3GPP TR 36.814 v1.5.1(2009-12), *Further Advancement for E-UTRA, Physical Layer Aspcts.*

[36] IEEE Std 802.16j at http://ieee802.org/16/pubs/80216j.html

[37] IEEE Std 802.16m at http://ieee802.org/16/tgm/

[38] Ericsson Research White Paper "LTE-Advanced: Evolving LTE towards IMT-Advanced," 2010.

[39] EARTH deliverable, "Most Promising Tracks of Green Network Technologies" 2011, at https://www.ict-earth.eu/publications/deliverables/deliverables.html.

[40] Yinan Qi, Reza Hoshyar, and Rahim Tafazolli, "A Novel Hybrid Relaying Scheme Using Multilevel Coding," in *Proc. of VTC-2010 Spring*, 2010, pp. 1-5.

[41] S. Cui, A. J. Goldsmith, and A. Bahai, "Energy-Constrained Modulation Optimization," *IEEE Trans. Wireless Commun.*, vol. 4, no. 5, pp. 2349-2360, Sep. 2005.

[42] Yinan Qi, *Single Relay Cooperative Transmission/Reception Techniques*, PhD thesis, University of Surrey, 2009.

[43] A. Wyner and J. Ziv, "The Rate-distortion Function for Source Coding with Side Information at the Decoder," *IEEE Trans. Inform. Theory*, vol. 22, pp. 1-10, Jan. 1976.

[44] A. Wyner, "The rate-distortion function for source coding with side information at the decoder-ii: General sources," *Inf. Control*, vol. 38, pp. 60-80, 1978.

[45] D. Slepian and J. K. Wolf, "Noiseless coding of correlated information sources," *IEEE Trans. Inform. Theory*, vol. 19, pp. 471-480, July 1973.

[46] Sergio D. Servetto, "Lattice quantization with side information," in *Proc. Data Compression Conf. (DCC'03)*, Snowbird, UT, Mar. 2003, pp. 510-519.

[47] Ram Zamir and Shlomo Shamai, "Nested linear / lattice codes for Wyner-Ziv Encoding," in *Proc. Information Theory Workshop (ITW)*, Killarney, Ireland, Jun. 1998, pp. 92-93.

[48] Zhixin Liu, S. Cheng, A.D. Liveris, and Zixiang Xiong, "Slepian-Wolf Coded Nested Lattice Quantization for Wyner-Ziv Coding: High-Rate Performance Analysis and Code Design," *IEEE Trans. Inform. Theory*, vol. 52, pp. 4358-4379, Oct. 2006.

[49] K. Singhal, J. Vlach, and M. Vlach, "Numerical inversion of multidimensional Laplace transform," in *Proc. of the IEEE*, vol. 63, no. 11, pp. 1627-1628, Nov. 1975.

[50] Larry L. Peterson and Bruce S. Davie, *Computer Networks: A Systems Approach*, Third Edition, London, Academic Press, 2003.

[51] S. Kallel and D. Haccoun, "Sequential decoding with ARQ and code combining: a robust hybrid FEC/ARQ system," *IEEE Trans. Commun.*, vol. 36, no. 7, pp. 773-780, Jul. 1988.

[52] D. N. Rowitch and L. B. Milstein, "On the performance of hybrid FEC/ARQ systems using rate compatible punctured turbo (RCPT) codes," *IEEE Trans. Commun.*, vol. 48, no. 6, pp. 948-959, Jun. 2000.

[53] F. Babich, G. Montorsi, and F. Vatta, "Some notes on rate-compatible punctured turbo codes (RCPTC) design," *IEEE Trans. Commun.*, vol. 52, no. 5, pp. 681-684, May 2004.

[54] J. Hagenauer, "Rate-compatible punctured convolutional codes (RCPC Codes) and their applications," *IEEE Trans. Commun.*, vol. 36, no. 4, pp. 389-400, Apr. 1988.

[55] S. Kallel and D. Haccoun, "Generalized type-II hybrid ARQ scheme using punctured convolutional coding," *IEEE Trans. Commun.*, vol. 38, no. 11, pp. 1938-1946, Nov. 1990.

[56] Giuseppe Caire and Daniela Tuninetti, "The Throughput of Hybrid-ARQ Protocols for the Gaussian Collision Channel," *IEEE Trans. Inform. Theory*, vol. 47, no. 5, pp. 1971-1988, Jul. 2001.

[57] Wang Rui, Lau, V.K.N., "Combined cross-layer design and HARQ for multiuser systems with outdated channel state information at transmitter (CSIT) in slow fading channels," *IEEE Trans. Wireless Commun.*, vol. 7, pp. 2771-2777, Jul. 2008.

[58] T. Tabet, S. Dusad, and R. Knopp, "Diversity-Multiplexing-Delay Trade-off in Half-Duplex ARQ Relay Channels," *IEEE Transactions on Inform. Theory*, vol. 53, no. 10, pp. 3797-3805, Oct. 2007.

[59] Lin Dai and K. Letaief, "Throughput maximization of ad-hoc wireless networks using adaptive cooperative diversity and truncated ARQ," *IEEE Transactions on Communications*, vol. 56, no. 11, pp. 1907-1918, Nov. 2008.

[60] Bin Zhao and M.C. Valenti, "Practical relay networks: a generalization of hybrid-ARQ," *IEEE Journal on Selected Areas in Communications (JSAC)*, vol. 23, no.1, pp. 7-18, Jan. 2005.

[61] M.N. Khormuji and E. G. Larsson, "Analytical Results on Block Length Optimization for Decode-and-forward Relaying with CSI Feedback," in *Proc. 8th IEEE Workshop on Signal Processing Advances in Wireless Communications*, Jun. 2007, Helsinki, Finland.

[62] S. Igor, S. Osvaldo, B. Yeheskel and Cho Myeon Yun, "On the Optimal Number of Hops in Linear Wireless Ad-Hoc Networks with Hybrid ARQ," in *Proc. of WiOPT*, Apr. 2008.

[63] R. Hoshyar, and R. Tafazolli, "Performance evaluation of HARQ schemes for cooperative regenerative relaying," in *Proc. of IEEE Int. Conf. on Commun.(ICC'09)*, Dresden, Germany, Jun. 2009.

[64] Technical Specification Group Radio Access Network, "Evolved universal terrestrial radio access (E0UTRA); LTE radio frequency (RF) system scenarios," 3rd Generation Partnership Project (3GPP), Tech. Rep. TS 36.942, 2008-2009.

[65] G. Auer et al., "D2.3: Energy Efficiency Analysis of the Reference Systems, Areas of Improvements and Target Breakdown," INFSO-ICT-247733 EARTH (Energy Aware Radio and NeTwork TecHnologies), Tech. Rep., Nov. 2010.

[66] EARTH deliverable 3.2 online at https://www.ict-earth.eu/publications/deliverables/deliverables.html

[67] J. Wagner, B. Rankov, and A. Wittneben, "Large N Analysis of Amplify-and-Forward MIMO Relay Channels with Correlated Rayleigh Fading," *IEEE Trans. Inf. Theory*, vol. 54, no. 12, pp. 5735–5746, Dec. 2008.

[68] F. Héliot, M. A. Imran, and R. Tafazolli, "Energy Efficiency Analysis of In-Building MIMO AF Communication," in *Proc. IWCMC 2011*, Istanbul, Turkey, Jul. 2011.

[69] G. Miao, N. Himayat, and G. Y. Li, "Energy-Efficient Link Adaptation in Frequency-Selective Channels," *IEEE Trans. Commun.*, vol. 58, no. 2, pp. 545–554, Feb. 2010.

[70] S. Verdu, "Spectral Efficiency in the Wideband Regime," *IEEE Trans. Inf. Theory*, vol. 48, no. 6, pp. 1319–1343, Jun. 2002.

Chapter 4

Cross-Layer Design and Optimization for Green Wireless Communications and Networking

*Ting Zhu, +Sheng Xiao and #Chang Zhou
*Binghamton University, tzhu@binghamton.edu
+University of Massachusetts Amherst, sxiao@ecs.umass.edu
#China Jiliang University, chowchang@163.com

Energy-efficient wireless communication is critical for long-term wireless network applications, such as military surveillance, habitat monitoring and infrastructure protection. Typically, energy in communication can be optimized through (i) physical-layer transmission rate scaling, (ii) link layer optimization for better connectivity, reliability, stability, (iii) network-layer enhancement for better forwarders and routes, (iv) application-layer improvements for both content-agnostic and content-centric data aggregation and inference. These solutions are highly diversified. In order to further improve energy efficiency, researchers proposed cross-layer design and optimization techniques. In this chapter, we first summarize energy efficient design at different layers, then discuss cross-layer design and optimization in energy static and dynamic networks.

4.1   Energy Efficient Design at Different Layers

For years, many researchers have been focusing on energy efficient design in wireless communications and networks at different layers. In this section, we briefly introduce: (i) energy efficient hardware platforms, (ii) energy efficient MAC, (iii) energy efficient networking and (iv) energy efficient applications.

4.1.1   Energy Efficient Hardware Platforms

As the building foundation for wireless communications and networks, numerous efforts have been devoted to develop energy efficient hardware platforms such as MicaZ [1], Telos [2], iMote [3], BTnode [4], Eyes [5], TinyNode [6], Sensinode [7], mPlatform [8] and IRIS [9]. To further support long-term applications, many new platforms are built to harvest ambient energy from the surrounding environment. Notable ones include Heliomote [10], Prometheus [11], Trio [12], AmbiMax [13] and PUMA [14].

According to the type of energy storage used, these platforms can be separated into three categories: (i) rechargeable battery-based platforms [15], (ii) designs combining ultra-capacitors and rechargeable batteries [11,12] and (iii) capacitor-based designs [16].

- In rechargeable battery designs, such as Heliomote [10,15], the energy harvesting panel is directly connected to its battery. As the primary energy storage device, the rechargeable battery is charged and discharged frequently, leading to low system lifetimes due to the physical limitation on the number of recharge cycles.

- In designs that combine ultra-capacitors and rechargeable batteries such as Prometheus [11], the solar energy is first stored in the primary energy buffer, which is one or multiple ultra-capacitors. The rechargeable batteries are then used as the secondary energy buffer. This design inherits both the advantages and limitations of batteries and capacitors. For example, it is difficult to predict remaining energy because of the inclusion of batteries, and the lifetime of energy storage sub-system is decided by the shelf time of batteries (in the order of a few years). Within this category, several similar systems have been built with a few enhanced features. For example, AmbiMax [14] harvests energy from multiple ambient power sources (e.g., solar and wind generators), and PUMA [14] uses a power routing switch to route multiple power sources to multiple sub-systems. The higher utilization of ambient power is achieved through a combination of MPPT and power defragmentation.

- Previous works have intentionally avoided capacitor-only design, citing the leakage issue [11]. To alleviate leakage, small capacitors are normally used, which makes secondary energy storage (batteries) necessary. However, the development of battery capacities is very slow and still leakage-prone. In addition, charging efficiency for batteries is comparatively low [17]. For example, according to the Natural Resources Defense Council [17], common battery chargers provide an efficiency between 6% and 40%. Different from previous works, TwinStar [16] investigates the frontiers of capacitor-only design and studies not only hardware designs but also related software control techniques to reduce the impact of energy leakage.

4.1.2 Energy Efficient OS

Operating system designs for wireless communications and networks have drawn a significant amount of attention in recent years. Many excellent operating systems have been developed, including such notable ones as TinyOS/T2 [18, 19], LiteOS [20], Contiki [21], SOS [22], Mantis [23], t-Kernel [24] and Nano-RK [25]. Popular TinyOS supports a single-thread mode with two-level scheduling. SOS [22] is also a single-thread, even-driven architecture with dynamically loaded modules and a common kernel. Differently, LiteOS [20], Contiki [21] and Mantis [23] provide multi-threading support with different features in the file system, dynamic memory and debugging supports. In the area of mobile computing, Odyssey [26] and ECOSystem [27] are proposed for laptops and PDAs. A few recent works provide resource-aware adaptions such as Pixie [28] and Eon [29] specifically for data-flow-oriented applications.

4.1.3 Energy Efficient MAC

MAC layer design has been a major focus in the community for supporting energy efficient networking. In general, existing MAC protocols can be categorized into two categories. One category is synchronous MAC protocols, including S-MAC [30], T-MAC [31], RMAC [32] and DW-MAC [33], which synchronize neighboring nodes in order to align their active or sleeping periods. The other category is asynchronous MAC protocols, including B-MAC [34], X-MAC [35], WiseMAC [36] and RI-MAC [37], which allow sensor nodes to work individually with their own working schedule through techniques such as Low-Power-Listening. More recently, some hybrid MAC protocols such as SCP-MAC [38], Z-MAC [39] and Funneling-MAC [40] are designed to take the advantages of two traditional approaches. Although these MAC protocols are effective to provide energy efficiency for single-hop communications, further optimizations and improvements for multi-hop communications within a network remain a new research frontier.

4.1.4   Energy Efficient Networking

Besides energy efficient hardware platforms, operating systems, and MAC protocol designs, energy efficient network layer design is also very important. Due to the growing gap between limited energy and increasing energy demand in long-term applications, researchers have started to realize the importance of low-duty-cycle networking. Dousse et al. provide a solid analysis of bounds of the delay for sending data from a node to a sink in the networks with completely uncoordinated node working schedules [41]. Lu et al. provide optimal forwarding solutions for duty-cycled networks with tree and ring topologies [42]. In [43], Lai and Paschalidis present both global and distributed algorithms to trade-off between energy consumption and communication latency in duty-cycled networks. More recently, Su et al. introduce two proactive minimum latency routing algorithms in intermittently connected sensor networks [44].

4.1.5   Energy Efficient Application

Despite a wide range of applications, energy efficiency has been a major focus for nearly all wireless network applications, ranging from military surveillance [45], disaster response [46, 47], habitat monitoring [48, 49], infrastructure protection [50–53], scientific exploration [54, 55], to participatory urban sensing [56–58]. With long-term operation requirements and limited energy supply, many of these applications shall operate under low duty-cycle.

Wireless sensor networks is one of the most important type of wireless networks. In wireless sensor networks research, sensing is an indispensable research component. With limited available energy, how to conduct energy efficient sensing operations is a challenging problem. In [59], the authors support full surveillance coverage based on an off-duty eligibility rule. In [60], surveillance coverage is achieved through probing. Kumar et al. [61] identify a critical bound for k-coverage in a network, assuming a node is randomly turned on with a certain probability. In [62], Kumar et al. investigate the k-barrier coverage problem, identifying the critical condition for weak k-barrier coverage. In [63, 64], the authors provide a theoretical analysis and simulation of the delay (or stealth distance) before a target is detected. Several algorithms are designed based on the concept of set cover. In [65], Cardei et al. propose two heuristic algorithms to identify a maximum number of set covers to monitor a set of static targets at known locations. In [66], Abrams et al. propose three approximation algorithms for a relaxed version of the previously defined SET K-COVER problem [67].

4.2 Cross-Layer Optimization in Energy Static Networks

In the previous section, we learned that in order to bridge the growing gap between lifetime requirement of sensor applications and the slow progress in battery capacity [62], it is critical to have an energy-efficient communication stack. A lot of energy-efficient communication protocols have been proposed.

On the other hand, wireless networks with intermittent receivers have caught unproportionately little attention, despite the known fact that communication energy is consumed mostly for being ready for potential incoming packets, a problem commonly referred to as idle listening. For example, the widely used Chipcon CC2420 [68] radio draws 19.7mA when receiving or idle listening, which is actually larger than 17.4mA when transmitting. More importantly, packet transmission time is usually very small (e.g., less than 1 millisecond to transmit a TinyOS packet using a CC2420 radio), while the duration of idle listening for reception can be orders of magnitude longer. For example, most environmental applications, such as Great Duck Island [48], Redwood Forest [49] and James Reserve [69], sample the environment at relatively low rates (on the order of minutes between samples). With a comparable current draw and a $3 \sim 4$ orders of magnitude longer duration waiting for reception, idle listening is a major energy drain that accounts for most energy in communication if it is not optimized. To reduce energy lost to idle listening, a low-duty-cycle network is formed by nodes that listen to the channel very briefly and shut down their radios most of the time (e.g., 99% or more). At any given time, this type of network is actually fragmented (partitioned) and network connectivity (topology) becomes intermittent. Uniquely, communication delay in low-duty cycle networks is dominated by sleep latency, the delay a sender waits for its receiver to wake up. Although low-duty-cycle networking is an ideal fit for many long-term unattended sensor applications, research has been lacking and predominantly focuses on physical and link-layer designs. To ensure packet reception at low-duty-cycle receivers, several pioneer researchers have proposed S-MAC [30], B-MAC [34], X-MAC [35], WiseMAC [36], SCP-MAC [38] and the 802.15.4 beacon-enable mode [70], which successfully reduce the amount of idle listening through techniques such as Low-Power-Listening (LPL) and/or Synchronous Channel Polling (SCP). These link-layer designs are effective, but further improvement becomes difficult without utilizing information about topology and multi-hop connectivity information at the network layer. For example, data reliability is commonly supported by link-layer protocols [30, 34, 35, 38] through retransmission to a same receiver if previous transmission fails. In a low-duty-cycle network, without the network layer re-routing capability, these link-layer protocols have to wait for an intended receiver to wake up again, introducing excessive sleep latency in the orders of seconds or possibly minutes.

Motivated by the insufficiency of link-layer designs, researchers have proposed cross-layer design and optimization in energy static and dynamic low-duty-cycle wireless networks. In this section, we introduce the cross-layer optimization in energy static low-duty-cycle wireless networks. The cross-layer design in energy dynamic low-duty-cycle wireless networks is introduced in Section 4.3.

4.2.1 Network Model

In energy static low-duty-cycle networks, sensor devices are normally powered by batteries. The available remaining energy does not change dynamically. The static low-duty-cycle wireless network supports mission-driven applications such as military surveillance [71, 72] with a specified network lifetime requirement and a fixed energy budget [45]. For example, a military strategic area shall be covered until a stronghold is established in two months. The duty-cycle of this type of network is a fixed ratio of the battery lifetime and network lifetime.

We assume a network with N sensor nodes. At a given point of time t, a sensor node is in either a wake-up or sleep state. When a node is in the wake-up state, it can sense and receive packets transmitted from neighboring nodes. When a node is in the sleep state, it turns off all function modules except a timer (for the purpose of waking itself up). In other words, a node can schedule itself to wake up and transmit a packet at any time, but can receive packets only when it is in its wake-up state. The rationale behind this network model is to reduce idle listening that dominates energy consumption of wireless sensor nodes. In a low-duty-cycle network, a sensor node specifies its working behaviors by a set of wake-up instances, during which a node can receive incoming packets. Each wake-up instance j at node i can be represented by a tuple (t_j^i, d_j^i), where t_j^i denotes the starting time of the wake-up instance and d_j^i denotes the corresponding duration of wake-up instance j. Let Γ_i symbolize the working schedule of node i, then we can have $\Gamma_i = \{(t_1^i, d_1^i), (t_2^i, d_2^i), ...\}$. Essentially, Γ_i stores all wake-up instances of node i during its lifetime. Since many sensor node working schedules [71, 73, 74] are periodic, it is sufficient to represent an infinity sequence of wake-up instances, using repeated subsequences with a period of T. For example, for a periodic working schedule $\{(1,1), (11,1), (21,1), ...\}$ where sensor nodes wake up every 10 units of time, we can represent this working schedule as (1,1) with $T = 10$. For a given Γ_i with a period of T, the listening duty cycle of the sensor node i is $\Sigma d_j^i / T$ where $0 \leq t_j^i \leq T$.

For a sensor node, its neighbors transit between wake-up and sleep states, and thus its neighborhood connectivity varies over time and becomes intermittent. Formally, for a given time t, we denote the network topology as

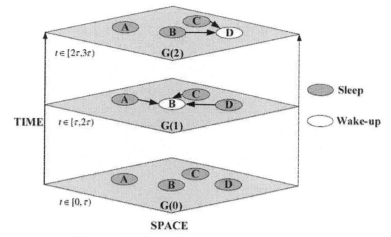

Figure 4.1: Low duty cycle network model.

$G(t) = (V, E(t))$, where V is a complete set of N sensor nodes within the network, and $E(t)$ is a set of directed edges at time t. An edge $e(i, j)$ belongs to $E(t)$ if and only if (i) node j is within the communication range of node i, and (ii) node j is awake and hence able to receive packets at time t. Essentially, $G(t)$ represents the potential traffic flow within the network at time t. Fig. 4.1 shows three instances of connectivity graph $G(t)$ along the time line, when node B and node D wake up at time t and $2t$, respectively. Clearly in such a network, nodes are not always ready to receive and a sender needs to wait for its receiver to wake up before transmitting.

In an energy static low-duty-cycle sensor network, each node i adopts a fixed working schedule $\Gamma_i = \{(t_1^i, d_1^i), (t_2^i, d_2^i), ...\}$ and announces Γ_i to its K-hop neighbors (normally, $K = 1$). Given its K-hop neighbors' working schedule information, the challenge here is how to design provably optimal algorithms/protocols in terms of delay, reliability and cost for pair-wise unicast, multicast and broadcast; a problem becomes very challenging when unreliable communication links among sensor nodes are taken into consideration.

4.2.2 Network Protocols

After introducing the network model, we briefly describe energy static low-duty-cycle network protocols in this section. We use DSF [75] as an example. DSF investigated the combined effect of sleep latency and unreliable communication links for source-to-sink communication, given one-hop neighbor's working schedule is known and fixed. The main idea of Dynamic Forwarding (DSF) is shown in Fig. 4.2. In traditional routing protocols, packets are forwarded either along a single fix-path (e.g., ETX [76], MintRoute [77], DSR [78] and AODV [79]) or multiple fix-paths (e.g., Directed Diffusion [80], Disjoint Multi-Path [81]). These protocols are very suitable for always-awake networks;

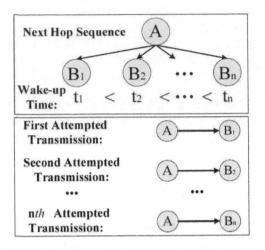

Figure 4.2: Dynamic switch-based forwarding (DSF).

however, they suffer $1 \sim 2$ orders of magnitude sleep latency in a low-duty-cycle network, when waiting for a fixed receiver to wake up again if a previous packet transmission is not successful. To minimize sleep latency, DSF utilizes multiple potential forwarding nodes at each hop. As shown in Fig. 4.2, for a given sink, each node maintains a sequence of forwarding nodes sorted in the order of the wakeup time associated with them. Packet transmission starts with the first node in the sequence. In case of failure, retransmission follows the sequence order until the packet is successfully received by one of the forwarding nodes. The key optimization problem in DSF is how to select a subset of forwarding nodes among all eligible nodes. For the first time, DSF reveals low-duty-cycle networks have fundamentally different properties from always-awake networks.

4.3 Cross-Layer Designs in Energy Dynamic Networks

For long-term unattended application such as habitat monitoring in remote areas, battery-powered sensors are not desirable because replacing batteries would add significantly to the cost of maintenance. To address this issue, researchers have designed various types of energy harvesting technologies to collect ambient energy from the environments. These include solar [12, 82], wind [13], kinetic [83], piezoelectric strain [83] and vibrational [84] energy. The dynamic change of environmental energy introduces new design challenges for wireless networks. This section discusses cross-layer designs for low-duty-cycle wireless networks with variable energy supply from the environment.

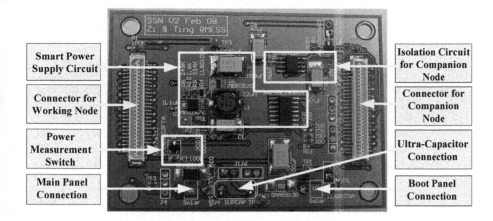

Figure 4.3: TwinStar power board.

4.3.1 Hardware and Communication

In [85], the TwinStar power board (shown in Fig. 4.3) is designed and introduced. It features a unique ultra-capacitor-only design. Different from existing rechargeable battery-based designs [82, 86], TwinStar design possesses a set of clear advantages: it has (i) more than 1 million recharge cycles, (ii) predictable remaining energy independent of discharge modes and (iii) very high charging and discharging efficiency. To support smart power management, as shown in Fig. 4.3, the TwinStar power board provides dual interfaces. First, a main interface supplies power to and receives/sends signals from/to a working platform (e.g., MicaZ, mPlatform). Second, an optional interface is designed for profiling purposes by attaching a companion node. This companion node is powered independently to eliminate the observer effect, i.e., energy consumed by profiling operations affects the accuracy of energy profiling.

With the TwinStar power board, Zhu et al. have conducted a series of experiments to investigate energy harvesting rates in outdoor, indoor and mobile environments. As shown in Fig. 4.4, 4.5 and 4.6, ambient energy varies noticeably over time and space. Since energy harvested is driven dynamically and unpredictably by the environment, in energy dynamic low-duty-cycle sensor networks, each node i maintains a changing working schedule $\Gamma_i = \{(t_1^i, d_1^i), (t_2^i, d_2^i), ...\}$ based on the energy availability. The key challenge in this part of research is to balance energy supply and demand with a provable performance stability and convergence.

To optimize performance in energy dynamic low-duty-cycle networks, Zhu et al. proposed energy-adaptive control in [16]. The delay in a low-duty-cycle network is affected by the amount of wake-ups duration at the receiving nodes (i.e., $\Sigma d_j^i / T$) and the layout of these wake-ups along a timeline (i.e., t_j^i). When energy from an environment is abundant, a node can wake up more, reducing the delay; on the other hand, if less energy is available for commu-

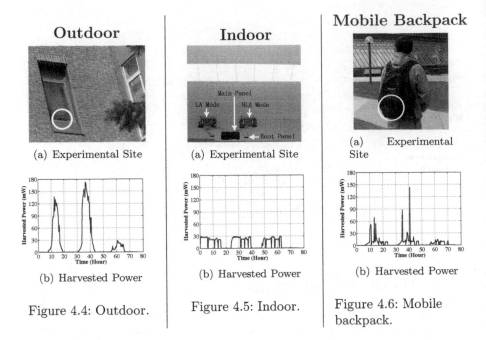

Outdoor

(a) Experimental Site

(b) Harvested Power

Figure 4.4: Outdoor.

Indoor

(a) Experimental Site

(b) Harvested Power

Figure 4.5: Indoor.

Mobile Backpack

(a)   Experimental Site

(b) Harvested Power

Figure 4.6: Mobile backpack.

nication, a node shall reduce the frequency or duration of wake-ups, leading to a longer delay. In contrast to previous heuristic algorithms [87, 88], the feedback-control-based approach proposed in [16] not only rigorously studied the system steady state behavior, but also transited overshoot and settling time.

One interesting phenomena of low-duty-cycle sensor networks is the separation between physical connectivity and logical connectivity. A low-duty-cycle network could be physically connected but logically partitioned. For example, if a node does not know when its neighbors wake up, it cannot communicate. With dynamic energy supply, energy adaptive control must adjust wake-up schedules, which introduces the possibility of logical partition. Incremental schedule adjustment can avoid the chance of logical partition but leads to a suboptimal timing of wake-ups.

In [89], an Energy Synchronized Communication (ESC) protocol is designed as a generic middle-ware service for supporting existing network protocols. The forwarding delay is modeled at low-duty-cycle sensor nodes. The basic idea is shown in Fig. 4.7. Let *predecessors* of Node A be the set of nodes that use Node A as forwarder (e.g., P_1, P_2, P_3 in the example) and *successors* of Node A be the set of nodes that Node A forwards to (e.g., S_1, S_2 in the example). Note that a node can be both a predecessor and successor of Node A, if it exchanges data bi-directionally. The delay experienced by Node A is modeled as the sum of the delay of both incoming and outgoing packets. To illustrate the main idea without loss of generality, let's consider the delay in

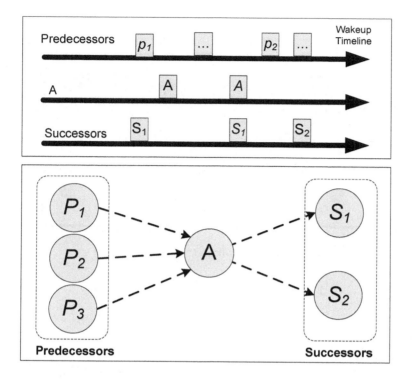

Figure 4.7: Cross-traffic delay modeling.

one link between P_i and A. Suppose the bi-directional link quality q denotes the success ratio of a round-trip transmission (DATA and ACK) between node A and node P_i. The probability that the packet reaches node A at its ith attempt can be expressed as $P(i) = (1 - q)^{i-1}q$. For a packet ready time t at node P_i, the expected transmission delay to reach node A is the sum of the product of probability that the packet reaches node A at its ith attempt and corresponding sleep latency $L_t^P(i)$ (where $i = 1, 2, \ldots$). $L_t^P(i)$ can be calculated according to the wake-up schedule of Node A. Consequently, the link delay between P_i and A can be formulated as: $D(t) = \sum_{i=1}^{\infty} P(i)L_t^P(i)$. Similarly, we can model the delay between A and one of the successor nodes S_i. To obtain the average delay, a weighted average can be used based on traffic rates at individual links.

The research in [89] reveals that cross-traffic delay through a duty-cycled node is determined only by the number of active instances at intervals, partitioned by active instances of predecessor and successor nodes. Evaluation result demonstrates that ESC can effectively reduce delay and increase delivery ratios, while synchronizing radio activity with available energy.

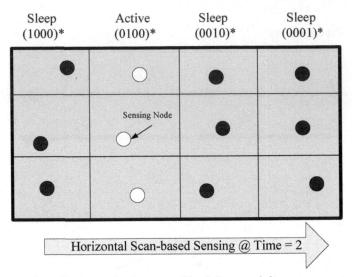

Figure 4.8: Cross-traffic delay modeling.

4.3.2   Sensing

Besides communication, Sensing is another major source of energy consumption for small sensor nodes. For example, a magnetic sensor draws 6.5 mA when active, which is as 26 times as much energy of that of a MSP430 microcontroller used in Telos motes. Although sensing is a very active research topic, almost all previous work [59,60,62–67,90,91] assumes battery-operated networks with a fixed energy budget. Clearly, the sensing problem becomes more realistic and challenging when non-uniform and dynamic energy budgets are considered.

Energy-efficient sensing extends system lifetime by reducing the working duty-cycle of sensing components and leveraging the redundant deployment of sensor nodes. In [92], Jason et al. model the low-duty-cycle sensing activity of node i with a tuple (S_i, R_i). S_i is a binary schedule in which 1 denotes the active state and 0 denotes the inactive state. R_i is the switching rate, which defines the reciprocal of the time duration a bit represents. For example, the tuple $((0010)^*, 2)$ denotes a repeated off-off-active-off schedule in which each bit represents 0.5 seconds. Theoretically, when the switching rate approaches infinity, schedule bits R_i can precisely characterize any on/off sensing activity of the node i, decided by any coverage algorithms. Consequently, the sensing activity of a sensor network N can be described as $\{(S_i, R_i)|i \in N\}$. For example, Fig. 4.8 shows the schedule bits of all nodes that perform a horizontal scan-based sensing. One of the key research problems addressed in [92] is how to schedule the sensing activity to minimize the worst-case area breach (i.e., the largest percentage of the area that a target can reach without being detected).

Bibliography

[1] CrossBow, "Crossbow micaz oem module," Available at http://www.xbow.com/Products/Product pdf files/Wireless pdf/MICAz OEM Edition Datasheet.pdf.

[2] J. Polastre, R. Szewczyk, and D. Culler, "Telos: enabling ultra-low power wireless research," in *Fourth International Symposium on Information Processing in Sensor Networks (IPSN 2005)*, pages 364–369, 2005.

[3] Intel, "Intel imote," in Available at http://www.intel.com/research/exploratory/motes.htm.

[4] J. Beutel, O. Kasten, F. Mattern, K. Romer, F. Siegemund, and L. Thiele, "Prototyping wireless sensor network applications with btnodes," in *Proceedings of the 1st European Workshop on Sensor Networks(EWSN 2004)*, pages 323–338, 2004.

[5] Eyes Project, "Eyes module," in Available at http://www.eyes.eu.org/.

[6] H. Dubois-Ferriere, R. Meier, L. Fabre, and P. Metrailler, "Tinynode: a comprehensive platform for wireless sensor network applications," in *The Fifth International Conference on Information Processing in Sensor Networks (IPSN 2006)*, pages 358–365, 2006.

[7] Sensinode Ltd., "Sensinode," in Available at http://www.sensinode.com.

[8] D. Lymberopoulos, N. B. Priyantha, and F. Zhao, "mplatform: a reconfigurable architecture and efficient data sharing mechanism for modular sensor nodes," in *IPSN 07: Proceedings of the 6th International Symposium on Information Processing in Sensor Networks*, pages 128–137, 2007.

[9] CrossBow, "Crossbow iris oem module," in Available at http://www.xbow.com/Products/Product pdf files/Wireless pdf/IRIS OEM Datasheet.pdf.

[10] K. Lin, J. Yu, J. Hsu, S. Zahedi, D. Lee, J. Friedman, A. Kansal, V. Raghunathan, and M. Srivastava, "Heliomote: enabling long-lived sensor networks through solar energy harvesting," in *SenSys '05*, 2005.

[11] X. Jiang, J. Polastre, and D. Culler, "Perpetual environmentally powered sensor networks," in *IPSN '05*, 2005.

[12] P. Dutta, J. Hui, J. Jeong, S. Kim, C. Sharp, J. Taneja, G. Tolle, K. Whitehouse, and D. Culler, "Trio: enabling sustainable and scalable outdoor wireless sensor network deployments," in *Proceedings of the Fifth International Conference on Information Processing in Sensor Networks (IPSN 2006)*, pages 407–415, 19-21 April, 2006.

[13] C. Park and P.H. Chou, "Ambimax: Autonomous energy harvesting platform for multi-supply wireless sensor nodes," in *3rd Annual IEEE Communications Society on Sensor and Ad Hoc Communications and Networks*, 1:168–177, 28-28, 2006.

[14] C. Park and P. Chou, "Power utility maximization for multiple-supply systems by a load-matching switch," *Proceedings of the 2004 International Symposium on Low Power Electronics and Design, 2004. ISLPED '04*, pp. 168–173, 2004.

[15] A. Kansal, J. Hsu, S. Zahedi, and M. B. Srivastava, "Power management in energy harvesting sensor networks," *TECS*, vol. 6, no. 4, 2007.

[16] T. Zhu, Z. Zhong, Y. Gu, T. He, and Z.-L. Zhang, "Leakage-aware energy synchronization for wireless sensor networks," in *MobiSys '09: Proceedings of the 7th Annual International Conference on Mobile Systems, Applications and Services*, 2009.

[17] Suzanne Foster, "The energy efficiency of common household battery charging systems: Results and implications," in *Natural Resources Defense Council*, http://www.efficientproducts.org/reports/bchargers/NRDC-Ecos_Battery_Charger_Efficiency.pdf.

[18] J. Hill, R. Szewczyk, A. Woo, S. Hollar, D. Culler, and K. S. J. Pister, "System architecture directions for networked sensors," in *ASPLOS '00*, 2000.

[19] P. Levis, D. Gay, V. Handziski, J. Hauer, M. T. Be. Greenstein, J. Huio, K. Klues, C. Sharp, R. Szewczyk, J. Polastre, P. Buonadonnao, L. Nachman, G. Tolleo, D. Cullero, and A. Wolisz, "T2: A second generation OS for embedded sensor networks," in *Technical Report TKN-05-007*, 2005.

[20] Q. Cao, T. Abdelzaher, J. Stankovic, and T. He, "LiteOS, A Unix-like operating system and programming platform for wireless sensor networks," in *IPSN'08*, 2008.

[21] A. Dunkels, B. Gronvall, and T. Voigt, "Contiki - A lightweight and flexible operating system for tiny networked sensors," in *LCN '04: Proceedings of the 29th Annual IEEE International Conference on Local Computer Networks*. Washington, DC, USA: IEEE Computer Society, 2004, pp. 455–462.

[22] C.-C. Han, R. Kumar, R. Shea, E. Kohler, and M. Srivastava, "A dynamic operating system for sensor nodes," in *MobiSys '05: Proceedings of the 3rd International Conference on Mobile Systems, Applications, and Services*. New York, NY, USA: ACM, 2005, pp. 163–176.

[23] S. Bhatti, J. Carlson, H. Dai, J. Deng, J. Rose, A. Sheth, B. Shucker, C. Gruenwald, A. Torgerson, and R. Han, "MANTIS OS: An embedded multithreaded operating system for wireless micro sensor platforms," *Mob. Netw. Appl.*, vol. 10, no. 4, pp. 563–579, 2005.

[24] L. Gu and J. A. Stankovic, "t-kernel: Providing reliable OS support to wireless sensor networks," in *SenSys '06: Proceedings of the 4th International Conference on Embedded Networked Sensor Systems*. New York, NY, USA: ACM, 2006, pp. 1–14.

[25] A. Eswaran, A. Rowe, and R. Rajkumar, "Nano-RK: An energy-aware resource-centric RTOS for sensor networks," in *RTSS '05: Proceedings of the 26th IEEE International Real-Time Systems Symposium*. Washington, DC, USA: IEEE Computer Society, 2005, pp. 256–265.

[26] J. Flinn and M. Satyanarayanan, "Managing battery lifetime with energy-aware adaptation," *ACM Trans. Comput. Syst.*, vol. 22, no. 2, pp. 137–179, 2004.

[27] H. Zeng, C.S. Ellis, and A.R. Lebeck, "Experiences in managing energy with ECOSystem," *Pervasive Computing, IEEE*, Jan.-March 2005.

[28] K. Lorincz, B. rong Chen, J. Waterman, G. Werner-Allen, and M. Welsh, "Pixie: An operating system for resource-aware programming of embedded sensors," in *In Proceedings of the Fifth Workshop on Embedded Networked Sensors (HotEmNets 2008)*, 2008.

[29] J. Sorber, A. Kostadinov, M. Garber, M. Brennan, et al., "Eon: A language and runtime system for perpetual systems," in *SenSys '07*, 2007.

[30] W. Ye, J. Heidemann, and D. Estrin, "An energy-efficient MAC protocol for wireless sensor networks," in *INFOCOM*, 2002.

[31] T. van Dam and K. Langendoen, "An adaptive energy-efficient mac protocol for wireless sensor networks," in *SenSys 03: Proceedings of the 1st International Conference on Embedded Networked Sensor Systems,* pages 171–180, 2003.

[32] S. Du, A. K. Saha, and D. B. Johnson, "Rmac: A routing-enhanced duty-cycle mac protocol for wireless sensor networks," in *INFOCOM*, 2007.

[33] Y. Sun, S. Du, O. Gurewitz, and D. B. Johnson, "Dw-mac: a low latency, energy efficient demand wakeup mac protocol for wireless sensor networks," in *MobiHoc 08: Proceedings of the 9th ACM International Symposium on Mobile ad hoc Networking and Computing*, pages 53–62, 2008.

[34] J. Polastre and D. Culler, "Versatile low power media access for wireless sensor networks," in *SenSys '04*, 2004.

[35] M. Buettner, G. V. Yee, E. Anderson, and R. Han, "X-mac: a short preamble mac protocol for duty cycled wireless sensor networks," in *SenSys*, 2006.

[36] A. El-Hoiydi and J.-D. Decotignie, "Wisemac: an ultra low power mac protocol for the downlink of infrastructure wireless sensor networks," in *IEEE Symposium on Computers and Communications*, 1:244–251, 2004.

[37] Y. Sun, O. Gurewitz, and D. B. Johnson, "Ri-mac: a receiver-initiated asynchronous duty cycle mac protocol for dynamic traffic loads in wireless sensor networks," in *SenSys 08: Proceedings of the 6th ACM Conference on Embedded Network Sensor Systems*, pages 1–14, 2008.

[38] W. Ye, F. Silva, and J. Heidemann, "Ultra-low duty cycle mac with scheduled channel polling," in *Proceedings of the 4th ACM Conference on Embedded Networked Sensor Systems (SenSys 2006)*, 2006.

[39] I. Rhee, A.Warrier, M. Aia, J.Min, and M. Sichitiu, "Z-mac: A hybrid mac for wireless sensor networks," in *IEEE/ACM Transactions on Networking*, 16(3):511–524, 2008.

[40] G.-S. Ahn, S. G. Hong, E. Miluzzo, A. T. Campbell, and F. Cuomo, "Funneling-mac: a localized,sink-oriented mac for boosting fidelity in sensor networks," in *SenSys*, 2006.

[41] O. Dousse, P. Mannersalo, and P. Thiran, "Latency of wireless sensor networks with uncoordinated power saving mechanisms," in *MobiHoc*, 2004.

[42] G. Lu, N. Sadagopan, B. Krishnamachari, and A. Goel, "Delay efficient sleep scheduling in wireless sensor networks," in *Proceedings of the 24th IEEE International Conference on Computer Communications (INFOCOM 2005)*, 2005.

[43] W. Lai and I. C. Paschalidis, "Sensor network minimal energy routing with latency guarantees," in *MobiHoc 07: Proceedings of the 8th ACM International Symposium on Mobile ad hoc Networking and Computing*, pages 199–208, 2007.

[44] L. Su, C. Liu, H. Song, and G. Cao, "Routing in intermittently connected sensor networks," in *Proceedings of the 16th IEEE International Conference on Network Protocols (ICNP 2008)*, 2008.

[45] A. Arora, P. Dutta, S. Bapat, V. Kulathumani, H. Zhang, V. Naik, V. Mittal, H. Cao, M. Demirbas,M. Gouda, Y. Choi, T. Herman, S. Kulkarni, U. Arumugam, M. Nesterenko, A. Vora, and M. Miyashita, "A line in the sand: A wireless sensor network for target detection, classification, and tracking," in Computer Networks (Elsevier), 46:605–634, 2004.

[46] J.-H. Huang, S. Amjad, and S. Mishra, "Cenwits: A sensor-based loosely coupled search and rescue system using witnesses," in *SenSys*, 2005.

[47] E. A. Basha, S. Ravela, and D. Rus, "Model-based monitoring for early warning flood detection," in *SenSys*, 2008.

[48] R. Szewczyk, A. Mainwaring, J. Anderson, and D. Culler, "An analysis of a large scale habit monitoring application," in *SenSys '04*, 2004.

[49] G. Tolle, J. Polastre, R. Szewczyk, N. Turner, K. Tu, S. Burgess, D. Gay, P. Buonadonna, W. Hong, T. Dawson, and D. Culler, "A macroscope in the redwoods," in *SenSys '05*, 2005.

[50] K. Chebrolu, B. Raman, N. Mishra, P. K. Valiveti, and R. Kumar, "Brimon: A sensor network system for railway bridge monitoring," in *MobiSys '08*, 2008.

[51] Y. Kim, T. Schmid, Z. M. Charbiwala, J. Friedman, and M. B. Srivastava, "Nawms: Nonintrusive autonomous water monitoring system," in *SenSys '08*, 2008.

[52] H. Song, S. Zhu, and G. Cao, "Svats: A sensor-network-based vehicle anti-theft system," in *Proceedings of the 27th IEEE International Conference on Computer Communications (INFOCOM 2008)*, 2008.

[53] N. Xu, S. Rangwala, K. K. Chintalapudi, D. Ganesan, A. Broad, R. Govindan, and D. Estrin, "A wireless sensor network for structural monitoring," in *Second ACM Conference on Embedded Networked Sensor Systems (SenSys 2004)*, 2004.

[54] L. Selavo, A. Wood, Q. Cao, T. Sookoor, H. Liu, A. Srinivasan, Y. Wu, W. Kang, J. Stankovic,D. Young, and J. Porter, "Luster: wireless sensor network for environmental research," in *SenSys '07*, 2007.

[55] P. Juang, H. Oki, Y. Wang, M. Martonosi, L. Peh, and D. Rubenstein, "Energy-efficient computing for wildlife tracking: Design tradeoffs and early experiences with zebranet," in *Proceedings of the International Conference on Architectural Support for Programming Languages and Operating Systems*, 2002.

[56] E. Miluzzo, N. D. Lane, K. Fodor, R. Peterson, H. Lu, M. Musolesi, S. B. Eisenman, X. Zheng, and A. T. Campbell, "Sensing meets mobile social networks: the design, implementation and evaluation of the cenceme application," in *SenSys '08*, 2008.

[57] P. Mohan, V. N. Padmanabhan, and R. Ramjee, "Nericell: rich monitoring of road and traffic conditions using mobile smartphones," in *SenSys '08*, 2008.

[58] S. B. Eisenman, E. Miluzzo, N. D. Lane, R. A. Peterson, G.-S. Ahn, and A. T. Campbell, "The bikenet mobile sensing system for cyclist experience mapping," in *Sensys*, 2007.

[59] D. Tian and N. Georganas, "A node scheduling scheme for energy conservation in large wireless sensor networks," *Wireless Communications and Mobile Computing Journal*, May 2003.

[60] F. Ye, G. Zhong, S. Lu, and L. Zhang, "PEAS: A robust energy conserving protocol for long-lived sensor networks," in *Proc. of International Conference on Distributed Computing Systems (ICDCS)*, May 2003.

[61] S. Kumar, T. H. Lai, and J. Balogh., "On k-Coverage in a mostly sleeping sensor network," in *Mobicom*, 2004.

[62] S. Kumar, T. H. Lai, and A. Arora, "Barrier coverage with wireless sensors," in *MobiCom '05*, 2005.

[63] C. Gui and P. Mohapatra, "Power conservation and quality of surveillance in target tracking sensor networks," in *MobiCom'04*, 2004.

[64] S. Ren, Q. Li, H. Wang, X. Chen, and X. Zhang, "Analyzing object tracking quality under probabilistic coverage in sensor networks," *ACM Mobile Computing and Communications Review*, vol. 9, no. 1, January 2005.

[65] M. Cardei, M. T. Thai, Y. Li, and W. Wu, "Energy-efficient target coverage in wireless sensor networks," in *IEEE INFOCOM*, 2005.

[66] Z. Abrams, A. Goel, and S. Plotkin, "Set K-Cover algorithms for energy efficient monitoring in wireless sensor networks," in *IEEE IPSN*, 2004.

[67] S. Slijepcevic and M. Potkonjak, "Power efficient organization of wireless sensor networks," in *IEEE ICC*, 2001.

[68] Texas Intruments, "2.4 ghz ieee 802.15.4 / zigbee-ready rf transceiver (rev. b)," in Available at http://focus.ti.com/docs/prod/folders/print/cc2420.html, 2007.

[69] A. Cerpa, J. Elson, D. Estrin, L. Girod, M. Hamilton, and J. Zhao, "Habitat monitoring: Application driver for wireless communications technology," in *Proc. of the 2001 ACM SIGCOMM Workshop on Data Communications in Latin America and the Caribbean*, April 2001.

[70] IEEE Computer Society, "IEEE computer society," in *IEEE Standard for Information technology Telecommunications and Information Exchange between Systems Local and Metropolitan Area Networks Specific Requirements:* Part 15.4: Available at http://standards.ieee.org/getieee802/download/802.15.4a-2007.pdf.

[71] C. Gui and P. Mohapatra, "Power conservation and quality of surveillance in target tracking sensor networks," in *Proceedings of 10th Annual ACM International Conference on Mobile Computing and Networking (MOBICOM 2004)*, 2004.

[72] T. He, S. Krishnamurthy, J. A. Stankovic, T. Abdelzaher, L. Luo, R. Stoleru, T. Yan, L. Gu, J. Hui, and B. Krogh, "Energy-efficient surveillance system using wireless sensor networks," in *MobiSys '04*, 2004.

[73] A. Chen, S. Kumar, and T. H. Lai, "Designing localized algorithms for barrier coverage," in *MOBICOM*, 2007.

[74] X. Wang, G. Xing, Y. Zhang, C. Lu, R. Pless, and C. Gill, "Integrated coverage and connectivity configuration in wireless sensor networks," in *Proceedings of the 1st International Conference on Embedded Networked Sensor Systems (Sensys 2003)*, 2003.

[75] Y. Gu and T. He, "Data forwarding in extremely low duty-cycle sensor networks with unreliable communication links," in *Proceedings of the 5th International Conference on Embedded Networked Sensor Systems (Sensys 2007)*, pages 321–334, 2007.

[76] D. Couto, D. Aguayo, J. Bicket, and R. Morris, "A high throughput pathmetric for multi-hop wireless routing," in *MobiCom 2003*, 2003.

[77] A. Woo, T. Tong, and D. Culler, "Taming the underlying challenges of reliable multihop routing in sensor networks," in *SenSys '03*, 2003.

[78] D. B. Johnson and D. A. Maltz, *Dynamic Source Routing in Ad Hoc Wireless Networks*, T. Imielinski and H. Korth, Eds. Kluwer Academic Publishers, 1996.

[79] C. E. Perkins and E. M. Royer, "Ad-hoc on demand distance vector routing," in *WMCSA*, 1999.

[80] C. Intanagonwiwat, R. Govindan, and D. Estrin, "Directed diffusion: A scalable and robust communication paradigm for sensor networks," in *the Sixth Annual International Conference on Mobile Computing and Networks*, 2000.

[81] D. Ganesan, R. Govindan, S. Shenker, and D. Estrin, "Highly resilient, energy efficient multipath routing in wireless sensor networks," *Mobile Computing and Communications Review*, vol. 1, no. 2, 2002.

[82] K. Lin, J. Yu, J. Hsu, S. Zahedi, D. Lee, J. Friedman, A. Kansal, V. Raghunathan, and M. Srivastava, "Heliomote: Enabling long-lived sensor networks through solar energy harvesting," in *Sensys '05*, 2005.

[83] J. Paradiso and M. Feldmeier, "A compact, wireless, self-powered push-button controller," in *Ubicomp '01*, 2001.

[84] S. Meninger, J.O. Mur-Miranda, R. Amirtharajah, A. Chandrakasan, and J. Lang, "Vibration-to-electric energy conversion," in *ISLPED '99*, 1999.

[85] Z. Zhong, D. Wang, and T. He, "Sensor node localization using uncontrolled events," in *International Conference on Distributed Computing Systems (ICDCS)*, 2008.

[86] X. Jiang, J. Polastre and D. Culler, "Perpetual environmentally powered sensor networks," in *Proceedings of the Fourth International Symposium on Information Processing in Sensor Networks (IPSN2005)*, 2005.

[87] J. Sorber, A. Kostadinov, M. Garber, M. Brennan, M. D. Corner, and E. D. Berger, "Eon: A language and runtime system for perpetual systems," in *Proceedings of the 5th International Conference on Embedded Networked Sensor Systems (Sensys 2007)*, pages 161–174, 2007.

[88] A. Lachenmann, P. J. Marron, D. Minder, and K. Rothermel, "Meeting lifetime goals with energy levels," in *Sensys*, 2007.

[89] Y. Gu, T. Zhu, and T. He, "ESC: Energy Synchronized Communication in sustainable sensor networks," in *Proc. of the 17th International Conference on Network Protocols (ICNP '09)*, October 2009.

[90] T. He, S. Krishnamurthy, L. Luo, T. Yan, L. Gu, R. Stoleru, G. Zhou, Q. Cao, P. Vicaire, J. A. Stankovic, T. F. Abdelzaher, J. Hui, and B. Krogh, "VigilNet: An integrated sensor network system for energy-efficient surveillance," *ACM Transactions on Sensor Networks*, vol. 2, no. 1, pp. 1–38, February 2006.

[91] T. Yan, Y. Gu, T. He, and J. A. Stankovic, "Design and optimization of distributed sensing coverage in wireless sensor networks," *ACM Transaction on Embedded Computing System (TECS)*, 2007.

[92] Y. Gu, J. Hwang, T. He, and D. H. Du, "uSense: A unified asymmetric sensing coverage architecture for wireless sensor networks," in *ICDCS '07*, 2007.

Chapter 5

Energy-Efficient Rate Adaptation in Long-Distance Wireless Mesh Networks

Zenghua Zhao, Zhibin Dou and Yantai Shu
Tianjin University, Tianjin, China. zenghua@tju.edu.cn

5.1   Introduction

The issue of green communications is compounded by the incredible growth of wireless communications in the developing regions which use wireless as a medium to leap frog past traditional wire-line technologies. In such regions, energy is a precious resource, some remote areas even rely on inefficient diesel generators for power, which will significantly grow the carbon footprint of wireless communication. Therefore the impact of saving energy for wireless communication is huge in the developing regions [1].

IEEE 802.11-based LDmesh (Long-Distance wireless mesh) networks are emerging as cost-effective options to provide Internet connectivity to rural areas with low population density in developing regions, to enable ICT (Information and Communication Technology) services. The primary cost gains arise from the use of low-cost and low-power single-board computers and high-volume low-cost off-the-shelf 802.11 wireless cards using unlicensed spectrum. The nodes are also lightweight and do not need expensive towers. These networks are very different from the short-range multi-hop urban mesh networks [2]. Unlike mesh networks, which use omnidirectional antennas to cater to short ranges (less than 1-2 km at most), LDmesh networks are comprised of point-to-point wireless links that use high-gain directional antennas (e.g., 24 dBi) with line of sight over long distances (10-100 km).

Furthermore, LDmesh networks also provide green communication solutions for rural residents due to low power requirements. The power consumption of a LDmesh router is around 15w, which is low enough to avoid diesel generators and enable local generation via solar or wind. The LDmesh network in developing regions is promising with its cost efficiency and low-power requirements.

Rate adaptation is an efficient approach to improve link throughput in multi-rate supported wireless networks such as 802.11 networks. Although there have been numerous adaptation algorithms [9-19], all of them are designed for 802.11 networks with short links and cannot work in LDmesh networks directly. First, most practical rate adaptation algorithms make decisions solely based on the ACK, which is sent upon successful delivery of a DATA packet. However, in LDmesh networks there is no ACK in the same time slot, since data are transmitted or received in separate time slots. Second, none of them takes energy efficiency into consideration, since they work in urban areas where power supply has never been a problem.

While rate adaption schemes have been studied extensively in the past, ERAA is the first designed for *long-distance* wireless links considering the energy-efficiency issues [7]. Over-provisioning transmit power will obtain higher throughput but consume more energy. ERAA achieves the best in both worlds, that is, it saves energy without compromising the network performance.

To this end, ERAA integrates several innovative techniques. First, it proposes an efficient probing algorithm to obtain the FDR (Frame Delivery

Ratio)-RSSI (Received Signal Strength Indicator) envelope mapping for each bit rate. FDR-RSSI is linear and remains invariant for a period of time so that it can be used to facilitate rate selection. Second, it presents an energy-efficient rate selection approach, which leverages the path loss information based on channel reciprocity theory[1] [20]. It selects bit rate and txpower (transmit power) according to the FDR-RSSI mapping, to maximize the efficient link throughput but at minimum energy consumption. Third, it provides a CNP-CUSUM (Continuous Non-Parameter-CUmulative Sum) technique to detect the distortion of FDR-RSSI that arises from external WiFi interference. Based on extensive measurements, beacon loss ratio has been found to be an accurate indicator of the external interference intensity.

CNP-CUSUM is verified using large real-world trace data. Extensive simulations were conducted to evaluate the performance of ERAA via QualNet 4.5 [21]. The simulation results show that ERAA can improve link throughput efficiently with minimum energy consumption.

The rest of the chapter is organized as follows. The background of LDmesh network is introduced in Section 5.2. The rate adaptation algorithms are overviewed briefly in Section 5.3. Section 5.4 describes ERAA in details, and Section 5.5 gives concluding remarks.

Symbols and notations used throughout the paper can be found in the Appendix.

5.2 Background: Long-Distance Wireless Mesh Networks

LDmesh network is an IEEE 802.11-based long-distance multi-hop wireless mesh network. Fig. 5.1 depicts a schematic of such networks. The nodes connect to others through point-to-point long-distance wireless links with high gain directional antennas. One or several nodes are gateways connecting to the wired Internet. The others are mesh routers which collect data from end devices as well as forward data from other nodes to the gateways through multiple hops. LDmesh networks have three prominent characteristics [4]: (a) multiple radios (i.e., interfaces) per node, (b) use of directional antennas with gains as high as 30 dBi, and (c) long-distance point-to-point links (from tens to hundreds of kilometers). These factors distinguish these networks from most networks considered in prior literature on multi-hop 802.11 networks. CSMA/CA-based 802.11 MAC protocol is not suited for LDmesh networks. It suffers from unnecessary contention resolution in a point-to-point link, delays due to the larger round-trip time, and even ACK timeouts on the long-distance links.

---

[1]If the role of the transmitter and receiver are instantaneously interchanged, the signal transfer function between them remains unchanged.

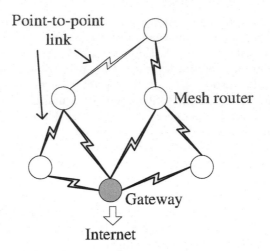

Figure 5.1: Long-distance wireless networks.

5.2.1   Inter-Link Interference Model

In an LDmesh network, each node has multiple point-to-point links, each with
a separate directional antenna, as shown in Figure 5.1. Although they use
directional antennas, the links at a node cannot really operate independently
due to the presence of side-lobes. This is also called the near field effect [3].
Because of this effect, when a node is transmitting along a link, it cannot
simultaneously receive along another link, since the transmitter will interfere
with the receiver at that node, as shown in Fig. 5.2.a. This is the inter-link
interference model in LDmesh networks.

However, it is possible to have synchronous operation (SynOp) where the
links at a node are all transmitting (SynTx), or all receiving (SynRx), as
depicted in Fig. 5.2.b and c.

5.2.2   2P MAC Protocol

TDMA-type MAC protocols have been proposed for LDmesh networks to
achieve high network performance, since CSMA/CA suffers poor performance
in such networks. 2P MAC is well-known among them. In 2P, the nodes trans-
mit/receive from each link simultaneously. 2P operates by switching each node
in the network between SynTx and SynRx phases (two phases, hence the
name 2P) alternatively. That is, when a node is in SynTx, its neighbors are in
SynRx, and vice versa. Further, when a node switches from SynRx to SynTx,
its neighbors switch from SynTx to SynRx, and vice versa. Obviously, 2P
operation requires the topology be bipartite.

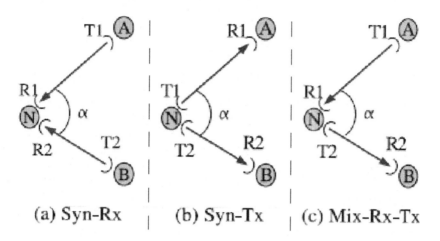

(a) Syn-Rx   (b) Syn-Tx   (c) Mix-Rx-Tx

Figure 5.2: Inter-link interference scenarios.

5.3 Overview: Rate Adaptation in Wireless Networks

Rate adaptation is a mechanism critical to the system performance by exploiting the multi-rate capability at the physical layer (PHY) in 802.11-based wireless networks, such as the widely deployed WLANs and the emerging mesh networks. The current 802.11 specifications mandate multiple transmission rates at the physical layer that use different modulation and coding schemes. For example, the 802.11b PHY supports four transmission rates (1, 2, 5.5, 11 Mbps), and the 802.11a PHY offers eight rates (6, 9, 12, 18, 24, 36, 48, 54 Mbps). The goal of rate adaptation is to maximize the transmission goodput at the receiver. It exploits the PHY multi-rate capability and enables the sender to select the best rate out of the mandated options based on the dynamic channel condition.

In recent years, a number of algorithms for rate adaptation [9-19] have been proposed in the literature. They can be classified into two categories: statistical based and PHY-metric based ones. The former uses either long-term or short-term statistics while the latter leverages PHY layer metrics such as RSSI (Received Signal Strength Indicator) to facilitate bit rate selection.

Statistical based algorithms. The first documented bit rate adaptation algorithm, ARF (Auto Rate Fallback) [8], keeps track of the current bit rate and the number of consecutive transmissions. Specifically, it increases the transmission rate after ten consecutive successful frame transmissions and lowers rates when encountering two successive transmission failures. Its extensions, such as Adaptive ARF [9], Fast-LA [11] and [10] try to reduce probing overhead by choosing adaptive success failure threshold for rate increase/decrease. Other derivatives, like LD-ARF [12] and CARA [13], seek to differentiate losses between wireless channel fading and frame collisions. Other statistical based algorithms, such as SampleRate [14] and RRAA [15], also require explicit feedback of transmission result of each frame.

PHY-metric based algorithms. RBAR [16] adopts RTS/CTS to obtain SNR (Signal-to-Noise Ratio) and selects data rates by looking up a predetermined SNR-data rates table. However, RTS/CTS is not supported in LDmesh networks with TDMA-type MAC. SoftRate [17] uses confidence information from the physical layer to estimate the prevailing channel BER (Bit Error Rate). The PHY layer coding used has to be implemented on software radios, which limits its prevalence on off-the-shelf 802.11 commodity hardware. Channel Busy Time (CBT) refers to the fraction of time for which the wireless channel is busy within a given interval. WOOF [18] adopts CBT as a congestion metric to reflect medium utilization, and incorporates it into SampleRate with congestion perception ability. It uses special hardware registers via reverse engineering.

They cannot work in LDmesh networks directly, since they need one-to-one acknowledgement frames from the receiver. However in such networks, TDMA-type MAC protocol is adopted, and data is transmitted and received in SynTX and SynRX time slots respectively, without feedback in one time slot.

In CHARM [19], the source station leverages channel reciprocity to obtain channel information (i.e., path loss and the predicted RSSI). Then it looks up a pre-determined RSSI-Bit Rates table to select a bit rate. CHARM shares some elements with ERAA, but there are significant differences. ERAA deploys CNP-CUSUM to process beacon loss ratio sequences, in order to detect the variations of interference intensity in time. When the variations exceed a threshold, a notification is sent to the sender to adjust the FDR-RSSI mapping. ERAA adapts not only bit rate but also transmission power to achieve energy efficiency. In contrast, CHARM takes the multi-rate retry features of the Atheros hardware, which is infeasible at TDMA-based long-distance links that prohibits timely one-to-one MAC ACK frames.

Besides, all above-mentioned schemes have no concerns about energy efficiency, which remains a significant problem for LDmesh networks. To the best of our knowledge, ERAA is the first to seek the energy efficiency and rate adaptation in LDmesh networks.

5.4   ERAA: Energy-Efficient Rate Adaptation Algorithm

In this section, we will first introduce the network model and operations, and then go to the details of ERAA.

5.4.1 Network Model and Operations

We consider a LDmesh network with several semi-urban long-distance wireless links (one end in urban, the other in rural). Semi-urban links are the usual link type in real world. They experience high channel dynamics due to time-variant channel fading and external WiFi interference. Therefore, these links have high loss rates ranging from 2% to as high as 80%, which can be highly asymmetric [22]. Moreover, the duration of loss bursts also vary from a transient high burst to a long burst lasting over 25-30 minutes.

We assume a simple TDMA-type MAC protocol used in the LDmesh network. It has the capability of providing bulk ACKs and PHY-metrics such as RSSI. The length of time slot is adaptive to the traffic load and the maximum value is 20ms. A marker frame is adopted for link synchronization. The sender ends its SynTx time slot with a marker frame, which notifies the receiver of its status change (from SynTx to SynRx). After correct reception of a marker frame, the receiver begins its SynTx time slot. In each SynTx time slot, bulk ACKs are piggybacked with the data frames. The ACKs are intended to acknowledge all correctly received data frames at the last SynRx time slot.

5.4.2 Design Overview

We now present the system design and the rationale behind our approach. ERAA is an energy-efficient rate adaptation algorithm designed for LDmesh networks, so it should address the following challenges:

- **How to determine the link quality accurately and quickly?** It is a primary problem for rate adaptation, and even harder in LDmesh networks. On one hand, the link quality is highly fluctuated thus difficult to capture. On the other hand, there is no one-to-one frame acknowledgment since the time is divided into SynTx and SynRx timeslot which makes feedback untimely.

- **How to identify the variations of external WiFi interference?** External WiFi sources working in the same/adjacent ISM bands are the key interference degrading the performance of LDmesh networks. They are the primary factors to cause link loss, therefore the variations of external WiFi interference make the link loss rate fluctuate together. This makes it harder to indicate the link quality. Moreover, because the strength of the external WiFi interference changes randomly, it is a great challenge to describe their variations.

- **How to save energy without compromising the link throughput?** Energy efficiency is a special requirement in LDmesh networks, since they are deployed in the wild areas with limited power supply. High txpower can increase RSSI thus to improve the link throughput; but over-provisioning txpower will consume a lot of energy. On

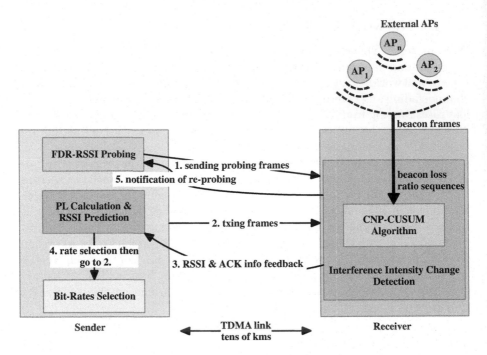

Figure 5.3: ERAA design framework.

the other hand, lower txpower can save energy, but lower RSSI will result in higher packet loss rate thus deteriorating link performance. It is challenging to make the best of both worlds.

To solve the above issues, ERAA adopts several novel techniques depicted in Fig. 5.3.

- **FDR-RSSI probing.**   FDR-RSSI is used to indicate the link quality, since their mapping is linear and keeps invariant in a period of time. The sender in its SynTx time slot sends probing frames to obtain the FDR-RSSI mapping. We designed a probing algorithm to reduce the probing overhead, i.e., the number of probing frames.

- **PL (Path Loss) calculation and RSSI prediction.**   We make use of receivers' feedback and channel reciprocity to obtain path loss in both directions of a link and predict RSSIs in the upcoming SynTx time slot, which enables us to estimate the link quality in an accurate manner.

- **Bit rate selection.**   We designed a bit rate selection algorithm to conserve energy without compromising the link throughput. To this end, an energy cost function is defined as the function of $rssi$, which represents energy consumption per bit at that $rssi$.

- **CNP-CUSUM algorithm.**   It is employed to detect the changes of the external WiFi interference, which may distort the FDR-RSSI mappings. We found that beacon loss ratio is capable of indicating the variations of the external interference. CNP-CUSUM algorithm is thus adopted to detect the changes of the interference based on the beacon loss ratio sequence. If changes are detected, the FDR-RSSI probing is triggered.

In general, ERAA consists of two stages, the probing stage and the adapting stage. The details of each component will be illustrated in the rest of this section.

5.4.3   The Probing Stage

In the probing stage, the sender transmits a group of probing frames at different bit rate and txpower to determine the relationship of FDR and RSSI, i.e., FDR-RSSI mappings.

FDR is the probability of the frame received successfully, and can be calculated as:

$$FDR = \frac{\text{\# of frames received successfully}}{\text{\# of frames sent}}.$$

RSSI is a good predictor of FDR when the multipath fading is negligible [19] (this is the case in LDmesh networks). The relationship of FDR and RSSI remains approximately invariant for a period of time; therefore, we can use FDR-RSSI as the link quality indicator.

In order to estimate FDR-RSSI mappings for a link, we transmit groups of probing frames at different bit rate and txpower. One technical difficulty in the above estimation is how to reduce the high probing overhead. Taking IEEE 802.11b MAC for example, there are 4 bit rates in total. For the wireless card used in our measurements, its txpower can be adjusted from 0 dBm to 24 dBm by a step size of 1 dBm. In this case, there are 100 (bit rate, txpower) combinations (4 bit rates × 25 steps/bit rate). If each probing frames group has 20 frames for each combination, then the sender should send 2000 probing frames to determine the FDR-RSSI mapping. Worse still, transmitting frames at lower bit rates significantly prolongs the transmission time, for example, approximately 11 times longer at 1 Mbps than that at 11 Mbps (and verified by our measurements), which is unacceptable and requires improvement.

Actually, at a certain txpower only one (bit rate, txpower) combination is meaningful that makes its effective throughput to be higher than any other bit rates. The effective throughput is defined as

$$ET_{tr}(rssi) = FDR_{tr}(rssi) \times tr, \qquad (5.1)$$

Algorithm 1 FDR-RSSI probing algorithm

1: **procedure** FDR_RSSI_PROBING(A, B, C, H, n, th)
2:     **if** $B \neq \phi$ **then**
3:         pair $a \leftarrow (max(A), max(B))$
4:         **if** $H(a) \notin H$ **then**
5:             probing FDR_a, $H(a) \leftarrow FDR_a$
6:         **if** $FDR_a >= th$ or $H = \phi$ **then**
7:             $D = \{d | d \in A \backslash \{txrate(a)\}, d \geq ET_a\}$
8:             $E = \{\text{pair } e : (x, y) | x \in A \backslash \{txrate(a)\}, y \in C, H(e) \in$
    $H, ET_e \geq ET_a\}$
9:             **if** $D \neq \phi$ **then**
10:                 **for all** $d \in D$ **do**
11:                     pair $z \leftarrow (d, txpwr(a))$
12:                     $G = \{\text{pair } g : (x, y) | x = d, y \in C, H(g) \in H\}$
13:                     **if** $G = \phi$ or $\exists g \in G, ET_g > ET_a$ **then**
14:                         probing FDR_z, $H(z) \leftarrow FDR_z$
15:                         $H \leftarrow H \backslash \{H(g) | g \in G\}$
16:                     **if** $H(z) \in H$ and $ET_z \geq ET_a$ **then**
17:                         $A \leftarrow A \backslash \{txrate(a)\}$
18:                         **break**
19:             **else**
20:                 $H \leftarrow H \backslash \{H(e) | e \in E\}$
21:                 **for all** $e \in E$ **do**
22:                     pair $f \leftarrow (txrate(e), txpwr(a))$
23:                     **if** $ET_a < ET_e$ **then**
24:                         probing FDR_f, $H(f) \leftarrow FDR_f$
25:                     **if** $H(f) \in H$ and $ET_f \geq ET_a$ **then**
26:                         $A \leftarrow A \backslash \{txrate(a)\}$
27:                         **break**
28:             **if** $D = \phi$ **then**
29:                 $B \leftarrow B \backslash \{txpwr(a)\}$
30:                 $C \leftarrow C + \{txpwr(a)\}$
31:         FDR_RSSI_PROBING(A, B, C, H, n, th)

---

where tr is the bit rate adopted. $FDR_{tr}(rssi)$ and $ET_{tr}(rssi)$ denote the frame deliver ratio and the effective throughput at the bit rate tr and RSSI $rssi$, respectively. One obvious but important observation is that at the same txpower, higher effective throughput cannot be obtained at lower bit rate even if FDR is almost 1. Consequently, it is unnecessary to try other combinations; thus the overhead is reduced. For example, if current combination (11 Mbps, 24 dBm) has yielded an effective throughput of 7 Mbps, then it is not necessary to try other combinations at the same txpower of 24 dBm, i.e., (1 Mbps, 24 dBm), (2 Mbps, 24 dBm) and (5.5 Mbps, 24 dBm), because 7 Mbps is already higher than any other bit rate among them.

In view of this, we design a FDR-RSSI probing algorithm that can obtain the FDR-RSSI mappings at minimum number of probing combinations, e.g., around 25 probing combinations in the above example. Algorithm 1 outlines our approach. In the algorithm, A is the set of bit rates and B is the set of adjustable txpower. C is the set of txpower probed and initially set as ϕ. H is the collection of FDR for each probed combination. n is the number of probing frames to be sent for each combination, and th is the customized threshold of FDR. If FDR is less than th, the probing algorithm terminates. In this situation, the corresponding RSSI is usually too low to obtain good link throughput. Therefore, it is unnecessary to probe any lower txpower. Finally, $txrate(a)$ and $txpwr(a)$ return the bit rate and txpower of probing combination a, respectively.

As shown in Algorithm 1, probing begins with combination of (maximum bit rate, maximum txpower). For each probing, it selects the maximum bit rate and maximum txpower from set A and B, respectively, sends probing frames at that combination and collects the corresponding FDR and effective throughput. If it is possible to achieve higher effective throughput, continue to probe. Otherwise algorithm terminates. All the (RSSI, FDR) in set H are the results. The piece-wise linear function of RSSI and FDR is obtained after this probing algorithm.

5.4.4 The Adapting Stage

When the FDR-RSSI relationship is determined in the probing stage, the algorithm enters the adapting stage to adapt its bit rate and txpower to achieve a tradeoff between energy efficiency and link throughput. There are three components involved in this stage, i.e., path loss calculation and RSSI prediction, bit rate selection and CNP-CUSUM.

5.4.4.1 Path Loss Calculation and RSSI Prediction

In order to select appropriate bit rate and txpower, the rate adaptation algorithm needs to estimate the RSSIs of frames sent at the specified settings. RSSI can be predicted by estimating the path loss at the forward direction (from the sender to the receiver). RSSI can be calculated as [15]:

$$RSSI = P_{tx} - PL + NF, \qquad (5.2)$$

where P_{tx} is the txpower of the sender, PL is the forward-direction path loss and NF is the noise floor detected by the wireless card.

Path loss calculation. According to the channel reciprocity theory, the path loss in the forward direction is approximately the same as that in the backward direction (from the receiver to the sender). The path loss in the backward direction can be obtained in the last time slot (i.e., SynRx time slot). In SynRx time slot, RSSI is obtained for each received frame from MAC layer. The txpower of the sender is piggybacked in the frames. Therefore, we can estimate the path loss in the backward direction using Eq. 5.2.

RSSI prediction.   It is straightforward to predict RSSI when we get the forward-direction path loss, i.e., submitting the path loss into Eq. 5.2. In other words, given a txpower, we can predict the RSSI of the frames sent at that txpower. Further we can estimate the FDR according to FDR-RSSI, then the effective throughput using Eq. 5.1 at different bit rate.

5.4.4.2   Energy-Efficient Bit Rate Selection

Energy consumption should be considered jointly with rate selection. According to FDR-RSSI mapping, when the FDR exceeds a certain threshold, a small improvement of FDR requires significant increase of RSSI. From Eq. 5.2, given path loss and noise floor, higher RSSI needs higher txpower thus more energy. In other words, much more energy consumption leads to little improvement of FDR above the threshold. Furthermore, it could be even worse in the presence of interference, as higher medium contention results in many more frame collisions, no matter how high the RSSI is. To avoid such circumstances, a cost function ERB (Energy Ratio per Bit) is introduced to represent energy consumed per bit at a *rssi*. Let *tr* be PHY bit rate and *tp* be txpower; we have:

$$
\begin{aligned}
ERB(rssi) &= \frac{EC(\Delta t)}{BS(\Delta t)} = \frac{\eta_{tr}(tp) \times tr \times \Delta t}{ET_{tr}(rssi) \times \Delta t} \\
&= \frac{\eta_{tr}(tp) \times tr \times \Delta t}{FDR_{tr}(rssi) \times tr \times \Delta t} \approx \frac{10^{k_{tr} \times (tp) + b_{tr}}}{FDR_{tr}(rssi)}
\end{aligned} \tag{5.3}
$$

where $EC(\Delta t)$ and $BC(\Delta t)$ are the total energy consumed and the total bits correctly received in time interval Δt, respectively. $\eta_{tr}(tp)$ is the energy utilization factor for transmitting 1 bit at the PHY rate *tr* and the txpower *tp*. It can be regarded as a linear function in log scale, as shown in Fig. 5.4, denoted as:

$$
Log(\eta_{tr}(tp)) = k_{tr} \times tp + b_{tr},
$$

which yields Eq. (5.3).

The next step is to determine the valid range of Δ_{rssi}. NIC (Network Interface Card) chips support only a limited range of txpower levels (conforming to FCC regulations), say, from tp_{min} to tp_{max}, where tp_{min} and tp_{max} are the minimum/maximum txpower supported, respectively. Since the *rssi* is predicted at txpower *tp*, the valid Δ_{rssi} value ranges from $(tp_{min} - tp)$ to $(tp_{max} - tp)$.

The range of $rssi + \Delta_{rssi}$ can be divided into several continuous subintervals such that at each subinterval, the effective throughput (ET_{tr}) is always

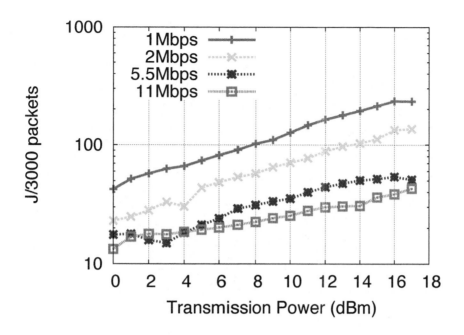

Figure 5.4: Energy consumption for transmitting 3000 1000-byte frames at different txpower and bit rate.

the highest one compared to any other tr', as follows:

$$RSSI_{tr} = \{rssi|ET_{tr}(rssi) > ET_{tr'}(rssi), \forall tr' \in R \backslash \{tr\}\},$$

where R is a set including all valid PHY rates. Each subinterval is continuous since $FDR_{tr}(rssi)$ is a monotonic function.

The sender tries to calculate each $ERB(rssi + \Delta_{rssi})$ with each discrete Δ_{rssi} decreasing from the maximum achievable value $(tp_{max} - tp)$ to the minimum value $(tp_{min} - tp)$ with corresponding tr that suffices $(rssi + \Delta_{rssi}) \in RSSI_{tr}$. Let $rssi^*$ denote the RSSI at which the minimum ERB is obtained as follows,

$$rssi^* = \operatorname*{argmin}_{rssi + \Delta_{rssi}} (ERB).$$

Therefore, to achieve energy efficiency, the sender should select the bit rate tr satisfying $rssi^* \in RSSI_{tr}$ and the corresponding tp, which is the optimal bit rate with minimum energy consumption.

5.4.4.3 CNP-CUSUM to Handle External Interference

Another aspect influencing the algorithm performance is external WiFi interference. The FDR-RSSI relationship could be severely distorted in the presence of external WiFi interference. In that case, a considerable rise in RSSI

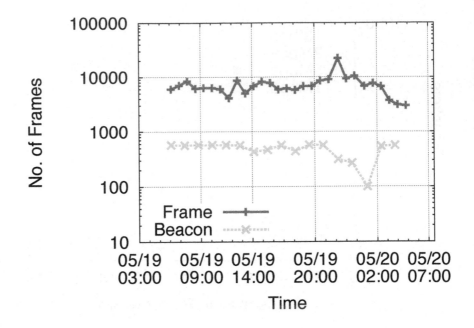

Figure 5.5: Number of frames (beacons) per minute received by a 802.11 router.

produces only little FDR improvement. To overcome this issue, the CNP-CUSUM algorithm leverages the beacon loss ratio of each external WiFi network observed by the receiver. It monitors the changes of the beacon loss ratio sequence, which means the interference intensity has been dramatically altered. When the changes are detected, the sender is required to re-probe FDR-RSSI mapping as stated in Section 5.4.3.

The beacon frames are periodically broadcasted by AP; any clients can receive them no matter whether the client is currently associated with the AP or not. The interval between two beacon transmissions is approximately 100ms.

The frame collision probability can be implicitly represented by beacon loss ratio. One may argue that channel fading also incurs fluctuations of beacon loss ratio. However, channel fading causes the ratio to fluctuate up and down around its average, which in a statistical way does not show drifts beyond its average. On the other hand, losses incurred by frame collision do express its drifting behavior since WiFi interference usually lasts tens of minutes [22] in view of people's behavior. Our outdoor experiments also show similar results. A mesh router is set to monitor mode to sniff all frames for nearly one day on the top of our teaching building at Tianjin University campus. We group frames and beacons from a specific AP every minute and plot the results in Fig. 5.5. It shows that beacon loss ratio is almost 0% (received 585 out of 600 beacons per minute) with network traffic of 5000–6000 frames/min while beacon losses are dramatically increased when traffic exceeds 10,000 frames/min from 22:00pm to 0:00am. Thus a rise/decline of contention level will result in a higher/lower beacon loss ratio statistically.

A key challenge lies in abrupt change detection of beacon loss ratio, since the change of beacon loss ratio reflects the medium contention level at the receiver. Fortunately, CUSUM test [23] is an effective technique to detect the distribution change of observations as quickly as possible. Its effectiveness on small drifts detection has been proved [24]. Here, we adopt a bilateral CUSUM[2] to identify the interference intensity changes.

Let $\{X_i | i = 0, 1, \cdots\}$ be a time sequence of beacon loss ratio. Since both wireless channel fading and interferences are random processes, we can consider $\{X_i\}$ being a stationary random process. Note that CUSUM algorithm has an assumption that the average value of the random sequence should be negative, and it becomes positive after change. Therefore, without losing any statistical properties, we transform the sequence $\{X_i\}$ into another random sequence $\{\overline{X}_i\}$ with a negative mean. Let $\overline{X}_i = X_i - a$ and $E(X_i) = c$, then $E(\overline{X}_i) = c - a$. In a given network environment, the parameter a is a constant. In this way, all the negative value of $\{\overline{X}_i\}$ will not be cumulated along the time. When drifts of beacon loss ratio take place, $\{\overline{X}_i\}$ will suddenly become large positive.

$$Z_i = \begin{cases} max(0, Z_{i-1} + X_i - a) & i > 0 \\ 0 & i = 0, \end{cases} \tag{5.4}$$

It is straightforward that Z_i is the maximum continuous increment until time i. A large Z_i is a strong indication of interference intensity rising. For the case of an opposite drifting (less intensive interference), it is equivalent to detect the increasing of the sequence $\{-X_i\}$; thus we have,

$$D_i = \begin{cases} min(0, D_{i-1} + X_i - b) & i > 0 \\ 0 & i = 0, \end{cases} \tag{5.5}$$

where b is larger than 0 and satisfies $E(X_i) - b > 0$. We determine the alarm time as follows,

$$\tau_1 = min\{i : i \geq 1, Z_i > h\} \tag{5.6}$$

$$\tau_2 = min\{i : i \geq 1, D_i < -h\}, \tag{5.7}$$

where h is the alarm threshold and it could be set to different values in Eq. (5.6) and Eq. (5.7). Thus, for a bilateral detection, the alarm time is $min\{\tau_1, \tau_2\}$, meaning that when $Z_i > h$ or $D_i < -h$, the system will give an alarm. Taking upward change detection (Eq. 5.4) for an example, assume the

---

[2]A bilateral CUSUM test is capable of detecting both the upward and downward changes of beacon loss ratio.

mean of beacon loss ratio after the alarm is K, and the sampling interval is Δ, then the detection delay is

$$D = \frac{h}{K - a} \times \Delta.$$

From the above equation, one can see that the detection delay D is a decreasing function of K but an increasing function of h. Furthermore, with an increase of the threshold h, the false alarm rate will become bigger; otherwise the false alarm time will become smaller. In order to achieve continuous detection of intensity change, after detecting the abnormal change, the algorithm needs to update a since the mean of beacon loss ratio is changed, and begins sampling with i reset to 0.

In the presence of multiple APs, a receiver applies the proposed CNP-CUSUM algorithm to each AP with carefully selected h and a, because, for example, an AP with lower RSSI is more vulnerable to channel fading, and a larger h and a should be selected. The change takes place if the majority of APs (greater than *threshold*) alarm, depicted as,

$$\sum_{ap \in AP} 1_{\{ap\ alarms\}} > threshold,$$

where AP denotes the set of interfering APs, and $1_{\{x\}}$ is the indicator of event x, defined as

$$1_{\{x\}} = \begin{cases} 1 & \text{event } x \text{ happens} \\ 0 & \text{otherwise.} \end{cases}$$

5.4.5 Performance Evaluation

We first verify CNP-CUSUM algorithm using trace data from test-bed experiment, then evaluate the performance of ERAA via extensive simulation in Qualnet 4.5. We compared the performance of ERAA with the fixed-bit-rate fixed-txpower mechanism. Detailed description of experimental setup for CNP-CUSUM verifications and simulation setup for ERAA evaluations can be found below.

The performance metrics link throughput and energy consumption ratio. The energy consumption ratio is defined as the ratio of total energy consumed (i.e., total frame sent × energy consumed per frame) to the total number of frames successfully received in a given time interval. We normalized the ratio in our algorithm.

5.4.5.1 CNP-CUSUM: Interference Intensity Change Detection

A. Experimental Setup

The CNP-CUSUM algorithm is verified using the trace collected from real experiments. In the experiments, a wireless node operating in monitor mode sniffs frames all day long in our lab. This choice is based on the fact that the lab is surrounded by many other 802.11b/g BSS (Basic Service Set) networks operating in various channels, most of which are in 3 orthogonal channels 1, 6 or 11. Besides, the network traffic shows obvious fluctuations at different times. We scanned the wireless channel and found that there were 8 APs in the vicinity.

For trace collecting, the sampling period is set to 1 second to generate the beacon loss ratio sequence. Moreover, for ease of plotting, the range of x axis is limited to 180 in Fig. 5.6 and Fig. 5.7, and 2500 in Fig. 5.8. Throughout the experiment result analysis, a number of randomly selected pieces of trace data are applied to our CNP-CUSUM algorithm. CNP-CUSUM keeps its effectiveness on all pieces of trace data. Due to limited space, we omit similar results obtained from other pieces of trace and pick one to present here.

We begin with the discussion of the upward drifts detection and then give the results for the bilateral case.

B. CNP-CUSUM Performance

The beacon loss ratio sequence $\{\overline{X}_i\}$ and the test statistics $\{Z_i\}$ are plotted in Fig. 5.6 and Fig. 5.7, respectively. The solid line in Fig. 5.6 is the beacon loss ratio, whereas the dashed line is its average. From the figures we observe that the loss ratio drifts by fluctuating rapidly after 80 samples. As shown in Fig. 5.7, the drift is detected when the CUSUM statistics exceed the given threshold $h = 50$. It also demonstrates that our CNP-CUSUM can detect the upward drifts in a timely manner.

The bilateral case is shown in Fig. 5.8. The straight line at y axis 10 and -10 are thresholds for $\{Z_i\}$ and $\{D_i\}$, respectively. The curved line represents the averaged beacon loss ratio sampled during 2500 seconds. It drifts upward at approximately 800, which is alarmed at asterisk "A". However, the drift does not stop increasing and begins to climb up at 920, which is detected at point 930, denoted as "B". In addition, downward drifts are also alarmed. Point "C" at 1090 alarms the decreasing of beacon loss ratio started at 1050.

5.4.5.2 Rate Adaptation Performance

A. Simulation Setup

In this scenario, a point-to-point 25km-long 802.11b WiFi link is set up operating on the aforementioned TDMA MAC with the maximum timeslot length of 20ms. Both nodes are equipped with directional antennas of 24 dBm gain. Txpower can be adjusted from 1 dBm to 25 dBm. The receiver is surrounded by two BSS networks with one AP for each. Every BSS network has several clients associated with the corresponding AP. This scenario represents a typical long-distance link with time-variant interference at the receiver side.

In addition, path loss model is two-ray model, fading model is the Rician model with parameter K of 3.8dB [25], and the shadowing model is constant with a mean of 4dB. There are three CBR flows with packet size 1000 bytes

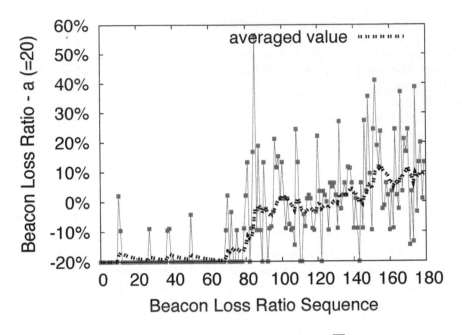

Figure 5.6: Beacon loss ratio sequence $\{\overline{X}_i\}$.

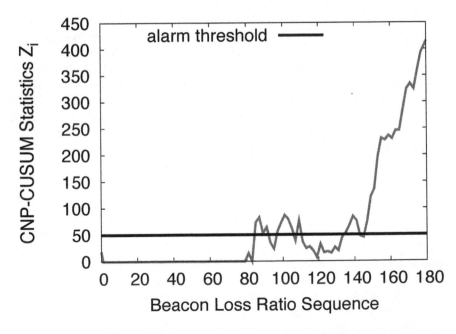

Figure 5.7: CNP-CUSUM sequence $\{Z_i\}$.

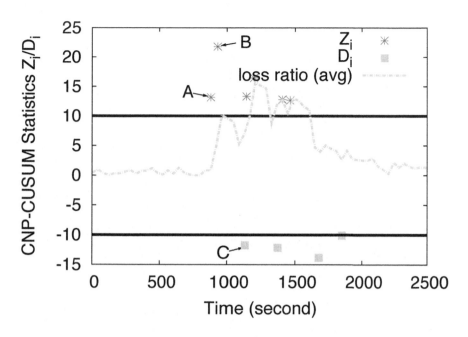

Figure 5.8: Bilateral drifts detection of beacon loss ratio.

in the scenario. One is for the point-to-point link with packet interval 1.5ms spanning the whole simulation time. The second is from a client station to another node in the same BSS network with a 5ms packet interval. It begins at 20s and terminates at 120s. At 70s, the third CBR traffic is injected into the other BSS network with a 2ms interval and lasts 100s.

We run 50 simulations with different random variable seeds and the result presented here is the average over all runs. Simulation time is set to 200s.

B. Simulation Results and Analysis

We compared the link throughput of ERAA with that of fixed schemes in Fig. 5.9. The sender sets its initial bit rate, txpower to 11 Mbps and 19 dBm, respectively. When the first interfering traffic CBR starts at 20s, the sender re-probes the FDR-RSSI mapping after receiving the notification of the interference intensity change detected by the receiver. After that, it chooses a higher txpower (21 dBm on the average) according to ERAA. This is in sharp contrast with the fixed scheme of (11 Mbps, 19 dBm) that exhibits significant throughput degradation, approximately 650 Kbps drop, when encountering the first interfering traffic. This situation is further exacerbated when another interfering traffic is injected into the network, which leads to the drop of link throughput to about 1650Kbps while ERAA roughly maintains its throughput by raising its txpower to about 23 dBm–24 dBm.

The deep throughput drop on the ERAA line corresponds to the FDR-

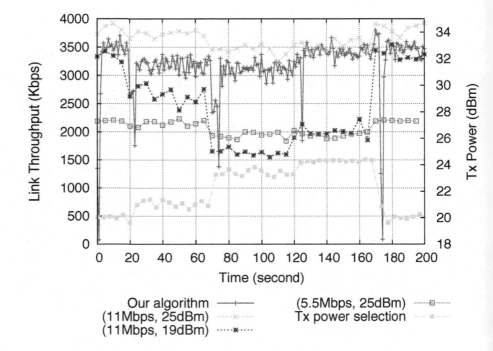

Figure 5.9: Link performance: ERAA versus fixed schemes.

Table 5.1: Probing Pairs for Interference Changes

	w/o Interference	CBR 1	CBR 2	CBR 3
11Mbps:	14~25dBm	14~25dBm	16~25dBm	16~25dBm
5.5Mbps:	9~15dBm	7~16dBm	10~18dBm	9~18dBm
2Mbps:	8, 11dBm	12dBm	16dBm	16dBm
1Mbps:	1~9dBm	N/A	N/A	N/A
Total:	30 pairs	21 pairs	20 pairs	21 pairs

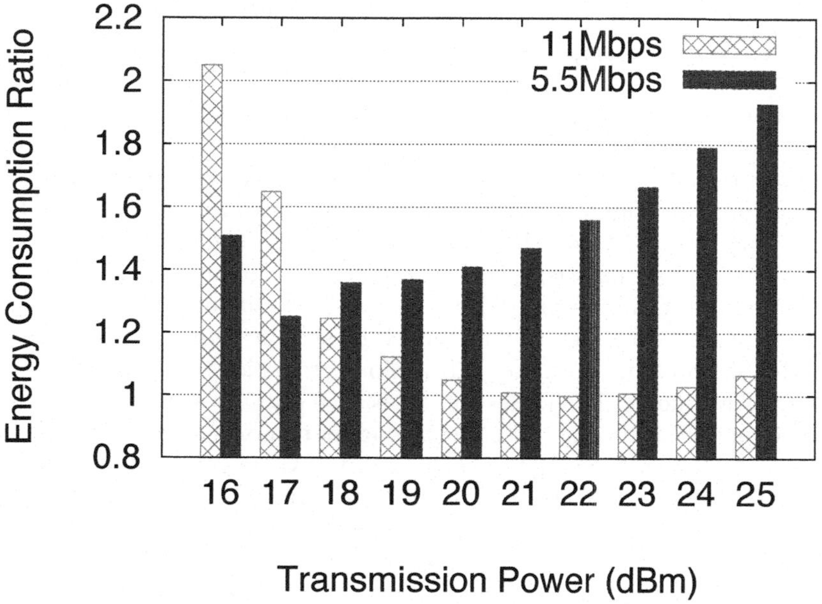

Figure 5.10: Energy consumption ratio.

RSSI re-probing, since re-probing requires transmitting packets at lower bit rates. However, by applying the algorithm stated in Section 5.4.3, the probing overhead is considerably reduced, as shown in Table 5.1. Moreover, it can be seen that the stronger the interference intensity is, the further the probing overhead is reduced. That is because lower bit rates prolong the transmission time which incurs a higher probability of frame error when encountering stronger interference. On the other hand, as depicted in Fig. 5.10, sending frames at the maximum txpower (25 dBm) outperforms ERAA slightly but leads to much more energy consumption. All fixed (bit rate, txpower) schemes have a higher energy consumption ratio except the one (11 Mbps, 22 dBm), which achieves the lowest ratio of 0.99956 among those schemes. However, it is only optimal for this network scenario since for links with different length, channel characteristics and interference intensity result in the diversity of optimal parameter settings. On the contrary, ERAA is self-adaptive and high energy efficiency in different interference environments. From the simulation results, we found that ERAA adopted 11 Mbps for majority of simulation time (285,790 out of 286,528 frames) and 5.5 Mbps for the remaining 738 frames. This is because for most of the time, raising txpower is sufficient to offer energy efficiency, which could avoid stepping down bit rates. Besides, Fig. 5.10 shows that reducing txpower at 11 Mbps before 22 dBm will improve the energy efficiency but degrade it after 22 dBm, since insufficient txpower incurs too severe frame losses.

5.5   Concluding Remarks

LDmesh networks have been deployed in developing regions of the world, providing low-cost low-power Internet access solutions for the residents scattered across rural regions. They use commercial off-the-shelf 802.11 wireless cards and high-gain directional antennas to reach long-distance range. Most of the routers work in wild areas with limited power supply (e.g., solar/wind power); therefore energy efficiency is a primary design paradigm in such networks. Unfortunately, the research work did not realize this issue until recently. ERAA has been proposed to seek saving energy in such networks. It is composed of FDR-RSSI probing, path loss calculation, predictions of RSSIs as well as txpower adjustments. It chooses bit rates in a way to achieve tradeoff between energy efficiency and effective throughput maximization. The receiver monitors the beacon loss ratio of each external AP in the neighborhood and detects the change of the interference intensity using CNP-CUSUM algorithm. If the change is above a threshold, the receiver notify the sender to trigger FDR-RSSI mapping process. Extensive simulation shows that ERAA achieves high energy efficiency while retaining considerable throughput gain by jointly selecting bit rates and txpower.

Although the LDmesh network is a promising green communication technology in developing regions, it still needs the design of the network in an energy-efficient way from bottom to top in view of the layered protocol stacks.

Acknowledgment

This research was supported in part by the National Natural Science Foundation of China under grant no. 61172063, and the Ministry of Education of China under grant no. 708024.

Appendix

a	A constant larger than c.
A	Set of available bit rates, e.g., 1, 2, 5, 5, 11 for 802.11b.
AP	Set of APs in the vicinity.
b	A constant value satisfying $E(X_i) - b > 0$.
b_{tr}	Intercept at bit rate tr.
B	Collection of adjustable txpower levels.
c	$E(X_i)$.
C	Set of txpower probed and initially is ϕ.
D_i, Z_i	CNP-CUSUM statistics.
ET_a	The effective throughput at pair $a = $ (txrate, txpower).
$E(X_i)$	Average value of sequence $\{X_i\}$.
$E(\overline{X}_i)$	Average value of sequence $\{\overline{X}_i\}$.
$FDR_{tr}(rssi)$	Return frame delivery ratio at bit rate tr and rssi $rssi$.
h	Alarm threshold for CNP-CUSUM statistics.
H	Collection of FDR for each probed (txrate, txpower) pair.
k_{tr}	Slope at bit rate tr.
K	Mean of beacon loss ratio after the alarm.
$max(T)$	Return the maximum element in set T.
n	Number of probing frames for each (txrate, txpower) pair.
NF	Noise floor.
$\overline{X}_i$	$X_i - a$
P_{tx}	Transmission power.
PL	Path loss.
R	Set of all valid PHY rates.
th	User-defined threshold to stop probing.
$threshold$	Threshold of alarming APs.
tp	Transmission power.
tp_{max}, tp_{min}	Maximum/minimum achievable tp.
tr	The bit rate.
$txpwr(p)$	Return the *txpower* of pair $p = $ (txrate, txpower).
$txrate(p)$	Return the *txrate* of pair $p = $ (txrate, txpower).

X_i	Beacon loss ratio of the ith sampling.
$1_{\{x\}}$	Indicator of event x.
$\eta_{tr}(tp)$	Energy utilization factor for transmitting 1 bit at bit rate tr and transmission power tp.
Δ	Sampling interval.
$EC(\Delta t)$	Total energy consumed in Δt.
$BS(\Delta t)$	Total bits correctly received in Δt.
Δ_{rssi}	Incremental/decreased amount of rssi.

Bibliography

[1] H. Sistek, "Green-tech base stations cut diesel usage by 80 percent," *in CNET News Green Tech*, 2008.

[2] I. Akyildiz, X. Wang, and W. Wang, "Wireless mesh networks: a survey," *Computer Networks*, vol. 47, issue 4, 15 March 2005, pp. 445-487.

[3] K. Chebrolu, B. Raman, and S. Sen, "Long-distance 802.11b links: performance measurements and experience," in *Proc. ACM MOBICOM*. ACM, New York, NY, USA, pp.74-85, 2006.

[4] B. Raman, and K. Chebrolu, "Design and evaluation of a new MAC protocol for long-distance 802.11 mesh networks," in *Proc. ACM MOBICOM*, Cologne, Germany, August 2005.

[5] S. Surana, R. Patra, S. Nedevschi, M. Ramos, L. Subramanian, Y. Ben-David, and E. Brewer, "Beyond pilots: keeping rural wireless networks alive," in *Proc. ACM NSDR*, August 18, 2008, Seattle, Washington, USA.

[6] The Akshaya E-Literacy Project. http://www.akshaya.net

[7] Z. Dou, Z. Zhao, Q. Jin, L. Zhang, Y. Shu, and O. Yang, "Energy-efficient rate adaptation for outdoor long distance WiFi links," in *Proc. workshop GNC'11*, April 10-15, Shanghai, China, 2011.

[8] L. Monteban, and A. Kamerman, "WaveLAN-II: a high-performance wireless LAN for the unlicensed band," *Bell Lab. Tech. J.*, vol. 2, no. 3, pp. 118-133, Aug. 1997.

[9] M. Lacage, M. H. Manshaei, and T. Turletti, "IEEE 802.11 rate adaptation: a practical approach," in *Proc. IEEE MSWiM*, Venice, Italy, October 2004.

[10] P. Chevillat, J. Jelitto, A. Noll Barreto, and H.L. Truong, "A dynamic link adaptation algorithm for IEEE 802.11a wireless LANs," in *Proc. IEEE ICC*, Anchorage, USA, May 2003.

[11] D. Qiao, and S. Choi, "Fast-responsive link adaptation for IEEE 802.11 WLANs," in *Proc. IEEE ICC*, Seoul, Korea, May 2005.

[12] Q. Pang, V. Leung, and Soung C. Liew, "A rate adaptation algorithm for IEEE 802.11 WLANs based on MAC-layer loss differentiation," in *Proc. IEEE BROADNETS*, Boston, USA, October 2005.

[13] S. Kim, S. Choi and D. Qiao, "CARA: collision-aware rate adaptation for IEEE 802.11 WLANs," in *Proc. IEEE INFOCOM*, Barcelona, Spain, April 2006.

[14] J. Bicket, "Bit-rate selection in wireless networks," *MIT Master Thesis*, 2005.

[15] S. H. Wong, H. Yang, S. Lu, and V. Bharghavan, "Robust rate adaptation for 802.11 wireless networks," in *Proc. ACM MOBICOM*, Los Angeles, USA, September 2006.

[16] G. Holland, N. Vaidya, and V. Bahl, "A rate-adaptive MAC protocol for multihop wireless networks," in *Proc. ACM MOBICOM*, Rome, Italy, July 2001.

[17] Mythili Vutukuru, Hari Balakrishnan and Kyle Jamieson, "Cross-layer wireless bit rate adaptation," in *Proc. ACM SIGCOMM*, Barcelona, Spain, August 2009.

[18] P. Acharya, A. Sharma, E. Belding, K. Almeroth and K. Papagiannak, "Congestion-aware rate adaptation in wireless networks: a measurement-driven approach," in *Proc. IEEE SECON*, San Francisco, USA, June 2008.

[19] G. Judd, X. Wang, and P. Steenkiste, "Efficient channel-aware rate adaptation in dynamic environments," in *Proc. ACM MOBISYS*, Breckenridge, USA, June 2008.

[20] C. Tai, "Complementary Reciprocity Theorems in Electromagnetic Theory," *IEEE Trans. Antennas Propagation*, vol. 40, pp. 675-681, 1992.

[21] QualNet simulator, http://www.scalable-networks.com/

[22] A. Sheth, S. Nedevschi, R. Patra, S. Surana, L. Subramanian, and E. Brewer, "Packet loss characterization in WiFi-based long distance networks," in *Proc. IEEE INFOCOM*, Anchorage, USA, May 2007.

[23] E. S. Page, "Continuous inspection schemes," *Biometrika*, vol. 41, pp. 100-115, 1954.

[24] M. Basseville and I. V. Nikiforov, "Detection of Abrupt Changes: Theory and Application," Prentice Hall, 1993.

[25] C. B. Dietrich Jr., K. Dietze, J. R. Nealy, and W. L. Stutzman, "Spatial, polarization, and pattern diversity for wireless handheld terminals," *IEEE Trans. Antennas Propagation*, vol. 49, pp. 1271-1281, 2001.

Part II

Green Wireline Communications and Networking

Chapter 6

Graph-Theoretic Algorithms for Energy Saving in IP Networks

Francesca Cuomo, Antonio Cianfrani and Marco Polverini
University of Rome Sapienza, francesca.cuomo@uniroma1.it,
cianfrani@infocom.uniroma1.it, polverini@net.infocom.uniroma1.it

Energy saving in the Internet can be achieved by controlling the topology of the network and by turning off, or by putting into sleep mode, network links or devices. Indeed actual Internet topologies have space to turn off some links and devices to reduce the energy consumed in off-peak periods, e.g., during the night. This should be achieved still guaranteeing specific properties of the resulting topology, such as the connectivity among terminals or the resulting path lengths, to cite a few. In this chapter we present different methodologies, based on the graph theory, that allow to identify in the network graph links that can be put in sleep mode. In the following we indicate that these links can be *switched off*. In practice this switching off corresponds to IP line cards of a router that are put into sleep mode. In this way these line cards consume a low amount of energy. We denote the family of the discussed algorithms as *Graph-based Energy Saving* (GES). Different graph properties can be used and algorithms based on these properties can be defined.

6.1 Elements of the Graph Theory Used in GES

In network science the graph theory is a key component to represent the network structure and to capture some important characteristics of it. Thanks to this theory we are able to model the topology of a network by a bidirectional graph $G = (\mathcal{N}, \mathcal{E})$ where $\mathcal{N}$ is the set of vertices and $\mathcal{E}$ is the set of edges. Let $N = |\mathcal{N}|$ and $E = |\mathcal{E}|$ be the cardinalities of sets $\mathcal{N}$ and $\mathcal{E}$, respectively. A vertex models a network node, e.g., a router, and an edge, also named link in the following, models the logical interconnection of two routers.

The *adjacency matrix* of a graph G, denoted with $A(G)$, is a $N \times N$ binary matrix with a generic element

$$a_{nm} = \begin{cases} 1 & \text{if } (n,m) \in \mathcal{E} \\ 0 & \text{otherwise} \end{cases} \tag{6.1}$$

where (n, m) indicates the link between node n and m. The adjacency matrix of a simple bidirectional graph is symmetric and has all diagonal elements equal to 0. The degree matrix $D(G)$ of G is a diagonal matrix with the generic element d_{nn} equal to the degree of node n, i.e., to the number of edges incident on n: $d_{nn} = \sum_{m=1}^{N} a_{nm} = \sum_{m=1}^{N} a_{mn}$.

By using this description we can identify three key components in a graph G:

- the *shortest path tree* of node n toward all the other nodes, denoted as $\mathcal{T}_n$;

- the *edge betweenness* of a link l, denoted as $\mathcal{B}_l$;

- the *algebraic connectivity* of G, denoted as $\mathcal{A}(G)$.

The first component $\mathcal{T}_n$, that can be computed for each network node n, is the subgraph of G that connects n with all other vertex by the shortest paths. This subgraph is a tree.

The second parameter $\mathcal{B}_l$ for the generic edge l in G is defined as the number of paths from all nodes to all other nodes that cross this edge. It hence measures how many times this edge is crossed by the network paths. Finally, $\mathcal{A}(G)$ measures the degree of connectivity of the graph and it is greater than 0 if and only if G is a connected graph.

It is important to highlight that $\mathcal{T}_n$ and $\mathcal{B}_l$ can be defined even if the network is modeled by a directed graph, i.e., with directed edges, while $\mathcal{A}(G)$ can be defined only if the graph is undirected, i.e., with undirected links.

By using one or a combination of these components we are able to define different GES algorithms that can be used to derive a set of links, denoted in the following as switching off list $\mathcal{SL}$, that can be put into sleep to achieve energy saving.

6.1.1 Shortest Path Tree

Given a connected, undirected graph G, a spanning tree of that graph is a subgraph that is a tree and connects all vertices together. A single graph can have many different spanning trees. We can also assign a weight to each edge, w, which measures the cost of the edge if it is included in the spanning tree. Edges with less weight are preferred. Also a weight to a spanning tree can be assigned by computing the sum of the weights of the edges in that spanning tree. The shortest path tree $\mathcal{T}_n$, also named minimum spanning tree or minimum weight spanning tree, is then a spanning tree originating from n with weight less than or equal to the weight of every other spanning tree.

6.1.2 Edge Betweenness

The *edge betweenness* $\mathcal{B}_l$ for the generic edge l in the graph is defined as the number of paths from all nodes to all other nodes that cross the edge. This parameter is directly derived by the node betweenness [4]. The edge betweenness $\mathcal{B}_l$ measures the intermediary role that a link l has in the network. In our algorithm we use a simplified version of edge betweenness derived by considering only the shortest paths between every pair of nodes. In this way we measure the role that link l has in the minimum spanning trees $\mathcal{T}_n$. To compute $\mathcal{B}_l$ we find the shortest paths for every node $n \in \mathcal{N}$, and for every shortest path we increase $\mathcal{B}$ values for those links composing such a path. More formally we can define $\mathcal{B}_l$ as follows:

$$\mathcal{B}_l = \sum_{n \in \mathcal{N}} b_l(n) \quad \text{where } b_l(n) = \begin{cases} 1 & \text{if } l \in \mathcal{T}_n \\ 0 & \text{otherwise} \end{cases} \quad (6.2)$$

The value of $\mathcal{B}$ is in the range $[0, N \cdot (N-1)]$. The minimum value is achieved when a link is not included in any shortest path, while the maximum value is achieved when a link is crossed by every shortest path. As an example in case of Fig. 6.1 the set of shortest paths are in Table 6.1. The resulting $\mathcal{B}$ for all the networks link is in Table 6.2.

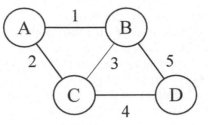

Figure 6.1: An example of graph with $\mathcal{A} = 2$ and $\mathcal{B}_1 = 4$.

6.1.3   Algebraic Connectivity

The algebraic connectivity $\mathcal{A}(G)$ of a graph $G(\mathcal{N}, \mathcal{E})$ is evaluated by using the *Laplacian matrix* $L(G)$ [4]. This matrix is equal to the difference between $D(G)$ and $A(G)$. The Laplacian matrix of a bidirectional graph is symmetric and all its row and column sums are equal to 0. The eigenspectrum of $L(G)$ is defined as the set of its N eigenvalues, denoted as $\lambda(G)$, that can be ordered from the smallest to the greatest, i.e., $\lambda_1(G) \leqslant \lambda_2(G) \leqslant ... \leqslant \lambda_{\mathcal{N}}(G)$. The eigenvalues of the Laplacian matrix measure the connectivity of the graph and the number of its connected components. The smallest eigenvalue of the Laplacian of a bidirectional graph G is equal to 0 (i.e., $\lambda_1(G) = 0$) and the number of eigenvalues equal to 0 is the number of connected components of G [7].

Consequently, $\lambda_2(G) = 0$ if and only if G is disconnected; $\mathcal{A}(G) = \lambda_2(G)$ is generally called *algebraic connectivity*.

$\mathcal{A}(G)$ can be used by a GES to decide links to be switched off, still keeping the resulting topology robust and resilient to failures. It is widely documented in the literature that $\mathcal{A}(G)$ encompasses several interesting characteristics of a network topology:

- it measures the stability and the robustness of complex network models [5]; a small perturbation in the network configuration will be attenuated

Table 6.1: Shortest Path Trees for the Graph of Fig. 6.1

n	Node A	Node B	Node C	Node D
Node A	-	1	2	1-5
Node B	1	-	3	5
Node C	2	3	-	4
Node D	5-1	5	4	-

Table 6.2: Edge Betweeness in Case of Fig. 6.1

Link ID	1	2	3	4	5
$\mathcal{B}$	4	2	2	2	4

back to the equilibrium with a rate proportional to $\mathcal{A}(G)$;

- the average distance between any two vertices of G is inversely proportional to $\mathcal{A}(G)$ [6];

- for any bidirectional graph G, the second eigenvalue of its Laplacian is upper bounded by its node connectivity which is equal to the minimum number of nodes whose deletion from G causes the graph to be disconnected or reduces it to a 1-node graph; $\mathcal{A}(G)$ is then a lower bound on the degree of robustness of the graph to node and edge failures;

- if a graph G^l is obtained from G by deleting the edge l, it has been proved that $\mathcal{A}(G^l) < \mathcal{A}(G)$; this means that the variation $\Delta^l = \mathcal{A}(G) - \mathcal{A}(G^l)$ measures the loss of connectivity produced by the removal of edge l.

By using the last property it is possible to evaluate the impact of an edge on the algebraic connectivity. Let l_1 and l_2 be two edges of graph G and let G^{l_1} and G^{l_2} be the graphs obtained from G by removing l_1 and l_2, respectively. In order to evaluate which of these two edges has a higher impact on the algebraic connectivity it is sufficient to compare Δ^{l_1} and Δ^{l_2}: we can state that l_2 has a higher impact on the algebraic connectivity compared with l_1 if $\Delta^{l_1} < \Delta^{l_2}$.

In the case of Fig. 6.1 we have

$$D(G) = \begin{pmatrix} 2 & 0 & 0 & 0 \\ 0 & 3 & 0 & 0 \\ 0 & 0 & 3 & 0 \\ 0 & 0 & 0 & 2 \end{pmatrix} \text{ and } A(G) = \begin{pmatrix} 0 & 1 & 1 & 0 \\ 1 & 0 & 1 & 1 \\ 1 & 1 & 0 & 1 \\ 0 & 1 & 1 & 0 \end{pmatrix}$$

the resulting Laplacian is then

$$L(G) = \begin{pmatrix} 2 & -1 & -1 & 0 \\ -1 & 3 & -1 & -1 \\ -1 & -1 & 3 & -1 \\ 0 & -1 & -1 & 2 \end{pmatrix}$$

and the eigenspectrum of $L(G)$ is

$$\lambda_1(G) = 0, \ \lambda_2(G) = 2, \ \lambda_3(G) = 4, \ \lambda_4(G) = 4$$

with a $\mathcal{A}(G) = 2$.

6.2   ESACON Algorithm

A first GES algorithm described in this chapter is ESACON (Energy Saving based on Algebraic CONnectivity) [3]. It models the Internet topology (i.e., the topology of an autonomous system in IP) with an undirected graph $G = (\mathcal{N}, \mathcal{E})$, where $\mathcal{N}$ is the set of routers and $\mathcal{E}$ is the set of bidirectional links connecting these routers.

ESACON is composed of two main steps:

1. the creation of an ordered list of links, denoted as $\mathcal{L}$;

2. the identification of a set of links to be switched off, denoted as $\mathcal{SL}$.

The pseudo-code in Algorithm 3 describes these two steps. As for step 1, the input is the network topology G and the output is an ordered list of the links composing the network, $\mathcal{L}$. The criterion used to order the network links is the algebraic connectivity: network links are ordered on the basis of their impact on the algebraic connectivity. The aim of ESACON is to include in the switching list $\mathcal{SL}$ those links that have a low impact on the network connectivity. The idea is that by enabling the sleeping mode of these links energy can be saved without compromising the network connectivity and the associated performance. ESACON associates to each bidirectional link $l \in \mathcal{E}$ the variation $\Delta^l = \mathcal{A}(G) - \mathcal{A}(G^l)$, where $\mathcal{A}(G^l)$ is the algebraic connectivity that the network would have if the bidirectional link l was switched off (lines 6-10 of Algorithm 3). We name $\mathcal{V}$ a vector containing the E bidirectional links and $\mathcal{L}$ is the ordered list obtained from $\mathcal{V}$ by sorting in increasing order the values Δ^l. Links at the top of the list entail a lower reduction of the algebraic connectivity compared with those links at the end of the list.

As for step 2, the input is the ordered list $\mathcal{L}$ and the output is the set of those links that can be switched off still keeping the connectivity over a given threshold. The set of these links is $\mathcal{SL}$. In order to evaluate the loss of connectivity produced by powering off some links, we introduce the *residual normalized algebraic connectivity* defined as:

$$\gamma(G') = \frac{\mathcal{A}(G')}{\mathcal{A}(G)} \cdot 100 \tag{6.3}$$

where $\mathcal{A}(G')$ is the algebraic connectivity of a reduced graph G', obtained by deleting some links from the graph G.

Step 2 has as many iterations as the number of bidirectional links of the network. At each iteration the algorithm tries to include a new link in $\mathcal{SL}$. At the generic iteration i, the algorithm checks if the link in i-th position of the list, indicated with $\mathcal{L}[i]$, can be switched off. To this aim it derives the algebraic connectivity $\mathcal{A}(G')$ of the graph $G' = (\mathcal{N}, \mathcal{E} - \mathcal{SL} - \mathcal{L}[i])$. G' is obtained from G by deleting the link $\mathcal{L}[i]$ of the current iteration and the links already in $\mathcal{SL}$ (line 15). ESACON verifies if the following connectivity

---

Algorithm 3 *ESACON*

1: **Input:** a network graph $G(\mathcal{N}, \mathcal{E})$;
2: /* STEP 1 */
3: $E = |\mathcal{E}|$;
4: $\mathcal{A}(G)$ is the algebraic connectivity of graph G;
5: $\mathcal{V} = zeros$;
6: **for** *each* $l \in \mathcal{E}$ **do**
7: $\quad \mathcal{V}[l] = l$:
8: $\quad G^l = (\mathcal{N}, \mathcal{E} - l)$;
9: $\quad \Delta^l = \mathcal{A}(G) - \mathcal{A}(G^l)$;
10: **end for**
11: $\mathcal{L} = $ sort $\mathcal{V}$ in increasing order based on the values Δ^l; /*Output of STEP 1*/
12: /* STEP 2 */
13: $\mathcal{SL} = \emptyset$;
14: **for** $i = 1, 2, ..., E$ **do**
15: $\quad G' = (\mathcal{N}, \mathcal{E} - \mathcal{SL} - \mathcal{L}[i])$;
16: $\quad \gamma(G') = \frac{\mathcal{A}(G')}{\mathcal{A}(G)} \cdot 100$;
17: $\quad$ **if** $\gamma(G') \geq \gamma_{th}$ **then**
18: $\quad\quad \mathcal{SL} = \mathcal{SL} \cup \mathcal{L}[i]$ /*include $\mathcal{L}[i]$ in the switching list*/;
19: $\quad$ **end if**
20: **end for**
21: **Outputs:** $\mathcal{SL}$ /*list of network links that can be switched off*/ and $G^{fin} = (\mathcal{N}, \mathcal{E} - \mathcal{SL})$ /*final network topology*/.

---

condition is verified (line 17):

$$\gamma(G') \geq \gamma_{th} \qquad (6.4)$$

being $\gamma_{th} > 0$ a value that imposes a degree of connectivity to the resulting network with respect to the initial one. If this is the case the link $\mathcal{L}[i]$ can be switched off and it is added to the set $\mathcal{SL}$ (lines 18-19); otherwise it remains active and the algorithm goes to the $(i+1)$-th iteration.

At the end of these steps ESACON identifies the largest number of links that can be switched off by keeping γ greater than a given threshold γ_{th}. The resulting graph, denoted as G^{fin} is:

$$G^{fin} = (\mathcal{N}, \mathcal{E} - \mathcal{SL}). \qquad (6.5)$$

6.3 ESTOP Algorithm

ESTOP is an evolution of ESACON where, besides the algebraic connectivity, also the edge betweenness $\mathcal{B}_l$ is considered.

This GES algorithm is composed of two steps:

1. computation of $\mathcal{B}$ for each link and creation of an ordered list of links $\mathcal{L}$;

2. identification of a set of links to be put into sleep mode (by controlling the connectivity of the resulting topology) and creation of a sleeping list ($\mathcal{SL}$).

The first two steps are described in Algorithm 4. As in the ESACON case the input of the algorithm is the network topology G and the output is an ordered list of the links composing the network, denoted as $\mathcal{L}$. The key difference with respect to ESACON is that the criterion used by ESTOP to order the network links is $\mathcal{B}$. Links are arranged in increasing order of $\mathcal{B}$, since the aim is to put into sleep mode those links that are less used in the paths between each pair of nodes. The algorithm computes, by using the Dijkstra algorithm, the shortest paths for every pair of nodes, and for each path it finds the links belonging to it and increases $\mathcal{B}$ associated with them (lines 6-10 of Algorithm 4). After that, on the basis of the $\mathcal{B}$ values, the ordered list $\mathcal{L}$ is created containing every bidirectional link of the network. Links at the top of the list are those less used in the shortest paths and their switch off entails a small number of paths to be recalculated. As for the second step, the input is the ordered list $\mathcal{L}$ and the output is a final topology G^{fin}, defined as in (6.5), and the set of links that can be put into sleep mode still keeping the connectivity over a given threshold. In order to evaluate the loss of connectivity produced by this switching off, we use also in this case the expression in (6.3) where $\mathcal{A}(G')$ is the algebraic connectivity of a reduced graph G', after the deletion of some links from G. Also for ESTOP the connectivity condition is the one in (6.4).

Step 2 has as many iterations as the number of bidirectional links of the network. At each iteration the algorithm tries to include a new link in $\mathcal{SL}$. At the generic iteration i, the algorithm checks if the link in the i-th position of the list $\mathcal{L}$, indicated with $\mathcal{L}[i]$, can be switched off. To do this the algebraic connectivity $\mathcal{A}(G')$ of the graph $G' = (\mathcal{N}, \mathcal{E} - \mathcal{SL} - \mathcal{L}[i])$ is computed. G' is obtained from G by deleting link $\mathcal{L}[i]$ of the current iteration and the links already in $\mathcal{SL}$ (line 17). The algorithm then verifies if $\gamma(G')$ is greater than γ_{th} (line 19). If this is the case link $\mathcal{L}[i]$ can be switched off and it is added to the set $\mathcal{SL}$ (lines 20-21), otherwise it remains active and the algorithm goes to the $(i + 1)$-th iteration.

6.4   EAR Algorithm

The Energy Aware Routing (EAR) algorithm is instead an example of GES using the shortest path trees $\mathcal{T}$ of the network graph [1, 2]. The main idea is to reduce the number of links used for routing packets, allowing two adjacent nodes to share the same shortest path tree, without involving other nodes. Let n and m be two adjacent nodes. These nodes use the routing tree $\mathcal{T}_n$ and $\mathcal{T}_m$, respectively, which are quite similar except for a small number of links.

Algorithm 4 *ESTOP*

1: **Input:** a network graph $G(\mathcal{N}, \mathcal{E})$;

2: /* *STEP 1* */

3: $\mathcal{B}$ = zeros; /*Array with betweenness values for every link */

4: $\mathcal{L}$ = zeros;

5: **for** *each* $n \in \mathcal{N}$ **do**

6:     Compute, by running Dijkstra, $\mathcal{T}_n$ = shortest path tree from n to all the other nodes;

7:     **for** *each edge* $l \in \mathcal{T}_n$ **do**

8:         $\mathcal{B}_l = \mathcal{B}_l + 1$;

9:     **end for**

10: **end for**

11: $\mathcal{L}$ = sort $\mathcal{E}$ in increasing order on the basis of $\mathcal{B}$; /*Output of STEP 1*/

12: /* *STEP 2* */

13: $\mathcal{SL} = \emptyset$;

14: **for** $i = 1, 2, ..., E$ **do**

15:     $G' = (\mathcal{N}, \mathcal{E} - \mathcal{SL} - \mathcal{L}[i])$;

16:     $\gamma(G') = \frac{\mathcal{A}(G')}{\mathcal{A}(G)} \cdot 100$;

17:     **if** $\gamma(G') \geq \gamma_{th}$ **then**

18:         $\mathcal{SL} = \mathcal{SL} \cup \mathcal{L}[i]$ /*include the link $\mathcal{L}[i]$ in the switching list/*;

19:     **end if**

20: **end for**

21: **Outputs:** $\mathcal{SL}$ /*list of network links that can be switched off*/ and $G^{fin} = (\mathcal{N}, \mathcal{E} - \mathcal{SL})$ /*final network topology/*.

So if n uses $\mathcal{T}_m$ instead of $\mathcal{T}_n$, a subset of paths used by n to route packets is modified; as a consequence some links outgoing from n can be switched off. We say that n uses a modified path tree. This sharing of tree is made possible thanks to an ad-hoc mechanism, called *exportation*, that allows a node, named Importer Router (IR), to get the $\mathcal{T}$ of an adjacent node, called Exporter Router (ER), without creating any link state variation in the network. More in detail, during an exportation the whole computational load is done by the IR, that recomputes Dijkstra algorithm using the ER as root. In this way the exportation is a local operation and no other nodes, except IR and ER, are made aware that some paths have changed. The consequence is that one or more links are excluded by the routing without any link state variation. To guarantee the correctness of routing paths, i.e., avoiding routing loops, and to limit excessive performance degradation, such as an increased delay, a set of restrictions is introduced:

1. a node can be an IR just one time;

2. a node involved in an exportation (i.e., a node with at least a path modified by an exportation) cannot be an ER.

If these rules are respected, it can be proved that all destinations remain reachable (no cycles appear during the EAR execution) and path lengthening is limited.

A fundamental step in EAR is how to find the ER-IR pairs that maximize the number of switched off links: this is called the EAR problem. By definition every directed network link is associated to a possible exportation, in which the tail is the IR and the head is the ER, so the number of possible exportations is limited by the number of network links. This gives rise to a combinatorial problem.

We model every exportation as a tupla $m_l = \langle l, \mathcal{S}_{m_l}, v_{m_l} \rangle$ denoted as *move*, where l is the link associated with m_l, $\mathcal{S}_{m_l}$ is the set of links that are switched off due to m_l and v_{m_l} is a vector representing the set of moves that are not compatible due to the rules defined above. The cardinality of the set of moves m_l is E.

If move m_i is still possible after the execution of move m_j, we said that m_i and m_j are compatible. Given a move m_i, its compatibility can be represented by means of the compatibility vector v_{m_i} of size E whose j-th element is defined as:

$$v_{m_i}[j] = \begin{cases} 1 & \text{if } m_i \text{ is compatible with } m_j \\ 0 & \text{otherwise} \end{cases} \tag{6.6}$$

The compatibility degree c_i of the move m_i represents the number of moves compatible with m_i:

$$c_{m_i} = \sum_{j=1}^{E} v_{m_i}[j]. \tag{6.7}$$

On the basis of (6.6) the graph of moves $G^m = (\mathcal{M}, \mathcal{I})$ can be defined. In this graph $\mathcal{M}$ is the set of moves and $\mathcal{I}$ is a $E \times E$ matrix where each row is the vector v_{m_i}. In G^m the i-th node represents the move m_i with a weight equal to the number of switched off links, i.e., the $|\mathcal{S}_{m_l}|$. In $G^m = (\mathcal{M}, \mathcal{I})$ there is an undirected edge between nodes i and j if $v_{m_i}[j] = 1$. Then, solving the EAR problem coincides with finding the maximum clique in G^m, that is an NP-hard problem. It is possible to define a greedy algorithm, named Max_Compatibility Heuristic; the heuristic is able to detect a set of compatible moves, trying to maximize the number of residual moves (see Algorithm 5).

The final solution, indicated as $\mathcal{C}$, is composed of a set of moves. For the $\mathcal{C}$ computations also the $|\mathcal{S}_{m_l}|$ are considered. Details on the implementation of this heuristic can be found in [2]. As in the other GES algorithms the final solution is a set of network links that can be put into sleep mode, $\mathcal{SL} = \bigcup \mathcal{S}_{m_l} \forall m_l \in \mathcal{C}$ and a $G^{fin} = (\mathcal{N}, \mathcal{E} - \mathcal{SL})$.

Algorithm 5 MaxCompatibility Heuristic

1: **Find** m_M s.t. $c_M = \max_{i \in E} c_{m_i}$
2: **for all** m_k s.t. $v_{m_M}[k] = 1$ **do**
3:    $temp_k = \{m_M, m_k\}$
4:    $v_{temp_k} = v_{m_M}$ AND v_{m_k}
5: **end for**
6: **for all** $temp_k \neq \emptyset$ **do**
7:   **while** $\sum_{j=1}^{E} v_{temp_k}[j] \neq 0$ **do**
8:     $m_l = \max\limits_{m_j \text{ s.t. } v_{temp_k}[j]=1} (v_{temp_k}^T * v_{m_j})$
9:     $temp_k = temp_k \cup m_l$
10:     $v_{temp_k} = v_{temp_k}$ AND v_{m_l}
11:   **end while**
12: **end for**
13: $\mathcal{C} = \max\limits_{temp_k \neq \emptyset} \sum\limits_{\substack{for\,all\,j\ \text{s.t.} \\ m_j \in temp_k}} |\mathcal{S}_{m_j}|$

6.5   A Practical Example of GES Application

To show in a practical case the behavior of the three GES algorithms described above we applied them to a network provided by SND-lib [9] composed of 17 nodes and 26 bidirectional links denoted as NOBEL-GERMANY. Fig. 6.2 reports the resulting topologies. In case of ESACON the links that are removed are the dotted ones if we set a $\gamma_{th} = 90\%$. In correspondence to each link the resulting normalized algebraic connectivity after its removal is reported (in the order indicated in Fig. 6.2a). It can be noticed that the removal of link 1 gives rise to a normalized algebraic connectivity equal to 99.82% and it decreases under 95% only after the removal of link 5. Under the same γ_{th} constraint, links removed by ESTOP are partially different from the ones selected by ESACON (see Fig. 6.2b) since they are selected under a the edge betweenness criterion (in Fig. 6.2b) the values of $\mathcal{B}_l$ are indicated for each link). Less links are switched off in this case even if they are more suitably selected as for other performance perspectives. Finally, the EAR case is reported in Fig. 6.2(c) where both importers and exporters are shown. This solution identifies links that are quite similar to the ESTOP case even if in general the number of switched off links is less than the other two solutions (some of them are switched off only in one direction as can be noticed by the arrows in Fig. 6.2c).

6.6　Performance Behavior of the GES Algorithms

To evaluate the performance of the different GES algorithms we use the following metrics:

δ: path length increase, in terms of number of hops;

$\Delta\%$: percentage of links that are switched off by the GES algorithm;

$\Delta\mathcal{P}\%$: percentage of power that can be saved by applying the GES algorithm;

$\rho\%$: mean traffic utilization, expressed as a percentage, of the links in final graph G^{fin}.

　　In the analysis we assumed that all links have an equal weight $w = 1$. The δ parameter measures the difference between the lengths of the shortest paths in G^{fin} and G. Indeed, due to the link switch off G^{fin} may present longer paths with respect to G.

　　The second parameter is defined as the ratio between the number of switched off links, i.e., $K = |\mathcal{SL}|$, and E (the cardinality of the initial set of links $\mathcal{E}$).

　　The amount of power saved using a GES is evaluated by computing the $\Delta\mathcal{P}\%$ parameter that is derived as the ratio between the power $\mathcal{P}_{fin}$(Watt) used in G^{fin} and the one used in G, $\mathcal{P}$. Power is computed on the basis of the capacity that is associated with each router line card and if this line card is in sleep mode or not. In all simulations we consider that a 2.5 Gbps module consumes 140 watts; therefore a link of capacity C_l Gbps has a power consumption equal to $\left(\frac{C_l}{2.5}\right) \cdot 140$ watts, while a link into sleeping mode consumes 20% of the total power (for fast recovery reasons, it cannot be totally turned off). In our $\Delta\mathcal{P}\%$ value we consider only the power used by the network links, i.e., by the network line cards.

　　Two different sets of real ISP networks are used to test the considered GES. The first set is composed of three topologies, provided by the Rocketfuel project [10]. The second set of network topologies, also with measurements

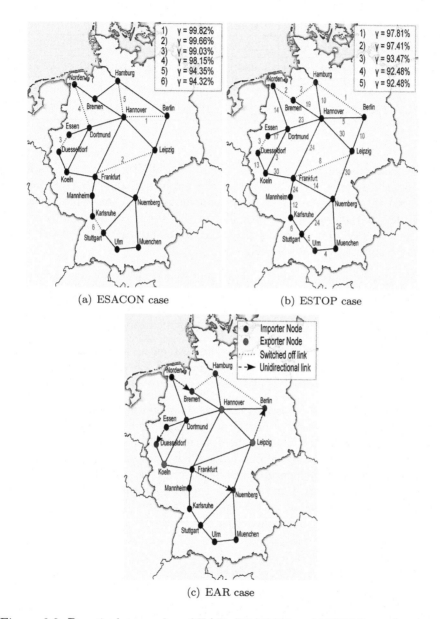

(a) ESACON case   (b) ESTOP case

(c) EAR case

Figure 6.2: Practical examples of EAR, ESACON and ESTOP application.

Table 6.3: Real Networks Used for the Performance Testing

Rocketfuel project [10]				SND-lib [9]			
	N	E	$\mathcal{P}$ *Watt*		N	E	$\mathcal{P}$ *Watt*
EBONE	159	614	162820	GERMANY50	50	176	25620
EXODUS	244	1080	350840	ABILENE	12	30	4200
ABOVENET	366	1936	685720	NOBEL- GERMANY	17	52	7280

on real traffic demands, is provided by SND-lib [9]. Table 6.3 reports for all networks the number of nodes N and the number of links E (this number is referred to the number of unidirectional links). Also the overall power $\mathcal{P}$ used in the initial topology is provided in this table. Since for all networks the actual link capacities are not provided, we assign to every network link a capacity computed in the following way:

1. we generate a traffic matrix supposing that each router generates traffic toward any other router; the composition is: 40% of high bit rate traffic, between 1 Mbit/s and 80 Mbit/s, and the remaining 60% at a low bit rate, about 1 Mbit/s (see guidelines provided in [8]).

2. we route all traffic flows on the shortest paths and then we assign to each link a number of capacity modules (of 2.5 Gbps each), such that the achieved utilization is less than or equal to 25% of the total capacity.

The first traffic pattern is also used in the performance analysis for evaluating $\rho\%$ that a key parameter to evaluate the resulting topology behavior when traffic is supported. In fact the switching off of some links may create some bottlenecks on the remaining links when some traffic is injected in the network. To this aim we measured the behavior of the proposed solutions also as for a traffic perspective and we used the $\rho\%$ parameter. After turning off the network links belonging to the set $\mathcal{SL}$, all traffic demands are routed in the new topology, G^{fin}, according to the shortest path criteria. After this new routing we can compute the $\rho\%$ as:

$$\rho\% = \frac{\sum_{\mathcal{E-SL}} \rho_l\%}{E - K} \tag{6.8}$$

where $\rho_l\%$ is the new traffic utilization of link l, with $l = 1, 2.., E - K$. In the following analysis we also discuss the impact on the performance when a maximum utilization of the network links is imposed, namely $\rho\%_{max}$.

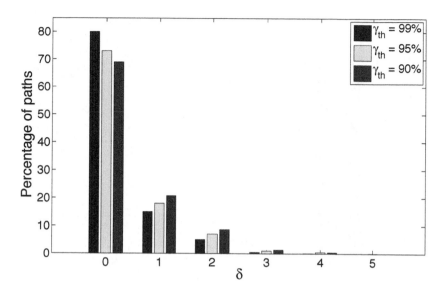

Figure 6.3: Distribution of the δ parameter in case of ESACON.

6.6.1 Path Length Increase

In this analysis we compute, by running Dijkstra, the shortest path trees $\mathcal{T}_n$ for all nodes in the initial topology G. For each $\mathcal{T}_n$ we measure the lengths of the associated paths. The same computation is then done for G^{fin}. For each network path we then are able to compute the path length increase δ measured in number of hops, after the application of the GES.

Fig. 6.3 depicts the distribution of the path length increase in case of ESACON for three different connectivity thresholds, $\gamma_{th} = 99\%$, $\gamma_{th} = 95\%$ and $\gamma_{th} = 90\%$ in case of the EXODUS network. The same result is reported in Fig. 6.4 in case of ESTOP. It can be observed that with both algorithms only few paths increase their lengths of more than one hop. This is an important result that may have an impact on the data transfer delay in the network. Moreover, by considering that the average distance between two vertexes of a graph is inversely proportional to the algebraic connectivity, we observe that, by choosing a connectivity threshold, we can also determine the maximum path length increase. For instance in case of ESACON and a $\gamma_{th} = 99\%$ the maximum value of δ is equal to 3, for $\gamma_{th} = 95\%$ is equal to 4 and only for $\gamma_{th} = 90\%$ reaches 5 hops. A similar behavior is measured in case of ESTOP.

By comparing Fig. 6.4 and Fig. 6.3 it can be observed that ESTOP has a better behavior than ESACON. This is due to the fact that ESTOP first switches off links crossed by a small number of shortest paths, while ESACON

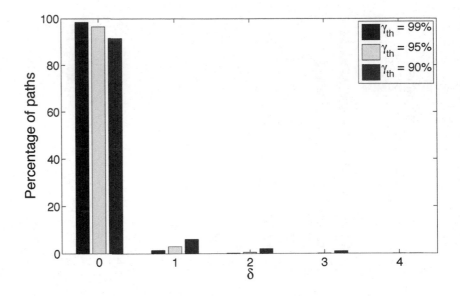

Figure 6.4: Distribution of the δ parameter in case of ESTOP.

works only on the algebraic connectivity.

As for EAR instead, the distribution of the path length increase is reported in Fig. 6.5. In this case it is to be noticed that, since we adopted equal weights on all links, the algorithm assures that a path cannot increase more than 2 hops. This property depends on the second rule introduced in EAR algorithm and can be intuitively explained considering that this rule guarantees that the paths of Exporter Routers remain the shortest ones. This result can be observed in Fig. 6.5.

6.6.2   Percentage of Links That Are Switched Off

We then observe the GES behavior in terms of $\Delta\%$ as a function of the connectivity threshold parameter. Fig. 6.6 reports the behavior of the $\Delta\%$ as a function of γ_{th} in case of ESACON and ESTOP. The lowest considered value of γ_{th} is 10%; this means that we are quite close to a disconnected network (achieved in case of $\gamma_{th} = 0$).

As the connectivity threshold increases, the number of links that can be switched off decreases. However, it can be noted that a good number of links can be switched off even if the connectivity threshold (and the related algebraic connectivity) does not reduce too much (e.g., 80%). This depends on the level of connectivity (meshed component) of the considered networks; in fact a high number of links can be deleted by these networks without reducing the

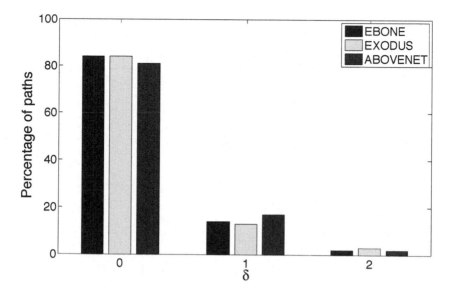

Figure 6.5: Distribution of the δ parameter in case of EAR.

minimum out degree, which is directly related to the algebraic connectivity. Furthermore, it can be noticed that the ESTOP and ESACON curves have an intersection point for a given γ_{th} that depends on the analyzed network. In all cases for low values of connectivity thresholds ESTOP has better performance, i.e., switches off more links; on the contrary for high values of γ_{th} ESACON is able to switch a greater number of links. This is due to the fact that ESACON is designed in order to maximize the algebraic connectivity, as a consequence until the G^{fin} remains quite similar to G (e.g., for high values of γ_{th}) the $\mathcal{SL}$ list is not too different from a list that is recomputed at each step on G'. On the other hand, for low values of γ_{th} the ESTOP method is preferable since it is based also on the betweenness.

The EAR behavior in terms of $\Delta\%$ is reported in Fig. 6.7 in case of three different networks. It can be observed that $\Delta\%$ depends on the network topology, varying from 20% to 30%. We also tested the algorithm under a $\rho\%_{max}$ constraint. This has been done to compare it also with ESACON and ESTOP. Among the three GES algorithms, ESTOP shows a more stable behavior with respect to the topologies while ESACON has a good behavior only in case of small topologies, i.e., GERMANY50. On the contrary EAR performs better in case of large networks (EBONE and EXODUS).

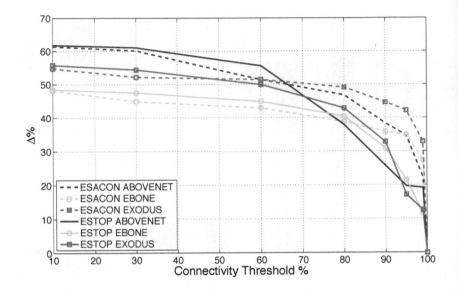

Figure 6.6: Percentage of links that are switched off by, $\Delta\%$, as function of connectivity threshold γ_{th} in case of ESACON and ESTOP.

6.6.3   Power Saving

Another result to be discussed is the power saved by the GES algorithms thanks to the switch off of the line cards. It is to be noticed that in the considered analysis we assume to put the line cards into sleep mode; in this way a line card consumes 20% of the total power of the working mode. Table 6.4 reports the $\Delta\mathcal{P}\%$ parameter in case of three networks for all the GES solutions. In the same table also the number of switched off links K is reported.

As can be observed, the percentage $\Delta\mathcal{P}\%$ in case of EAR applied to small networks as the ones considered here is almost the maximum that can be achieved on the basis of the intrinsic mechanism of the algorithm. In fact EAR intrinsically aims at reducing the perturbations of the shortest path trees. In small networks the shortest path trees are very similar and the number of links that can be switched off thanks to the exportation-importation mechanism are few. On the contrary, both ESACON and ESTOP are able to increase $\Delta\mathcal{P}\%$ (around 30% in some cases).

6.6.4   Impact in Terms of Traffic Utilization

Fig. 6.8 analyzes the utilization resulting in G^{fin} when the three GES are run on the EBONE network. The traffic is generated as described at the beginning of Section 6.6. The figure plots the utilization on each link of this network with

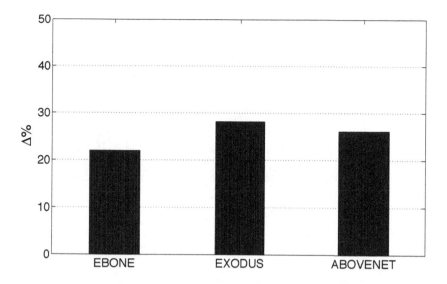

Figure 6.7: Percentage of links that are switched off by, $\Delta\%$, in case of EAR.

a $\Delta\%$ around 24% that is the gain that EAR achieves when a $\rho\%_{max} = 70\%$ is imposed. The distributions of the $\rho_l\%$ in the three cases is instead reported in Fig. 6.9. We can notice that, even if some bottlenecks arise, all the solutions behave quite well when traffic is supported. In general all the solutions are able to constrain the $\rho_l\%$ on a big amount of links under 40% as can be noticed in Fig. 6.9. Moreover, in case of ESTOP and ESACON it is possible to reduce the bottlenecks by acting on the connectivity threshold. As for EAR it is instead possible to control the overall number of exportations.

ESTOP is also able to avoid peaks of utilizations above 60% while ESACON on one link reaches 80%. This is the reason why in Table 6.5 ESACON under a $\rho\%_{max} = 50\%$ switches off only few links. As a conclusion, ESTOP and EAR have a better impact on $\rho\%$ than ESACON due to the fact that they have less peaks and traffic demands are almost uniformly distributed among the residual links.

6.7 Practical Implementation of the GES Algorithms

In this section implementation issues related to the proposed algorithms are discussed. In a pure IP network a link-state routing protocol, such as OSPF or IS-IS, is used. A link-state routing protocol assures that each IP router

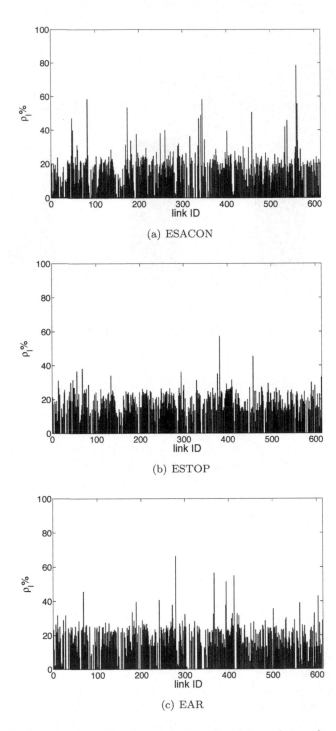

(a) ESACON

(b) ESTOP

(c) EAR

Figure 6.8: Utilization of each link in EBONE by applying the three GES.

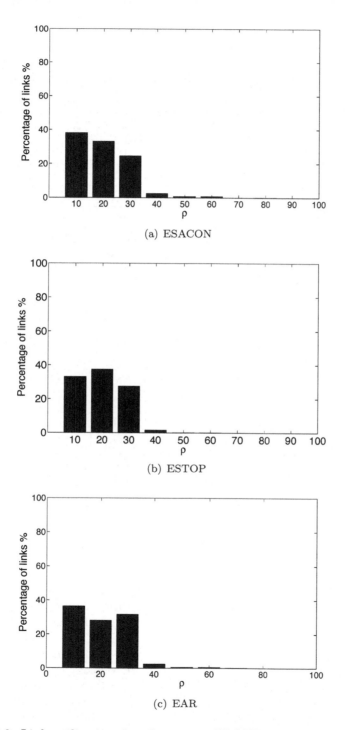

(a) ESACON

(b) ESTOP

(c) EAR

Figure 6.9: Link utilization distribution in EBONE by applying the three GES.

Table 6.4: Power Saving

		GERMANY50		ABILENE		NOBEL-GERMANY	
		K	$\Delta P\%$	K	$\Delta P\%$	K	$\Delta P\%$
EAR	$\rho\%_{max} = 30\%$	10	4%	2	5%	8	12%
	$\rho\%_{max} = 50\%$	10	4%	2	5%	8	12%
	$\rho\%_{max} = 70\%$	10	4%	2	5%	8	12%
ESACON	$\rho\%_{max} = 30\%$	24	10%	8	21%	20	31%
	$\rho\%_{max} = 50\%$	66	29%	8	21%	20	31%
	$\rho\%_{max} = 70\%$	72	32%	8	21%	20	31%
ESTOP	$\rho\%_{max} = 30\%$	22	10%	8	21%	20	31%
	$\rho\%_{max} = 50\%$	66	29%	8	21%	20	31%
	$\rho\%_{max} = 70\%$	74	33%	8	21%	20	31%

Table 6.5: Percentage of Links That Are Switched Off, $\Delta\%$, by the Three GES Algorithms under a $\rho\%_{max}$ Constraint

	EBONE		GERMANY50		EXODUS	
$\rho\%_{max}$	50%	70%	50%	70%	50%	70%
ESACON	3.91	6.19	36.36	38.64	11.67	20.56
ESTOP	21.5	37.13	37.5	40.91	32.78	38.7
EAR	22.64	23.62	5.68	5.68	27.59	28.06

knows the network topology; in this way it is possible to perform a topology based algorithm with no need to introduce extra information. In the following a distributed implementation of ESACON and ESTOP algorithms, and a centralized implementation of EAR algorithm, are discussed.

EASCON and ESTOP can be implemented in a distributed manner. In each router a new module, defined Green Algorithm module, performing the algorithms of Sections 6.2 and 6.3 has to be introduced. The Green Algorithm module executes the algorithm, i.e., ESACON or ESTOP, and detects network links that can be put to sleep. The algorithm execution can be performed independently on each router, because of network topology knowledge achieved via the routing protocol. After the algorithm execution each router is able to evaluate if one or more of its links can be put to sleep. In this case the router generates an OSPF control message, i.e., a Link State Advertisement

(LSA), to inform other routers that one or more links are not more available to route traffic. Finally, the router interface connected to the sleeping link can be put to sleep so that real energy saving is obtained. A possible disadvantage of these solutions is that some transitional instabilities can arise during the traffic routing. This due to the fact that they it takes some time to have all the LSA circulating and this can bring some packets losses.

EAR algorithm can be implemented in a centralized manner. First of all a network router is elected as EAR coordinator. The EAR coordinator performs two main operations:

1. it executes the EAR algorithm and so detects routers' role;

2. it generates special control messages, referred to as Importer Messages, to inform importer routers about their role and the corresponding exporter to be selected.

Each importer, after the reception of the Importer Message, executes the modified Dijkstra algorithm and modifies its paths. Finally all importer interfaces not used to route traffic can be put to sleep. It can be noticed that in EAR possible packet losses due to path recalculation are avoided, since, during the exportation process, the old shortest path tree can be used by the IR to route its packets.

6.8 Conclusions

In this chapter we presented graph-theoretic bases solutions that can be adopted in an IP network for energy saving purposes. The common idea of these solutions is to reduce the number of links, e.g., router line cards, that are used in the network during the off-peak period. To this aim different properties of the graph that models the network are used. ESACON in an algorithm based only on the algebraic connectivity. By controlling the impact of the removal of some links on the algebraic connectivity ESACON derives a list of links that can be switched off. ESTOP combines the algebraic connectivity also with the edge betweenness. This latter parameter allows to cut from the network graph links that are crossed only by few paths. The resulting graph algebraic connectivity is then used to control that the network remains connected and that its connection degree is above a suitable threshold. Finally EAR uses the shortest path trees and decides to allow to some nodes to share their trees in order to remove some links from the routing paths. Table 6.6 summarizes the key components that are used by these three algorithms and presents if these can be implemented in a centralized or distributed fashion.

As for the performance behavior of the solutions in this chapter we tested them on real ISP network topologies available in the literature and we measured some interesting performance metrics. As for ESACON we showed that even if the connectivity threshold is kept quite high (around 80%-90%), the

Table 6.6: Summary of Graph-Based Energy Saving

	Graph components used in the algorithm			Network implementation	
	Shortest path tree $\mathcal{T}$	Edge betweenesses $\mathcal{B}$	Connectivity $\mathcal{A}$	Centralized	Distributed
ESACON	-	-	√	-	√
ESTOP	√	√	√	-	√
EAR	√	-	-	√	-

algorithm is able to switch off a great number of links without compromising some other network performance like the path length increase. On the other hand, if we associate to the algebraic connectivity also the edge betweenness, as done by ESTOP, we are able to more suitably select links to be switched off. In this case the resulting performance enhances, as shown in case of all the ESTOP performance metrics. Finally, an accurate selection of the importers and exporters nodes in order to have the shortest path sharing can give rise to a satisfactory power saving without affecting too much the routing protocols adopted in the network.

Bibliography

[1] A. Cianfrani, V. Eramo, M. Listanti, M. Marazza, and E. Vittorini. An energy saving routing algorithm for a green ospf protocol. In *INFOCOM IEEE Conference on Computer Communications Workshops, 2010*, pages 1–5, March 2010.

[2] A. Cianfrani, V. Eramo, M. Listanti, and M. Polverini. An OSPF enhancement for energy saving in IP networks. In *Computer Communications Workshops (INFOCOM WKSHPS), 2011 IEEE Conference on*, pages 325–330, April 2011.

[3] F. Cuomo, A. Abbagnale, A. Cianfrani, and M. Polverini. Keeping the connectivity and saving the energy in the internet. In *Computer Communications Workshops (INFOCOM WKSHPS), 2011 IEEE Conference on*, pages 319–324, April 2011.

[4] Chris Godsil and Gordon Royle. *Algebraic Graph Theory*. Springer-Verlag, New York, April 2001.

[5] Almerima Jamakovic and Steve Uhlig. On the relationships between topological measures in real-world networks. *Networks and Heterogeneous Media*, 3:345–359, 2008.

[6] Bojan Mohar. Eigenvalues, diameter, and mean distance in graphs. *Graphs and Combinatorics*, 7(1):53–64, 1991.

[7] Bojan Mohar. The laplacian spectrum of graphs. In *Graph Theory, Combinatorics, and Applications*, pages 871–898. Wiley, 1991.

[8] A. Nucci, A. Sridharan, and N. Taft. The problem of synthetically generating ip traffic matrices: initial recommendations. In *SIGCOMM Comput. Commun. Rev.*, volume 35, pages 19–32, July 2005.

[9] S. Orlowski, R. Wessaly, M. Pioro, and A. Tomaszewski. Sndlib 1.0 survivable network design library. *Netw.*, 55:276–286, May 2010.

[10] Neil Spring, Ratul Mahajan, David Wetherall, and Thomas Anderson. Measuring ISP topologies with rocketfuel. *IEEE/ACM Trans. Netw.*, 12(1):2–16, February 2004.

Bibliography

Chapter 7

Architectural Design of an Energy-Efficient Router

*Chengchen Hu, +Bin Liu, †Mingui Zhang, ‡Beichuan Zhang and §Xiaojun Wang
*MoE KLINNS Lab, Department of Computer Science and Technology, Xi'an Jiaotong University, chengchenhu@xjtu.edu.cn
+Department of Computer Science and Technology, Tsinghua University, liub@tsnghua.edu.cn
†Huawei Inc., zhangmingui@huawei.com
‡Department of Computer Science, Arizona University, bzhang@arizona.edu
§School of Electronic Engineering, Dublin City University, xiaojun.wang@dcu.ie

The tremendous success of the Internet has made it a ubiquitous infrastructure nowadays comprising an enormous number of hardware components to deliver a variety of services to end users. However, the concern of energy consumption in today's Internet infrastructure is highlighted in [2]: in 2007, the annual energy consumption of the Internet terminal devices and networking and transmission equipment is about 350 billion kWh in the US (9.4% of the national electricity consumption), and about 868 billion kWh globally (5.3% of the global electricity consumption). In the most recent years, with the aim of building a more sustainable society, research efforts have been made to look into the feasibility and benefits of applying energy-efficient techniques in Information and Communication Technology (ICT) systems in order to obtain the best tradeoff between energy cost and system performance.

In fact, many power management solutions are available and have been well applied to Personal Computer (PC) systems and different types of battery-operated portable devices, e.g., dynamic voltage scaling technology in Central Processing Unit (CPU) of portable devices, energy-saving sleep modes in PC operating systems, and power-aware routing protocols in Wireless Sensor Networks (WSNs). However, the application of these techniques in the Internet-scale systems has not been fully considered yet. Some pioneering exploitations of saving energy across the Internet are discussed [8] which investigated the network protocols design with energy-saving considerations and device power consumption optimization techniques against expected performance. The previous studies have confirmed the feasibility and benefits of engineering the next generation energy-saving Internet infrastructure and pointed out a set of research directions [1, 8, 10]. In this chapter, we concentrate on the exploration of power/energy-saving mechanisms through the design of Internet transmission equipment, e.g., routers. By revisiting the characteristics of the Internet behaviors and the modular architecture of routers, this chapter suggests the approach for engineering an energy-efficient Internet from three different perspectives and discusses the imposed technical challenges. To address the challenges and seize the energy-saving opportunities, a new conceptual router model/architecture as the guide to design and implement power efficient router, as well as the Internet, is pursued.

7.1   Opportunities and Challenges

Several intrinsic characteristics of Internet behaviors imply the opportunities of applying energy-saving techniques in routers toward an energy-efficient Internet.

Low average link utilization and significant path redundancy. Over-provisioning of the link bandwidth is common in network planning to harness burst traffic. It is stated that the average utilization of backbone links is less than 30% [18]. Besides, massive link redundancies are also built to sustain the network during potential failures. Although low link utilization and massive link redundancy improve the network resilience, they greatly hurt the energy efficiency of Internet. The links are operated at full rate all the time while they are highly under-utilized most of the time. Our first opportunity to save energy is shifting and aggregating traffic from lightly utilized links. In this way, some of the routers (or line cards in a router) can be shut down to reduce the power instead of being in operation all the time.

Variable Internet traffic demand. It is now well known that the Internet traffic demand exhibits fluctuations over time due to end user behavior, temporary link failures, anomalies and so forth [5]. The dynamics of data traffic rate allows us to incorporate energy control mechanisms in the router design, e.g., adapting router processing speed based on the detected traffic rate or queue length. The processors can be decelerated to reduce the energy consumption for incoming traffic with low data rate.

Energy-efficient packet processing. At present, the packet processing mechanisms inside a router are mainly driven by the pursuit of desired performance, with little consideration of energy consumption. From the energy efficiency perspective, there remains a large potential to optimize the design of router packet processing components through various techniques, e.g., reducing the peak power of the processor using architectural approaches.

Although the aforementioned opportunities are promising to minimize the energy consumption in current Internet infrastructure, they also impose the following questions:

How to properly aggregate traffic and route packets? With traffic aggregation and shift in mind, the end-to-end routing paths of individual flows needs to be determined which do not necessarily follow the conventional shortest path first routing paradigm, e.g., Open Shortest Path First (OSPF) protocol. Mathematically, it can be formulated as an optimization problem, which maximizes the energy-savings by identifying as many idle links as possible to be put into sleep mode while guaranteeing the expected network performance. The nature of Internet traffic dynamics over time, e.g., variable traffic demand, imposes additional complexity on finding the solution and needs to be taken into account.

How to manage routers under energy saving states to achieve performance-energy tradeoff. Generally, off-the-shelf routers can only work in one of two operational states: on and off, which cannot be flexibly configured or switched to cope with traffic fluctuations. To meet the needs of energy consumption reduction, we suggest that multiple router operational states (e.g., energy saving state) with fast switching ability among them should be supported. Advanced strategies determining when to switch and to which state should be adopted to achieve the best tradeoff between the performance and energy efficiency.

How to reduce peak power with performance guarantees. It is also quite challenging to reduce the router's peak power without harming the system performance in the worst case. If the router could freely switch working states to reduce the average power, it is impossible to lower the peak power, which is related to the processing in the worst case, without changes on the router architecture. Unique features of network processing should be utilized and elaborate architectural design of router functions should be investigated to take this challenging task.

7.2   Architecture

The concept of Green Reconfigurable Router (i.e., GRecRouter) is presented in [10], which aims to contribute to the creation of the next generation energy-efficient Internet infrastructure. Through the enhancement of the router architectural design, it is expected to reduce both average power and peak power consumption during network operation.

GRecRouter is designed in such a way that its settings (e.g., routing path, clock frequency, supply voltage) are "reconfigurable" based on its awareness of traffic rate fluctuations. In details, this can be interpreted from two aspects as follows.

- At a large time-scale, it has been known for a long time that the Internet traffic exhibits strong daily and weekly patterns and the behavior remains unchanged over years [5]. Take the enterprise network as an example. The traffic volume in daytime could be about ten times more than that in night time, and the similar difference is also observed between workdays and weekends. The time-of-day/time-of-week effect makes the link load vary slowly, but with a large magnitude. Bear this in mind. GRecRouter first manages the energy consumption of the Internet in a macro time scale by periodically aggregating traffic during lightly loaded periods, and in contrast distributing traffic in heavily loaded periods. The realization of this control mechanism needs modifications of the underlying routing protocols and the mechanism of routing path selection.

- At a small time-scale, the flow rate varies in a smaller magnitude but more frequently (compared with time-of-day/time-of-week effect). According to this, GRecRouter also adopts an energy control mechanism operating in a micro time scale, which adaptively tunes the processing rate of the function blocks inside individual routers based on the detected link utilization, traffic rate or queue length. To minimize the energy consumption, fast but more energy-consuming states are suggested to be used under heavy traffic scenarios, while slow but energy efficient states are suitable for light traffic scenarios.

With the elaborate design of functional blocks inside the GRecRouter using architectural advances, the router's peak power could also be reduced. The energy-efficient architectural designs for implementing main router functions in GRecRouter, i.e., routing lookup and packet queuing, are presented in later sections.

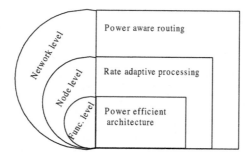

Figure 7.1: GRecRouter controls power dissipation in three levels.

To summarize, the design of GRecRouter exhibits many desirable and unique features compared with conventional routers, as shown in Fig. 7.1. The major features are discussed as follows:

- At the network level, power-aware routing is applied to determine the end-to-end routing paths and forward the packets from the source to the destination with minimal energy consumption. It may change packet routes during quiet periods for traffic aggregation and inform the idle devices to work in sleep mode. It should be noted that the expected performance, e.g., Quality of Service (QoS), needs to be guaranteed with the energy-saving consideration.

- At the node level, rate adaptive processing should be activated inside individual routers. Different function blocks could be flexibly configured to be operated in a specific state, e.g., on, off or low energy-consumption states (with slower clock frequency or/and lower supply voltage) according to the network traffic load.

- At the function block level, each functional block should be designed with energy-efficiency in mind so as to reduce the peak power of the router. Many architectural design techniques could be employed as appropriate, e.g., caching, clock gating and processing separation.

The GRecRouter is presented as a conceptual architectural model which is not limited to any specific deployment. The analysis and discussions on a reference implementation of GRecRouter based routers are presented in detail in the following sections.

7.3 Power Aware Routing through Green Traffic Engineering

Today's wide-area networks usually have redundant and over-provisioned links, resulting in low link utilization most of the time. Take Abilene, a large

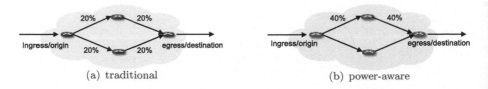

(a) traditional   (b) power-aware

Figure 7.2: Different traffic engineering goals.

US education backbone, as an example. It is shown that the average link utilization is only about 2%; the maximum fluctuates mostly between 10% and 20%; and only one rare event pushes the maximum over 50% [20]. High path redundancy and low link utilization combined also provide a unique opportunity for power-aware traffic engineering. Traditional traffic engineering spreads the traffic evenly in a network (Fig. 7.2a), trying to minimize the chance of congestion induced by traffic bursts. However, in power-aware traffic engineering (Fig. 7.2b), one can free some links by moving their traffic onto other links, so that the links without traffic can go to sleep for an extended period of time. This should result in more power saving than pure opportunistic link sleeping because the sleep mode is much less likely to be interrupted by traffic.

To generalize the basic idea illustrated in Fig. 7.2b, we develop the GreenTE model, which, given the network topology and traffic matrix, finds a routing solution (i.e., the links to be used and the traffic volume to be carried on each link) that maximizes the power saving from turning off line cards as well as satisfying performance constraints including link utilization and packet delay.

7.3.1   The General Problem Formulation

We model the network as a directed graph $G = (V, E)$, where V is the set of nodes (i.e., routers) and E is the set of links. A port can be put to sleep if there is no traffic on the link, and a line card can be put to sleep if all its ports are asleep. Let M be the set of line cards in the network. For a single line card $m \in M$, its base power consumption is B_m, its set of ports is S_m, and each port $l \in S_m$ consumes power P_l. Then the power saving from turning of one port is P_l, and the power saving from turning off the entire line card is $B_m + \sum_{l \in S_m} P_l$. The objective is to find a routing that maximizes the total power saving in the network. This general power-aware traffic engineering problem can be formulated based on the Multi-Commodity Flow model as follows. Please see Table 7.1 for the notation used in this chapter.

Table 7.1: Summary of Notation Used in This Chapter

Notation	Meaning
S_m	Set of links connected to line card m
P_l	Power consumption of the port connected to link l
B_m	Base power consumption of line card m
x_l	1 if link l is sleeping, 0 otherwise
y_m	1 if line card m is sleeping, 0 otherwise
$f_l^{s,t}$	Traffic demand from s to t that is routed through link l
H_l	l's head node
T_l	l's tail node
I_l^v	1 if v is the head node of link l, 0 otherwise
O_l^v	1 if v is the tail node of link l, 0 otherwise
$D_{s,t}$	Traffic demand from s to t
C_l	Capacity of link l
u_l	Utilization of link l
$r(l)$	Reverse link of l
k	Number of candidate paths for each origin-destination (OD) pair
U_T	Threshold for the maximum link utilization (MLU)
$Q_i^{s,t}(l)$	1 if the ith candidate path from s to t contains link l, 0 otherwise
$\alpha_i^{s,t}$	Ratio of traffic demand from s to t that is routed through the ith candidate path

$$\text{maximize} \quad \sum_{l \in E} P_l x_l + \sum_{m \in M} B_m y_m \tag{7.1}$$

$$\text{s.t.} \quad \sum_{l \in E} f_l^{s,t} O_l^i - \sum_{l \in E} f_l^{s,t} I_l^i =$$

$$\begin{cases} D_{s,t}, & i = t \\ -D_{s,t}, & i = s \\ 0, & i \neq s,t \end{cases} \quad , \quad s,t,i \in V, s \neq t \tag{7.2}$$

$$|S_m| y_m \leq \sum_{l \in S_m} x_l \tag{7.3}$$

$$u_l = \frac{1}{C_l} \sum_{s,t \in V, s \neq t} f_l^{s,t}, \quad l \in E \tag{7.4}$$

$$x_l = x_{r(l)}, \quad l \in E \tag{7.5}$$

$$x_l + u_l \leq 1, \quad l \in E \tag{7.6}$$

Equation 7.1 computes the objective that maximizes the total power saving in the network. Equation 7.2 states the flow conservation constraints. Let $|S_m|$

be the cardinality of S_m; then Equation 7.3 ensures that a line card is put to sleep only when all its ports are asleep. Equation 7.4 calculates the link utilization. Equation 7.5 ensures that links are put to sleep in pairs, i.e., there is no inbound traffic nor outbound traffic. Equation 7.6 states that a link can be put to sleep only if there is no traffic on it, and when it is on, it does not carry traffic more than its capacity. Solving this problem gives which links to be turned off, and how much traffic each remaining link should carry.

The binary (integer) variables x_l and y_m that denote the power state of link l and line card m make the model a Mixed Integer Programming (MIP) problem. Generally speaking, MIP problems are NP-Hard, thus its computation time for medium and large networks is a concern. This model, though it maximizes power saving in the network, does not consider some practical constraints. For example, packet delay could be much longer than that of current shortest path routing, and links may operate at unacceptably high link utilization, making them vulnerable to any traffic bursts.

7.3.2 A Practical Heuristic

To consider the practical constraints and reduce computation time, we refine the problem formulation as follows.

$$\text{maximize} \tag{7.7}$$

$$\sum_{l \in E} P_l x_l + \sum_{m \in M} B_m y_m \tag{7.8}$$

$$\text{s.t.} \tag{7.9}$$

$$f_l^{s,t} = \sum_{0 \le i < k} Q_i^{s,t}(l) D_{s,t} \alpha_i^{s,t}, s, t \in V, l \in E, s \ne t \tag{7.10}$$

$$\sum_{0 \le i < k} \alpha_i^{s,t} = 1, \quad s, t \in V, s \ne t \tag{7.11}$$

$$|S_m| y_m \le \sum_{l \in S_m} x_l \tag{7.12}$$

$$u_l = \frac{1}{C_l} \sum_{s,t \in V, s \ne t} f_l^{s,t}, \quad l \in E \tag{7.13}$$

$$x_l = x_{r(l)}, \quad l \in E \tag{7.14}$$

$$x_l + u_l \le 1, \quad l \in E \tag{7.15}$$

$$u_l \le U_T, \quad l \in E \tag{7.16}$$

One change is the addition of the bound on maximum link utilization (MLU) in a network. Equation 7.16 states that MLU must be no greater than a configured threshold U_T. In this chapter, we use 50% as the default value of U_T.

Another change is the use of the k-shortest paths as the candidate paths instead of searching the solution in all possible paths. Equations 7.10 and 7.11 are equivalent to the flow conservation constraints under this change. It reduces overall computation time as well as adding path length as another constraint. The general model introduced in the previous subsection considers all possible paths for each origin-destination (OD) pair, making the search space extremely large. To reduce search space and computation time, for each OD pair, we pre-compute its set of k-shortest paths and only search solutions within this set. Since the k-shortest paths are pre-computed with network topology as the only input, they do not change with the traffic matrix and the computation does not add run-time overhead. Note that when k is set to be large enough, we can actually consider all possible paths for each OD pair, which will give the maximal power saving under the MLU constraint. However, the computation time increases with the value of k; therefore there is a tradeoff between the precision of the heuristic and the computation time. Our evaluation later will show that a reasonably large k can achieve near optimal results.

Searching solutions only within k-shortest paths also avoids very long paths. In practice, network operators can have their own definitions of link delays and path lengths, and choose the set of k candidate paths accordingly. In this chapter we add up link propagation delays to get path lengths, and consider two different constraints in selecting the k-shortest path. One is that any candidate path should not be longer than the diameter of the network, i.e., the shortest path between the farthest pair of nodes in the topology. The other is that between any OD pair, the path length of the candidate should not be greater than twice that of the shortest path. Depending on how the candidate paths are chosen, in this chapter, we will evaluate three different combinations:

- *basic*: The candidate paths are the k-shortest paths. MLU bound is applied.

- *basic+nd*: The candidate paths are k-shortest paths which also satisfy the network diameter constraint. MLU bound is applied.

- *basic+e2e*: The candidate paths are k-shortest paths which also conform to the OD-pair end-to-end delay constraint. MLU bound is applied.

With these changes, the GreenTE model now has practical constraints on link utilization and path length, and also can be solved within a reasonable time.

Fig. 7.3 shows that the power saving potential grows as the value of k increases. However, increasing k also increases the computation time. When k is large enough (20 in this example), increasing k only improves the power

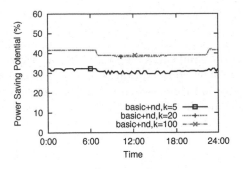

Figure 7.3: Power saving potential of GÉANT with different k values.

saving potential by a small amount. Therefore, GreenTE is able to achieve near optimal power savings as long as k is reasonably large.

As mentioned previously, the computation time becomes unacceptable for large topologies such as Sprint and AT&T. We use CPLEX [12] as the computation tool and in order to limit the computation time, we force CPLEX to stop after 300 seconds. Table 7.2 shows that the computation time for near optimal results within the k-shortest paths solution space increases dramatically as value of k grows. When $k = 20$, we can obtain about 96% of the optimal power saving potential when we limit the computation time to be 300 seconds. More details about GreenTE can be found in [20].

Table 7.2: Power Saving Potential of AT&T under *basic+e2e* with Traffic That Has 21% MLU under OSPF

k value	Computation Time	Status	Power Saving Potential
5	65s	Optimal	11.90%
10	5747s	Optimal	17.54%
20	100892s	Optimal	19.79%
20	300s	Non-optimal	18.99%

7.4   Rate Adaptive Processing inside Routers

7.4.1   Dynamic Voltage and Frequency Scaling

Among the factors determining the dynamic power, voltage and frequency have a far-reaching impact. The dynamic power is proportional to the frequency and the square of supply voltage. The frequency also influences the supply voltage: maintaining a higher clock frequency may mean maintaining a higher supply voltage. Therefore, it has a cubic impact on power dissipation combining the voltage and frequency. Although voltage and frequency have

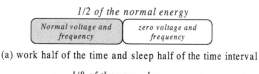

(a) work half of the time and sleep half of the time interval

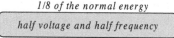

(b) processing is stretched to the whole time interval

Figure 7.4: Benefits of DVFS on energy savings.

significant leverage on energy savings, it may degrade the system performance due to the decrease of clock frequency. If one could recognize the periods when absolute guarantees or stringent requirements of network performance are not required, the energy can be greatly saved by reducing the voltage and frequency. Power-aware technique managing the supply voltage and clock frequency, known as Dynamic Voltage and Frequency Scaling (DVFS) has been extensively studied in the area of microprocessor (e.g., CPU) design. However, little research work has been reported on the application of DVFS technique in the context of router design.

Consider a router operation scenario: a router completes the processing of all the packets within a time interval, and the energy will be wasted in the remaining idle time periods. A coarse granularity DVFS method is proposed in [7] for Ethernet devices, which attempts to identify the inactive periods of the links and puts the associated devices into sleep in such periods. As shown in Fig. 7.4a, if the processing could be completed by the middle of the interval, half of the energy could be saved in an ideal case.[1] The key question behind this approach is in deciding when to turn off a link and for how long. Without the detailed knowledge of incoming traffic pattern, this approach has to compromise with network performance (e.g., packet delay, loss) on energy consumption reduction.

Rather than simply pausing the packet processing in idle periods, more energy is expected to be saved if the packet processing is stretched to a time interval by scaling the frequency and voltage, as illustrated in Fig. 7.4b which is with the same traffic load as in Fig. 7.4a. It reduces the clock frequency by half and stretches the processing time covering the whole time interval. Meanwhile, with halved clock frequency, the voltage could also be scaled down. Suppose that the voltage could also be halved, then the total energy needed is reduced to 1/8 of its normal consumption considering the quadratic impact of voltage on power dissipation. It is obvious that more gain will be achieved if stretching the processing into a larger interval, yet resulting in a longer packet delay. Therefore, the energy saving can be optimized subject to the constraint

---

[1]Suppose zero switching time to sleep state and zero energy consumption in sleep state.

of maximum tolerable delay.

7.4.2 Adaptive Link Rate Interface

Take Ethernet as an example, a network 1Gbps interface consumes 4W more than a 100Mbps interface and a 10Gbps interface consumes about 10 to 20W. In addition, the impact of link utilization on the power consumption is considered minimal [6]. Therefore, the interface data rate should be dynamically tailored with flexible configuration according to the link utilization. This approach is known as Adaptive Link Rate (ALR), which was proposed in [6] aiming to reduce energy consumption of full-duplex Ethernet networks. The proposed ALR solution only works for Ethernet and more research efforts are still needed to extend ALR to be applied for other network interface types. In addition, switching the interface rate with finer granularity steps (instead of only three steps suggested in [6]) could further enhance the energy consumption efficiency. In this case, two issues need to be addressed for the enhancement of ALR.

First, it can work properly only under the condition that the interfaces of the two end nodes of a link operate with the same data rate. Certain mechanism or protocol is required to negotiate the interface rate between these two nodes. The process of negotiation and switching of different interface rates is expected to be quick so as not to introduce additional packet loss and delay. A two-way handshake procedure could be implemented: receiver (RX) or transmitter (TX) sends a request message of rate change which contains the information of desired rate; upon the receipt of the request, the other node responses with a message with ACK to agree on the change action or NACK to decline the request.

Second, a policy is required to determine when to change the link rate and what the target rate is. The operational states, as well as the interface rate of each state, can be predetermined. The queue length of RX/TX can be the indicator to trigger the transition of the states: the increase of queue length leads to a transition to a state with faster rate, whereas the decrease of queue length results in a lower rate. The system is first operated with the idle state before switching to another rate to prevent potential problems caused by glitches [13].

7.4.3 A Multi-Frequency Scaling Prototype

In this section, we describe a Multi-Frequency Scaling (MFS) scheme for energy conservation of network devices like routers and switches [16]. The frequency of components in a network device is scaled dynamically according to the real-time workload. Based on this scheme, we first present a prototype of this scheme in the data path of a general IPv4 router on a real hardware platform — NetFPGA. Experimental results show excellent energy savings at the cost of a tolerable latency under various ranges of traffic loads and indicate the feasibility and possibility of deploying this mechanism into real network devices for energy saving.

In essence, there are two performance metrics for a rate adaptive scheme of network devices, and we present, analyze and evaluate the MFS scheme according to them.

- *Energy saving:* The amount of energy saving is linear to the reduction in rate [16]. Therefore, we use the *rate reduction* to measure the efficiency of energy saving. In particular, we use the average rate reduction which is more meaningful than instantaneous energy saving.

- *Packet delay:* Generally, more time working on lower frequency, longer delay suffers. Moreover, variation of packet delay brought by frequency switching may have an impact on routing protocols that update routing based on packet delay. Therefore, the rate adaptive scheme must incur limited increase in packet delay. The packet delay consists of *queuing delay* which is used as the metric for delay in this chapter, and a constant amount of time for forwarding and switching.

Basically, a rate adaptive scheme needs to strike a balance between energy saving and packet delay even with highly dynamic traffic loads. In the following sections, we investigate, both in theory and in data result, the proposed MFS scheme with the above metrics. We show that it is feasible to achieve excellent energy saving with tolerable cost in traffic delay using multi-frequency.

Generally, a network device works at two states: *active* state and *idle* state. When the device is actively processing traffic, it works at the active state. When there is no traffic for processing but the device is still powered on, it works at the idle state. Thus, the energy consumption of a general network device could be modeled as:

$$E = P_a T_a + P_i T_i, \tag{7.17}$$

where T_a and T_i denote the time spent in active state and idle state, respectively, P_a and P_i represent the power consumption in each mode. For both active and idle states, there is a static portion of power draw which is independent of the device's operating frequency, while the rest of the power draw indeed depends on the operating frequency:

$$P_a(r) = C + f(r),$$
$$P_i(r) = C + \beta f(r), \tag{7.18}$$

where $f(r)$ reflects the dynamic portion of energy consumption working at frequency r, and C denotes the static portion of energy consumption. The parameter β represents the relative magnitudes of routine work incurred even in the absence of packets to the work incurred when actively processing packets [17]. In general, the dynamic portion of energy consumption depends on the operating frequency and voltage of the network device:

$$f(r) \propto r V_{dd}^2, \tag{7.19}$$

where V_{dd} is the voltage of the device [19]. Hence, energy consumption could be reduced for both states by scaling the device's operating frequency according to its workload.

It is observed that buffers are commonly adopted in current design of network devices as queue for processing jobs. Fig. 7.5 shows an example of this; the Framer and Packet Processor in the line card are connected by a buffer. Intuitively, the actual buffer occupancy is considered as an indicator for real-time traffic load. Our MFS mechanism uses this indicator for frequency switch. We assume that the hardware supports working at several different rates. The operating rate of devices is dynamically switched according to the buffer occupancy. A range of thresholds is set to evaluate the relative workload of the devices.

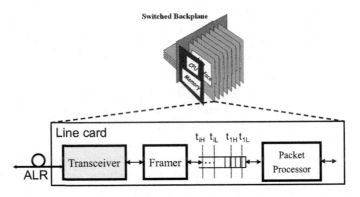

Figure 7.5: Example of components connected by buffers in the line card.

To avoid rate oscillation of single threshold, Dual-Threshold scheme is proposed, which uses both high threshold t_H and low threshold t_L for rate switch, as illustrated in Fig. 7.6a. When component works at low rate mode, the buffer occupancy (denoted as Q_Length in Fig. 7.6) would rise with the increase of traffic load. The transition from low rate to high rate is induced when the buffer occupancy exceeds t_H. On the other hand, when component operates at high speed mode, the queue length would drop while the traffic load decreases. Similarly, a transition from high rate to low rate is caused when the queue length drops below t_L.

Now, we give a detailed description of the implementation of MFS on NetFPGA.

The NetFPGA is a line-rate, flexible and open platform for gigabit-rate network switching and routing, which enables students and researchers to build high-performance networking systems using Field-Programmable Gate Array (FPGA) hardware. The core data processing functions are implemented in a modular style, enabling researchers to design and build their own functional components independently without modifying original codes [15]. Here, we use NetFPGA-1G card for hardware prototype, which contains one Xil-

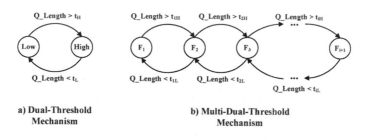

a) **Dual-Threshold Mechanism**

b) **Multi-Dual-Threshold Mechanism**

Figure 7.6: State machine for buffer occupancy based frequency scaling mechanism.

inx VirtexII-Pro 50 FPGA, The core clock of NetFPGA-1G card could be configured to operate at either 125 MHz or 62.5 MHz. Two SRAMs work synchronously with this core FPGA. In our implementation, the core clock rate is set at 125 MHz, which is the default setting of NetFPGA-1G card.

The *Rate Adapter* (*i.e.*, *CLK Adapter* in Fig. 7.7) is the main component for rate adaptation. The CLK Adapter reads the FPGA's core clock as input, and generates a range of exponentially distributed frequencies based on the input, by simply using a clock cycle counter. The buffer is used for frequency switch decision. Frequency transition is achieved by switching to another frequency according to the input clock, and setting it as the output. In our implementation in NetFPGA, the frequency switch does not involve much overhead, which suggests the frequency switch has negligible impact on packet delay.

The output is used for driving the modules connected to the output of buffer, as shown in Fig. 7.7. To avoid misalignment between input and output clock frequencies, D flip-flop is adopted for buffering the generated output clock. Thus, the actual frequency transition happens one clock cycle after the decision of transition is made.

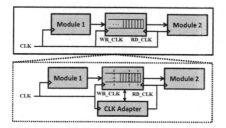

Figure 7.7: Adaptive frequency scaling module.

In the implementation, the CLK Adapters are embedded between *Input Arbiter* and *Output Port Lookup*, and into some sub-modules of *Output Queues*. Six frequencies distribute from 3.096 MHz to 125 MHz in an exponential order of two. The core processing rates are accordingly from 250

Mbps to 8 Gbps. Since not all components in NetFPGA support operating at a frequency different to the default value, it is difficult to directly measure the reduction in energy saving of components supporting frequency scaling. Energy conservation could only be estimated using Equation 7.19 in Section 7.3. However, the rate reduction could be measured by exploiting the *Register System* of NetFPGA. The register interfaces allow software running on host system to send data to and receive data from the hardware modules. A few registers are used for recording operating time of each frequency in the prototype, so that we could estimate average rate reduction.

The power saving percentages of dynamic portion of energy consumption of modules supporting MFS in our prototype are estimated by using (7.19), and are presented in Fig. 7.8. The reason that the minimum saving percentage is around 50% is also because the full processing rate is 8 Gbps. Since frequency only has influence on the dynamic power consumption, the total energy saving may be less than the results presented in Fig. 7.8. However, the dynamic power consumption is much higher than the static power consumption generally, when there is no rate adaptation in hardware. Thus, MFS would achieve great energy saving in modules supporting it, especially when the average utilization is low. Based on the results presented above, we draw the conclusion that the MFS mechanism is feasible and effective for energy conservation in real systems.

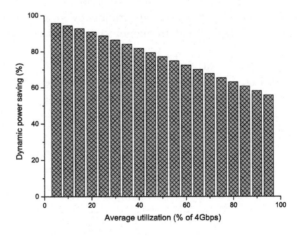

Figure 7.8: Dynamic power savings of modules supporting MFS in MFS Router.

7.5 Power Efficient Architectures for Router Functions

7.5.1 Routing Lookup

In general, two major categories of routing lookup algorithms are used based on Classless Inter-Domain Routing (CIDR) IP addresses: Ternary Content Addressable Memory (TCAM) based schemes and trie-based algorithms. TCAM is a fully associative memory and is capable of returning the matching result within a single memory access. Therefore, it dominates today's high-end routers. Although the lookup speed of trie-based mechanisms using Dynamic Random Access Memory (DRAM) or Static Random Access Memory (SRAM) is slower than TCAM-based method, its cost and power consumption are significantly lower. Estimation shows that the power consumption per bit of TCAM is about 150 times more than that of SRAM. Therefore, it would be desirable to reduce the energy consumption of TCAM-based mechanisms while keeping its high performance.

The power consumption of TCAM is high since it is a fully parallel device. When the search key is input, all entries are triggered to perform the matching operations. Current high-density TCAM devices consume as much as 12-15 watts per chip when all the entries are enabled for search. However, the power consumption of TCAM can be reduced. We observe that not all the entries need to be triggered simultaneously during a lookup operation. The entries that really need to be searched are the ones matching the input key. Thus, one way to make TCAM-based method power efficient is to prevent the non-essential entries from being triggered during a lookup operation. There are two ways to decrease the number of entries triggered during a lookup operation. The first one is to store the entries in multiple small chips instead of a single large one. The second way is to partition one TCAM into small blocks and trigger only one of them during a lookup operation. Exploring these two energy-saving possibilities, multiple small TCAMs instead of only one large TCAM is utilized and the entries inside each TCAM are divided into several blocks. In this way, only one block on one TCAM is activated for routing lookup and so the energy can be saved. The division of routing table so as to fit into the TCAM blocks is the key to this problem, which should evenly allocate the route entries among the TCAM chip, balance the lookup traffic load among the TCAMs and avoid possible prefix conflicts. A detailed mechanism can be found in our previous work [21].

7.5.2 Packet Classifier

Packet classification involves matching several fields extracted from a packet header to a set of pre-defined rules in order to determine the follow-up action to be performed on the packet by networking devices. As one of the key functions of a router/switch, packet classification is widely used in applications such as policy-based routing, QoS, Virtual Private Networks (VPN), network security (e.g., firewall, and intrusion detection) and sophisticated traffic billing. Transmission line speeds keep increasing. For example, the state-of-the-art line rate of OC-768 (40 Gbps), used in optical fiber backbone networks, demands a worst case processing speed of 125 million packets per second. In addition to meeting the ever-increasing line rates, growing attention has been paid to reducing the power consumption in the design of next generation networking devices, which make it desirable for the packet classifier to consume as little energy as possible.

The most commonly used header fields in packet classification form a 5-tuple which includes source IP address and destination IP address for prefix matching, source port and destination port for range matching, and protocol number for exact or wildcard matching. The typical size of a rule-set ranges from hundreds to millions of rules, depending on the application and location of router/switch. Due to the complexity of packet classification, TCAM is frequently used, which can guarantee millions of searches per second by examining all rules simultaneously. However, the high power consumption caused by the parallel comparison in TCAM makes it less likely to be incorporated in the design of future power-efficient networking equipment.

We aim to design a packet classifier that meets the challenges of both high speed and low power consumption [13]. In order to achieve this goal, we choose to implement a decision tree-based packet classification algorithm (HyperCuts) in hardware, so that TCAM is replaced with energy efficient memory devices such as SRAM or DRAM. This type of algorithm recursively cuts the hyperspace represented by the rule-set into smaller hyper-boxes called Regions along some selected dimensions, which form a decision tree, until the number of rules contained in the resulting hyper-boxes is smaller than a predefined threshold. When a packet is received, the classifier traverses down the decision tree, based on the value of header fields, until a leaf node is found and searched for the matching rules. The structure and parameters of the packet classifier are carefully selected to make sure that only a small number of memory accesses are needed in order to guarantee the throughput.

The major parts of the packet classifier consist of a Tree Traverser and Leaf Searcher as shown in Fig. 7.9. Each internal node contains the cutting scheme to be performed on the hyperspace represented by this node, and the starting address of each child node. Other information, such as whether the child node is a non-empty leaf, is also coded. In each leaf node, rules are stored in the order of priority, along with their IDs. When the cutting scheme is fetched and interpreted, the Region corresponding to the header fields is selected, i.e., the starting address of this child node will be read out from memory and used for the next round of tree traversing. If a non-empty leaf is encountered, the control is passed to Leaf Searcher. The rules are compared against packet header fields one by one, until either a matching rule is found or all of the rules have been examined.

Power consumption of packet classifier can be further reduced by exploiting

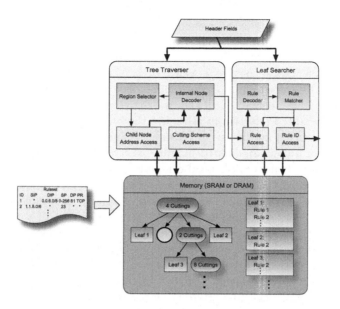

Figure 7.9: Reduced power vs. the number of queues.

the fluctuation in network traffic. It has been observed that high level of variation is common in network traffic. For example, traffic volume during the night might be less than one third of the peak day rate. This phenomenon implies that packet classifier does not have to run always at its highest possible clock speed. When the packet arrival rate is low, the packet classifier can reduce its working frequency to an appropriate level in order to save energy. In our project, we use a header buffer to monitor the change in network traffic, in which the number of outstanding packet headers waiting for classification works as an indicator of the frequency required for the packet classifier.

Our packet classifier can be easily implemented in hardware (FPGA or ASIC), working as an individual component on a line card or integrated into a network processor. The experiments performed on sample implementations in Altera FPGA and ASIC show that our packet classifier can achieve packet classification throughput of up to OC-768 line rate with significantly less power consumption than TCAM implementations.

7.5.3 Packet Queuing

The queuing process in the buffer is introduced in the routers to temporarily store the packets which cannot be processed immediately due to network congestion. Per-flow queuing mechanism is suggested to achieve advanced (QoS) guarantees by segregating individual flows into dedicated queues. Commercial products adopt "brute-force" implementation of per-flow queuing by deploying a separate physical queue for each in-progress flow.

The authors in [14] observe that the number of active flows in a packet scheduler is measured typically in hundreds even though there may be tens of thousands of flows in progress. With such a recognition, Dynamic Queue Sharing (DQS) mechanism, which provides dedicated queues for only simultaneous active flows instead of all the in-progress flows, is proposed in [9, 11]. DQS is demonstrated effective in reducing the required physical queues from millions to hundreds, and hence to reduce the energy consumption.

If packets of a flow are queued in the router buffer, it is considered as an active flow and a dedicated physical queue is assigned to it. In order to keep all the arrival packets of a certain flow to be in one physical queue, an Active Flow Mapping (AFM) table is introduced to record the mapping information between active flows and physical queues. When a packet of a flow arrives, DQS checks if the flow is in the AFM table. If yes, the packet is pushed into the corresponding queue; otherwise, a queue allocator creates a new queue to buffer the packets from this flow and correspondingly a new mapping entry between the flow and the queue is added to the AFM table. Once a physical queue becomes empty, the queue and its mapping to the flow will be withdrawn immediately. Instead of maintaining all the states of the flows, DQS only needs to manage a small number of physical queues and an AFM table. To achieve more efficient operations on the AFM table, the AFM table is first divided into a number of small AFM sub-tables by using hashing operations and each sub-table is organized through linked-list or binary sorting tree.

Since fewer queues are required, the peak power of packet queuing is reduced. We implement prototypes of DQS and brute-force per-flow packet queuing in Field Programmable Gate Array (FPGA)[2] in order to check the power consumption. With the help of Stratix PowerPlay Early Power Estimator, we are able to evaluate the power consumption of the prototypes. Considering the power consumption of the internal logic and on-chip memory to support 8000 flows, the estimator's results indicate that the power consumption with DQS is about 23.3% less than the brute-force approach. If one aims to support more flows, more gain on energy-saving can be achieved through DQS. Fig. 7.10 depicts how much power can be reduced by employing DQS compared with brute-force method.

7.5.4 Traffic Manager

Within a line card of a high-end router/switch, the traffic manager (TM) stores the arrival packets in its off-chip memory (e.g., a DDR-3 or RLDDR-2 memory). Even though only a small portion of the memory is used for queuing most of the time, TM cannot turn off the remaining portion of the memory to reduce power consumption.

Previous research has reported that the buffer size in Internet core router

---

[2]We use ALTERA Stratix EP1S80F1508C5 for implementation.

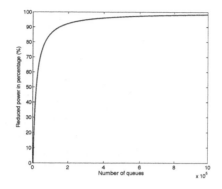

Figure 7.10: Reduced power vs. the number of queues.

can be very small [3] and the buffer size can be reduced to 20-50 packets if we are willing to sacrifice a fraction of link capacity, and if there is a large ratio between the speed of core and access links. Therefore the TM can utilize a small-size on-chip memory to buffer the packets most of the time, and activate the off-chip memory to store the packets only when the on-chip memory is about to overflow. In such a design, the off-chip memory and its corresponding memory controller are turned off by clock gating when there is no or light contention. With this idea in mind, we design and implement a *dynamic on-chip and off-chip scheduling scheme*, named *DPM*, to reduce both the peak and average dynamic power consumptions of the TM function.

In general, the packets need to be queued in both ingress and egress. For illustration, we first describe the architecture of the TM in the ingress direction as shown in Fig. 7.11. There are four key modules: segmentation, per-flow manager, packet manager and scheduler. Given the switch, fabrics can be optimized for switching fixed-sized data units (cells), and the packets arriving at the traffic manager are usually variable-length packets. Segmentation module will segment the packets into fixed length cells for further processing.[3] In order to provide advanced QoS, the arrival packets will be queued in a per-flow manner and the packets from a same flow are queued in one dedicated queue. Per-flow manager takes charge of packets to be written into the right queues, i.e., to determine in which queue an incoming packet will be kept in the buffer. The scheduler is responsible for reading out the packets from a certain queue in the buffer.

The TM's working flow-chart is described as follows. A packet from the upstream module[4] of TM is first segmented into fix-sized data units. Meanwhile, the flow ID of the packet is sent to the per-flow manager, where the

---

[3]The packets can also be segmented before their arrival at TM. In this case, the traffic is fixed-length cells.

[4]It is usually a network processor in commercial products.

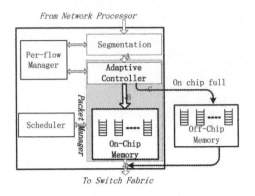

Figure 7.11: A schematic system of TM (Ingress direction).

address of the queue to store the packet is determined. And then the packets along with the address of the queue are sent to DPM (Point A in Fig. 1), where the packet management is conducted. DPM operates two kinds of packet queuing: on-chip memory queuing and off-chip memory queuing. If the adaptive controller (AC) in DPM detects that there is enough room in on-chip memory to hold the incoming packets, the packets are stored in the on-chip memory (Point B) directly. Otherwise, the packets will be kept in the off-chip memory (Point C). The detailed schemes to determine the switching between the on-chip memory and the off-chip memory will be described later. In the on-chip memory queuing, it adopts the aforementioned DQS [9] technique to let all the packets from the same flow be stored in the same physical queue so as to improve QoS, as mentioned above. In the off-chip memory queuing, we can either adopt DQS technique to support per-flow queuing or utilize the class information of the flow to implement class-based queuing management. Scheduler determines which flow queue will be sent out from its head of line packet to the *Switch Fabric* (Point D).

The switching mechanism between the on-chip memory and off-chip memory is the key part of the system design and is described as follows.

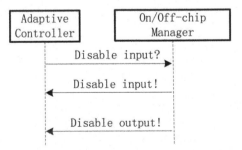

Figure 7.12: Scheduling between adaptive controller and on/off chip manager.

- **Switching from on-chip to off-chip** The adaptive controller described in Fig. 7.11 always queries the free memory size in the on-chip memory. As illustrated in Fig. 7.12, if the free size is smaller than T_{on} (T_{on} is a threshold for on-chip free memory size), the controller will send a "disable input?" message to the on/off-chip manager to attempt to disable the input direction of the on-chip memory. When the manager gets the message, it will not disable its input direction of on-chip memory until it has received a whole packet. And then the manager sends a "disable input!" message to controller so as to enable the input direction of the off-chip memory. When all the packets in the on-chip memory are scheduled out, the manager sends a "disable output!" message to controller and the on-chip memory is disabled.

- **Switching from off-chip to on-chip** The adaptive controller keeps querying the un-free memory size in the off-chip memory. If the un-free size is smaller than T_{off}, the DPM starts the switching operation. Similar to switching from on-chip to off-chip, the on/off chip manager will first disable the input of the off-chip memory and enable the input of the on-chip, and then enable the output of the on-chip memory when the off-chip memory is empty.

If the on-chip memory size is M_{on}, and the size of the packets buffered in the packet manager oscillates around $M_{on} - T_{on}$, DPM will switch frequently between the on-chip memory and the off-chip memory. To prevent the frequent switching, the threshold T_{off} for the off-chip memory should be smaller than $M_{on} - T_{on}$.

In this section, we have presented an energy-efficient packet management architecture named DPM. DPM can be utilized to construct the TM chip in a router with less peak and average dynamic power consumption while providing improved delay performance. Compared with the packet management schemes in traditional commercial TM chips, real prototype testing shows DPM has several advantages: first, the dynamic power of DPM is reduced by 60% as demonstrated through prototype testing with a small-sized on-chip memory; second, the peak power of DPM is reduced by 27.9%; third, DPM facilitates a reduced average delay, while the required total memory and the registers are both reduced accordingly. Please refer to [4] for the details about the experiments and evaluations.

7.6 Summary

A great deal of research has contributed to the energy efficiency on battery-operated devices in the area of wireless communications, but only recently has the energy consumption for underlying network infrastructures started to get more attention since we begin to be aware of the significant fraction of energy consumed by the Internet over the entire energy consumed.

Router is the main equipment that is used to construct the Internet. We argue that the architecture of the routers to construct the Internet can be improved so as to fully utilize the chances to save the energy related to transmission equipment. By exploring the means to control the energy consumption of routers, the Internet could become energy efficient. GRecRouter presented in this chapter seeks opportunities and means to notably reduce the power dissipation at network level, node level and function level.

Bibliography

[1] Raffaele Bolla, Roberto Bruschi, Franco Davoli, and Andrea Ranieri. Energy-aware performance optimization for next-generation green network equipment. In *PRESTO '09: Proceedings of the 2nd ACM SIG-COMM Workshop on Programmable Routers for Extensible Services of Tomorrow*, pages 49–54, New York, NY, USA, 2009. ACM.

[2] David Sarokin. Energy use of Internet, 2007, available at http://uclue.com/index.php?xq=724.

[3] M. Enachescu, Y. Ganjali, A. Goel, N. McKeown, and T. Roughgarden. Routers with very small buffers. In *INFOCOM 2006*, pages 1 –11, April 2006.

[4] Jindou Fan, Chengchen Hu, Keqiang He, Junchen Jiang, and Bin Liu. Reducing power of traffic manager in routers via dynamic on/off-chip scheduling. In *INFOCOM 2012*, Berkeley, CA, USA, Mar. 2012.

[5] C. Fraleigh, S. Moon, B. Lyles, C. Cotton, M. Khan, D. Moll, R. Rockell, T. Seely, and S.C. Diot. Packet-level traffic measurements from the sprint ip backbone. *Network, IEEE*, 17(6):6–16, Nov.-Dec. 2003.

[6] Chamara Gunaratne, Kenneth Christensen, Bruce Nordman, and Stephen Suen. Reducing the energy consumption of ethernet with adaptive link rate (alr). *IEEE Trans. Comput.*, 57(4):448–461, 2008.

[7] M. Gupta and S. Singh. Using low-power modes for energy conservation in ethernet lans. In *INFOCOM 2007*, pages 2451–2455, May 2007.

[8] Maruti Gupta and Suresh Singh. Greening of the Internet. In *SIGCOMM '03: Proceedings of the 2003 Conference on Applications, Technologies, Architectures, and Protocols for Computer Communications*, pages 19–26, New York, NY, USA, 2003.

[9] Chengchen Hu, Yi Tang, Xuefei Chen, and Bin Liu. Per-flow queueing by dynamic queue sharing. In *IEEE INFOCOM 2007*, pages 1613 –1621, May 2007.

[10] Chengchen Hu, Chunming Wu, Wei Xiong, Binqiang Wang, Jiangxing Wu, and Ming Jiang. On the design of green reconfigurable router toward energy efficient internet. *Communications Magazine, IEEE*, 49(6):83–87, June 2011.

[11] Chengchen Hu, Tang Yi, Kai Chen, and Bin Liu. Dynamic queuing sharing mechanism for per-flow qos control. *IET Communications*, 4(4):472–483, 2010.

[12] IBM. Ibm ilog cplex optimizer. http://www-01.ibm.com/software/integration/optimization/cplex-optimizer/.

[13] Alan Kennedy, Xiaojun Wang, Zhen Liu, and Bin Liu. Low power architecture for high speed packet classification. In *ANCS '08: Proceedings of the 4th ACM/IEEE Symposium on Architectures for Networking and Communications Systems*, pages 131–140, New York, NY, USA, 2008.

[14] A. Kortebi, L. Muscariello, S. Oueslati, and J. Roberts. Evaluating the number of active flows in a scheduler realizing fair statistical bandwidth sharing. In *SIGMETRICS '05: Proceedings of the 2005 ACM SIGMETRICS International Conference on Measurement and Modeling of Computer Systems*, pages 217–228, New York, NY, USA, 2005.

[15] J.W. Lockwood, N. McKeown, G. Watson, G. Gibb, P. Hartke, J. Naous, R. Raghuraman, and Jianying Luo. Netfpga–an open platform for gigabit-rate network switching and routing. In *IEEE International Conference on Microelectronic Systems Education, 2007*.

[16] Wei Meng, Chengchen Hu Yi Wang, Keqiang He, and Bin Liu. Greening the internet using multi-frequency scaling schemes. In *IEEE AINA 2012*, Fukuoka, Japan, Mar. 2012.

[17] Sergiu Nedevschi, Lucian Popa, Gianluca Iannaccone, Sylvia Ratnasamy, and David Wetherall. Reducing network energy consumption via sleeping and rate-adaptation. In *Proc. of NSDI, 2008*, pages 323–336, Berkeley, CA, USA. USENIX Association.

[18] Sergiu Nedevschi, Lucian Popa, Gianluca Iannaccone, Sylvia Ratnasamy, and David Wetherall. Reducing network energy consumption via sleeping and rate-adaptation. In *NSDI'08: Proceedings of the 5th USENIX Symposium on Networked Systems Design and Implementation*, pages 323–336, Berkeley, CA, USA, 2008. USENIX Association.

[19] Bo Zhai, David Blaauw, Dennis Sylvester, and Krisztian Flautner. Theoretical and practical limits of dynamic voltage scaling. In *DAC 04: Proceedings of the 41st Annual Conference on Design Automation*, 2004.

[20] Mingui Zhang, Cheng Yi, Bin Liu, and Beichuan Zhang. Greente: Power-aware traffic engineering. In *18th IEEE International Conference on Network Protocols (ICNP 2010)*, pages 21–30, Oct. 2010.

[21] Kai Zheng, Chengchen Hu, Hongbin Lu, and Bin Liu. A tcam-based distributed parallel ip lookup scheme and performance analysis. *IEEE/ACM Trans. Netw.*, 14(4):863–875, 2006.

Chapter 8

The Impact of Renewable Energy Sources in the CO_2 Emissions of Converged Optical Network and Cloud Infrastructures

*Markos P. Anastasopoulos, *Anna Tzanakaki and +Dimitra Simeonidou
*Athens Information Technology Center, {mark,atza}@ait.gr
+University of Essex, dsimeo@essex.ac.uk

This chapter focuses on converged optical network and IT infrastructures suitable to support cloud services. More specifically, the concept of Virtual Infrastructures (VIs), over one or more interconnected Physical Infrastructures

(PIs) comprising both network and IT resources, is considered. Taking into account the energy consumption levels associated with the ICT today and the expansion of the Internet, energy efficient infrastructures with reduced CO_2 emissions become critical. To address this, a hybrid energy power supply system for the high energy consuming IT resources is adopted. In this system conventional and renewable energy sources are cooperating to produce the necessary power for the IT equipment to operate and support the required services. The reduction in CO_2 emissions is further increased by applying energy aware planning of VIs over the converged PI. To quantify the benefits of the proposed approach a Mixed Integer Linear Programming model suitable for planning VIs is proposed and developed. This model takes into account multi-period and multi-service considerations over an integrated hybrid-solar powered IT and optical network infrastructure and aims at minimizing the CO_2 emissions of the planned VIs. Our modelling results indicate significant reduction of the overall CO_2 emissions that varies between 10-50% for different levels of demand requests.

8.1   Introduction

In response to the increased volume and requirements of existing and upcoming applications such as UHD IPTV, 3D gaming, virtual worlds, etc., and the limitations of existing large scale computer networks, the current trend is to support on demand delivery of infrastructures, applications and business processes in a commonly used, secure, scalable and computer based environment over the Internet for a fee [1]. This concept, known as **cloud computing**, is a natural evolution of traditional data centers (DCs) in which computing resources are provided as standard-based Web services adopting a pricing model where customers are charged based on their utilization of computational resources, storage and transfer of data [2]. This introduces a new business model and facilitates new opportunities for a variety of business sectors increasing the sustainability and efficiency in the utilization of available resources.

In cloud computing, subscription-based access to infrastructures, platforms or applications may be provided to users referred to as Infrastructure as a Service (IaaS), Platform as a Service (PaaS), Software as a Service (SaaS), respectively [2]. In IaaS, computation, storage and communication resources are offered on demand to users and are charged on a pay-per-use basis.

On the other hand, cloud computing services need to be supported by specific IT resources in DCs that may be remote and geographically distributed between themselves and the end users, requiring connectivity, through a very high capacity and increased flexibility and dynamicity transport network. A strong candidate to support these needs is optical networking due to its carrier-grade attributes, its abundant capacity, long reach transmission capabilities and recent technology advancements including dynamic control planes, elastic technologies, etc. In this context, an infrastructure comprising converged optical network and IT resources supporting the IaaS model introduces a paradigm shift in the way the infrastructure design and operation are performed, as optimal solutions require to take a holistic view of the infrastructure and jointly consider both types of resources their details, specificities and interplay.

In order to maximize the utilization and efficiency of infrastructures supporting converged network and IT resources, in support of cloud services, it has been recently proposed to partition the network into logical domains, which reduces the signalling overhead and generates more scalable solutions. A tool for hierarchically organizing domains is that of virtualization. Infrastructure virtualization aims at reducing the size of information exchanged between domains through topology aggregation and logical topology and resource creation and also maximizing the utilization and efficiency of infrastructures. The concept of Virtual Infrastructures (VIs) facilitates sharing of physical resources among various virtual operators introducing a new business model that suits well the nature and characteristics of the Future Internet and enables new exploitation opportunities for the underlying physical infrastructures. Through the adoption of VI solutions, optical network and IT resources can be deployed and managed as logical services, rather than physical resources. This results in resizable and on demand computing and network capacity that may be provided on a pay-as-you-go basis as a web service, known in the literature as elastic cloud. However, it should be noted that supporting the paradigm of elastic cloud introduces challenges to network designers who need to have accurate knowledge of both network and IT resources. More specifically underestimating the network and IT resources may lead to inability to satisfy end-users' requirements, whereas an overestimation may lead to increased operational and capital expenditures (OpEx and CapEx).

An important consideration that needs to be taken into account when designing and operating converged optical network and cloud infrastructures is their sustainability in the context of the required energy consumption and the associated CO_2 emissions. This is a key consideration, as ICT is responsible for about 4% of all primary energy today worldwide, and this percentage is expected to double by 2020 [3]. In this respect an additional benefit that optical network technology is offering compared to its conventional optoelectronic counterparts is improved energy efficiency [4]. It is evident that the level of energy efficiency that can be achieved through the use of optical networking is also very dependent on specific architectural approaches followed [5], [6], [7], technology choices made [8] as well as the selection of algorithms and provisioning schemes [9]. Recognizing the importance of these parameters several approaches have been proposed in the literature aiming at reducing the energy consumption of network infrastructures [5] - [9]. On the other hand, significant efforts are currently focused on minimizing the energy consumption of data centers and the associated computational resources [10].The concept of the use of renewable energy sources (green energy) such as solar, wind, geothermal or biomass is also becoming a promising approach toward the minimization of CO_2 emissions for the operation of this type of resource [11], [12].

In this chapter it is proposed to minimize the CO_2 emissions of converged optical network and cloud infrastructures by introducing a hybrid energy power supply system for the high energy consuming IT servers. In these systems, conventional and renewable energy sources (solar) are merged to produce the necessary power for an IT server to operate. In addition, in these infrastructures deploying the concept of virtualization, energy efficiency can be further addressed at both the VI planning and operation phase [13]. This chapter focuses on planning energy efficient VIs. In general, VI planning is responsible to generate dynamically reconfigurable virtual networks satisfying the end users' or VI provider's requirements and meeting any specific needs such as energy efficiency. Through this process the least energy consuming VIs that can support the required services are identified, in terms of both topology and resources. In the optimization process involved, joined consideration of the energy consumption of the converged network and IT resources is performed. As IT resources require very high levels of power for their operation and their conventional operating window is commonly not optimized for energy efficiency, allocating IT resources in an energy-aware manner interconnected through a low energy-consuming optical network can potentially offer significant energy savings and hence reduction in CO_2 emissions. This is particularly relevant for the cases that the IT resources are powered through conventional energy sources with high CO_2 emissions.

To quantify the reduction of the overall CO_2 emissions of the converged infrastructure a Mixed Integer Linear Programming (MILP) model, suitable for the planning of VIs, is proposed. This model takes into account multi-period and multi-service considerations over an integrated hybrid-solar powered IT and optical network infrastructure. The proposed scheme aims at reducing the non-renewable energy (non-green) consumption in the IT servers based on a simple model for the performance of solar panels that takes into account both the instantaneous variation of solar intensity and the locations of the IT servers. So far, existing energy aware VI planning schemes ([13], [14], [15], [16]) do not take into account time variability of services or the impact of renewable energy sources in the IT servers. However, this may lead to an inefficient utilization of the network and IT resources and, therefore, increased CO_2 emissions. Detailed modeling results indicate a clear benefit achieved through the proposed approach that varies between 10-50% reduction in terms of CO_2 emissions depending on the demand request volume that the VIs are expected to support.

The rest of the chapter is organized as follows. In Section 8.2, a brief description of the energy consumption models for both optical network and IT elements is given. Then, the energy-aware VI planning process, taking into account various optimization and design scenarios, is presented in Section 8.3. The performance of the proposed VI planning process is examined in terms of non-renewable energy consumption and utilization of physical resources in

Section 8.4. Comparisons with other similar schemes presented in the literature are also part of the performance analysis. Finally, Section 8.5 concludes the chapter.

8.2 Energy Consumption Models for the Physical Infrastructure

The estimation of the energy consumption of the physical infrastructure (PI) is highly sensitive to the architecture used and the specific technology deployed. In Sections 8.2.1 and 8.2.2, a detailed description of the energy consumption models for both optical network and IT resources is presented.

As already discussed in the introduction section, in this work a hybrid energy power supply system for the IT resources is assumed, in which conventional and renewable energy sources (solar) are merged to produce the necessary power for the IT servers to operate. To quantify the benefits of this type of approach, a renewable energy model based on solar radiation that is used to provide power for the operation of the IT servers in the data centers is adopted and presented.

8.2.1 Optical Network Elements

The estimation of energy consumption of the PI network resources is very much dependent on the network architecture employed and the specific technology choices made. The current chapter is focusing on optical network technologies based on wavelength division multiplexing (WDM) utilizing Optical Cross-Connect (OXC) nodes to perform switching and facilitate routing at the optical layer. The overall network power consumption model is based on the power-dissipating (active) elements of the network that can be classified as switching nodes (OXC nodes), and transmission line related elements. More specifically the OXCs assumed are based on the Central Switch architecture using Micro-Electrical Mechanical Systems (MEMS), while for the fiber links a model comprising a sequence of alternating single mode fiber (SMF) and dispersion compensating fiber (DCF) spans together with optical amplifiers based on Erbium Doped Fiber Amplifier (EDFA) technology to compensate for the losses are allocated at the end of each transmission span (Fig. 8.1). The details of these models are described in [4], [17], [18], [19], with the only difference being that unlike [17] the current work assumes wavelength conversion capability available at the OXC nodes.

Fig. 8.2 illustrates the OXC architecture assumed in this work. Each OXC node comprises a set of active and passive elements. The passive elements are: the multiplexers (MUX) and de-multiplexers (DEMUX), while the active elements indicated in Fig. 8.2 in gray color include: the photonic switching matrix, one Erbium-Doped Fiber Amplifier (EDFA) per input fiber port, one

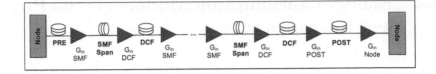

Figure 8.1: Link model.

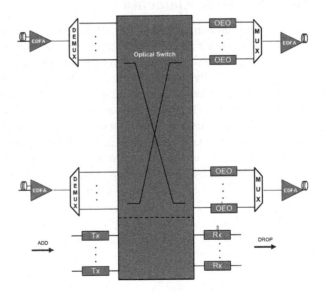

Figure 8.2: OXC architecture.

Table 8.1: Network Equipment Power Consumption Figures

Symbol	Description	Power
$P_{transponder}$	O/E/O: Line-side WDM Transponder (10G)	6W
$P_{Tx/Rx}$	E/O,O/E: Transmitter or Receiver	3.5W
P_{Edfa}	EDFA	13W

Erbium-Doped Fiber Amplifier (EDFA) per output fiber port, one Optical-Electrical-Optical (OEO) transponder per output wavelength port and one transmitter (Tx) -receiver (Rx) pair per lightpath located at the add and drop ports of the OXC. It has been assumed that every network node is equipped with full wavelength conversion capability, supported by the OEO transponders available on each wavelength port of the OXCs. Therefore, this OXC architecture is overcoming the wavelength continuity constraint present in non-wavelength converting solutions. It is also assumed that the add/drop capability of the node is 50% of the through (bypass) traffic.

The total network power consumption is determined by the power consumption of the individual OXCs and fiber links comprising the optical network. Table 8.1 lists the power consumption values (in watts) that have been assumed for the power-dissipating (active) devices to be used for the calculation of the overall network power consumption. These are typical consumption figures originating mostly from the literature and datasheet surveys. Additionally to power consumption due to active devices, we also incorporate power dissipation due to cooling in the power estimation mode used and a 100% power overhead due to cooling is assumed [20]. It should be noted that all energy consumption calculations account only for the transport equipment of the optical network; the power dissipated by electronic circuits, such as control boards of OXCs or hardware implementing protocol functionality is not considered. For more detailed information regarding the formulas describing the power consumption models of the OXCs the reader is referred to [21].

8.2.2 Energy Consumption Models for Computing Resources

The physical IT infrastructures are considered to be data centers housing computer systems and associated components. Data centers are primarily used to process data (servers) and store data (storage equipment). The collection of this processing and storage equipment is referred to as the "IT equipment." A large part of the energy consumption of a data center resides in the IT equipment, i.e., processing and storage equipment and the cooling system [22].

In this chapter a hybrid energy power supply system for the IT resources is assumed, in which conventional and renewable energy sources (solar) are co-operating to produce the necessary power for the IT equipment to operate and

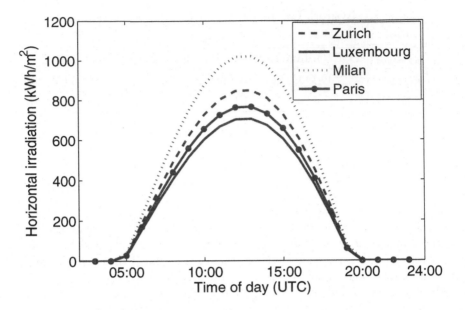

Figure 8.3: Solar energy in different IT locations.

support the required services. To describe mathematically this approach and quantify its benefits two different energy models are adopted: one deriving the power required by the IT resources when utilizing conventional energy sources and one renewable energy model based on solar radiation.

For the conventional energy sources a linear power consumption model that mainly concentrates on the power consumption associated with the CPU load of IT resources is assumed. More specifically p_s is referred to the CPU resources and E_s to the energy consumption for utilizing a portion $u_s = p_s/p_s^{max}$ of the maximum CPU capabilities, p_s^{max}, of server s. For simplicity, the following linear energy consumption model has been adopted [22], [23]:

$$E_s\left(u_s\right) = P_{idle}^s + \left(P_{busy}^s - P_{idle}^s\right)u_s \qquad (8.1)$$

where P_{idle}^s and P_{busy}^s is the energy consumption of IT server s at idle state and under full load, respectively. For further details regarding the technical specifications of the IT servers the reader is referred to [24]. Note that 50% overhead in energy consumption is assumed addressing the cooling associated power dissipation.

To describe the power supply of the IT servers by a dual system, that uses both conventional electrical energy sources and alternative energy sources based on solar radiation, it is assumed that each IT server is directly connected to a single closely located photovoltaic power station. The performance of the solar powered subsystem depends on the solar radiation that is received at the

Earth's surface and is mainly affected by atmospheric conditions, the location of the solar panels, the solar declination angle, the time of day, etc. [25]. Insolation (measured in W/m^2) is described by the following formula [25]:

$$I = S cos(Z) \tag{8.2}$$

where S is the solar insolation under clear sky conditions on a surface perpendicular to incoming solar radiation and Z the Zenith Angle that is the angle from the zenith (point directly overhead) to the Sun's position in the sky. Z is given by the following equation [25]:

$$Z = cos^{-1}(sin\phi sin\delta + cos\phi cos\delta cos H) \tag{8.3}$$

where ϕ is the latitude of the location of the IT servers, δ is the solar declination angle that takes values in the range $[-23.45^o, 23.45^o]$ and is calculated by

$$\delta = sin^{-1} \left\{ sin\,(23.45^o)\,sin \left[\frac{360}{365}(d-81) \right] \right\} \tag{8.4}$$

where d is the day of the year with Jan 1 as $d = 1$. The declination is zero at the equinoxes (March 22 and September 22), positive during the northern hemisphere summer and negative during the northern hemisphere winter [26]. The declination takes its maximum value (23.45^o) on June 22 (summer solstice in the northern hemisphere) and its minimum (-23.45^o) on December 22 (winter solstice in the northern hemisphere) [26]. Finally, H is the angle of radiation due to time of day. A typical numerical example is shown in Fig. 8.3 where the hourly variation of the measured daily solar radiation at the equinox is depicted for two different IT server locations. As expected, the performance of the solar powered system is maximized at noon, while before sunrise and after sunset the radiated power is zero. In practical systems, this issue is addressed by storing the energy produced by solar cells during the daytime in batteries to supply the IT servers when the generated power is not sufficient during the night time. In this chapter, energy storage elements are not included in the formulations and it is assumed that during the night time IT servers are powered by conventional non-renewable energy sources.

8.3 Energy Aware VI Planning

VI planning is the process of using historical operating data and estimated future virtual resource (VR) requirements, to determine the optimal design of the VI. The objective of VI planning is to identify the topology and determine the virtual resources required to implement a dynamically reconfigurable VI based on both optical network and IT resources. This VI will not only meet customers' specific needs, but will also satisfy the virtual infrastructure providers' (VIP's) requirements for minimum energy consumption in our case.

The Energy Aware VI planning problem is formulated through a Mixed Integer Linear Programming (MILP) model that aims at minimizing jointly the total non-renewable energy that is consumed by optical network components including WDM transponders, amplifiers, switches and IT resources.

8.3.1   VI Problem Formulation

The VI planning problem is formulated using a network that is composed of one resource layer that contains the physical infrastructure and will produce as an output the virtual infrastructure layer illustrated in Fig. 8.1. The physical infrastructure is described through an eleven-node topology corresponding to the Pan-European optical network in which randomly selected nodes generate demands d $(d = 1, 2, \ldots, D)$ that belong to service class i $(i = 1, 2, \ldots, I)$ to be served by a set of IT servers s $(s = 1, 2, \ldots, S)$. Each service class is associated with a particular delay priority in which strict or more relaxed delay boundaries are assigned to delay sensitive (DS) or delay tolerant (DT) services, respectively. In the current study, it is assumed that high priority services have to be served at a strict time frame while in case of low priority this constraint is relaxed. For example, critical data could be sent for processing and storage to an IT server at a specific time or day while data of low importance could be served anytime within a week.

In order to incorporate demand volumes which are time dependent t is introduced to capture time variability. Furthermore, it is assumed that the granularity of demands is the wavelength. The IT locations (demand destinations) at which the services will be handled, are not specified and are of no importance to the services themselves. To mathematically formulate this issue the binary variable a_{dsi} is introduced to indicate whether demand d belonging to service class i is assigned to IT server or not. This variable takes value equal to 1 if and only if demand d is processed on server s. Moreover, it is assumed that each demand can be assigned only to one IT server:

$$\sum_s a_{dsi} = 1, \quad d = 1, 2, \ldots, D, \quad i = 1, 2, \ldots, I \qquad (8.5)$$

Furthermore, for each demand d belonging to service class i at time t, its demand volume h_{dti} is realized by means of a number of lightpaths assigned to paths of the VI. This demand volume has to be served within a specific time frame, i.e., $t_{di}^m \leq t \leq t_{di}^M$ where t_{di}^m is the arrival time of the new service request and t_{di}^M the upper bound of service time. Now, let $p_s = 1, 2, \ldots, P_{dsti}$ be the candidate path list in the VI for the lightpaths required to support demand d in time t for service i at server s, and x_{dpti} the non-negative number of lightpaths allocated to path p for demand d that belongs to service i at time

t. The following demand constraints should be satisfied in the VI [27]:

$$\sum_s \sum_p a_{dsi} \sum_{t_{di}^m \le t \le t_{di}^M} x_{dpti} = \sum_{t \le T} h_{dti}, \quad d = 1, 2, \ldots, D,$$
$$i = 1, 2, \ldots, I, \quad t = 1, 2, \ldots, T$$
$$(8.6)$$

Summing up the lightpaths through each link e ($e = 1, 2, \ldots, E$) of the VI we can determine the required link capacity y_{et} for link e required at time t [28], [29]:

$$\sum_d \sum_i \sum_s \sum_p \delta_{edpti} x_{dpti} \le y_{et}, \quad e = 1, 2, \ldots, E, \quad t = 1, 2, \ldots T \quad (8.7)$$

where δ_{edpti} is a binary variable defined as follows [28], [29]:

$$\delta_{edpti} = \begin{cases} 1, & \text{if link } e \text{ of VI belongs to path } p \text{ realizing} \\ & \text{demand } d \text{ of service class } i \text{ at time } t \text{ at server } s \\ 0, & \text{otherwise} \end{cases} \quad (8.8)$$

Using the same rationale, the capacity of each link e at time t in the VI is allocated by identifying the required lightpaths in the PI. The resulting PI lightpath z determine the load of each link g ($g = 1, 2, \ldots, G$) of the PI, and hence its capacity u_{gt}. Assuming that $q = 1, 2, \ldots, Q$ is used for denoting the PI's candidate path list realizing link e, then, the following demand constraint for link e should be satisfied [28], [29]:

$$\sum_q z_{eqt} = y_{et}, \quad e = 1, 2, \ldots, E \quad (8.9)$$

where the sum is taken over all paths q on the routing list Q_e of link e. Introducing the link-path incidence coefficients for the PI

$$\gamma_{geqt} = \begin{cases} 1, & \text{if link } g \text{ of PI belongs to path } q \text{ realizing link } e \text{ of VI at time } t \\ 0, & \text{otherwise} \end{cases}$$
$$(8.10)$$

the general formula specifying the PI capacity constraint can be stated as :

$$\sum_e \sum_q \gamma_{geqt} z_{eqt} \le u_{gt}, \quad g = 1, 2, \ldots, G, \quad t = 1, 2, \ldots, T \quad (8.11)$$

where G is the total number of links in the PI and the summation for each link g is taken over all lightpaths in the PI layer.

Apart from link capacity constraints (8.7) and (8.11) for the VI and PI, respectively, the total demands that are assigned to each server should not exceed its capacity p_s, $s = 1, 2, \ldots, S$. The latter capacity corresponds to the underlying physical resources, such as CPU, memory, disk storage, etc. The inequality specifying servers' capacity constraints is given by

$$\sum_d \sum_i \sum_p a_{dsi} c_{dsi} (x_{dpti}) \le p_s, \quad s = 1, 2, \ldots, S, \quad t = 1, 2, \ldots, T \quad (8.12)$$

where the summation is taken over all demands that arrive at server s and $c_{dsi}(x_{dpti})$ is a parameter specifying the computational requirements for demand d of service class i on server s. In practice, this parameter is determined by the set of relevant benchmarks for computer systems provided by the Standard Performance Evaluation Corporation (SPEC) [30].

The objective of the current problem formulation is to minimize the total cost that is related to the CO_2 emissions that are produced by the power-dissipating (active) elements of the resulting network configuration. This cost consists of the following components:

a) k_{gt} that is the cost of the capacity u_{gt} of link g of the PI at time t. It consists of the energy consumed by each lightpath due to transmission and reception of the optical signal, optical amplification at each fiber span and switching, and

b) E_{st} that is the energy used for processing u_{st} wavelengths in the IT server s at time t. Since the IT servers are powered by a hybrid energy supply system, it is assumed that the impact of renewable sources on CO_2 emissions is negligible.

In this context, minimum energy consuming VIs are obtained by minimizing the following cost function

$$F = \sum_t \left(\sum_g k_g u_{gt} + \sum_s \left(E_{st}(u_{st}) - E_{st}^R(u_{st}) \right) \right) \qquad (8.13)$$

where $E_{st}^R(u_{st})$ is the energy used in the IT server s at time t from renewable sources to process u_{st} wavelengths.

8.4 Numerical Results

To investigate the energy efficiency of the proposed VI design scheme, the multilayer network architecture illustrated in Fig. 8.5 is considered: the lower layer depicts the PI and the layer above depicts the VI. For the PI the COST239 Pan-European reference topology [31] has been used in which four randomly selected nodes generate demands to be served by two IT servers located in Paris, Milan, Luxembourg and Zurich. For simplicity, the granularity of service duration for both DT and DS services is the hour since according to the pricing policy of Amazon's Elastic Compute Cloud (EC2) even if a partial instance-hour will be consumed this will be billed as a full hour [32]. The major difference between DT and DS services is that the former has to be assigned to an IT server at the requested point in time while the latter could be scheduled anytime within a day. Furthermore, we assume a single fiber per link, 40 wavelengths per fiber, and wavelength channels of 10Gb/s each. It is also assumed that each IT server can process up to 1Tb/s of and its power

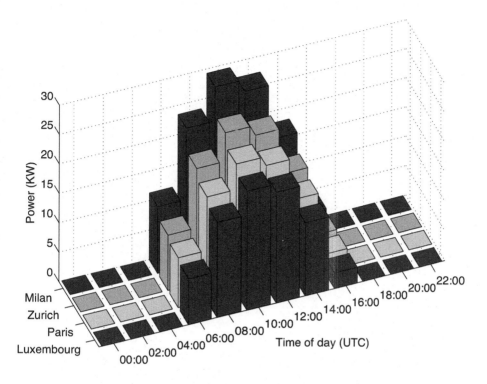

Figure 8.4: Solar energy output power per IT server.

consumption ranges from 6.6 to 13.2KW, under idle and full load, respectively
[8]. The necessary input power is provided by a hybrid energy power supply
system in which renewable energy is produced by photovoltaic panels based
on crystalline silicon with efficiency of 29% and $100m^2$ surface [33]. In the
numerical evaluations it is assumed that the power systems are not equipped
with energy storage elements, i.e., batteries.

An example of the optimal VI topology design for a scenario in which
4 source nodes that are located in London, Vienna, Copenhagen and Paris
generate demands equal to 40 wavelengths each, is depicted in Fig. 8.5. In the
proposed numerical example, it is assumed that each service has duration of
6 hours with the demands originating from London and Copenhagen to be
DS while those originating from Paris and Vienna to be DT. All demands
arrive at 02:00 UTC requesting instant access to the necessary network and
IT resources. However, DT services could be scheduled anytime within a day.
From Figs. 8.5a-b, where the VI is planned without considering renewable
energy sources, it is observed that the DS services are scheduled in the time
interval 22:00-04:00 UTC while DT services are scheduled in the time interval
04:00-10:00 UTC. As depicted in Fig. 8.5a, the associated VI topology for
the first two time periods consists of 2 virtual links and 3 virtual nodes,

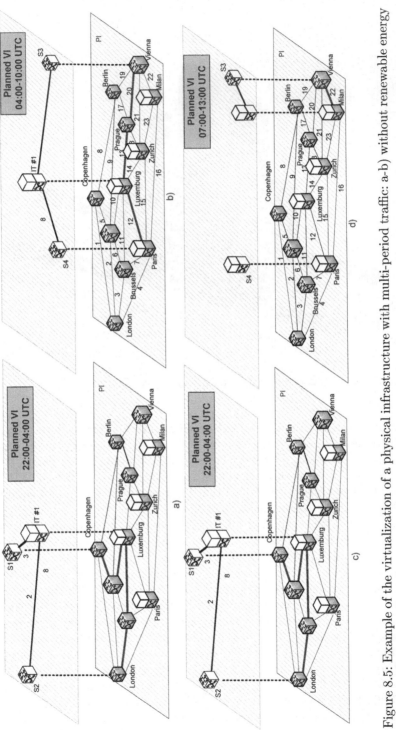

Figure 8.5: Example of the virtualization of a physical infrastructure with multi-period traffic: a-b) without renewable energy sources, c-d)with renewable energy sources.

while all demands corresponding to DS services are routed to the IT server in Luxemburg. For the following time periods the planned VI topology consists of 2 virtual links and 2 virtual nodes. Again all demands corresponding to DT services are forwarded to the same IT server. It is clear that during the entire planning horizon of the VIs, every time instant only one IT server is active while the other is switched off. On the other hand, if the VI is planned considering renewable energy sources, more IT servers will be included in the planned infrastructures. The above is illustrated in Fig. 8.5d where the planned VI consists of 3 virtual nodes and two active IT servers.

In Fig. 8.6a, the performance of the proposed VI planning scheme with solar powered IT servers and multi-time traffic considerations is compared to

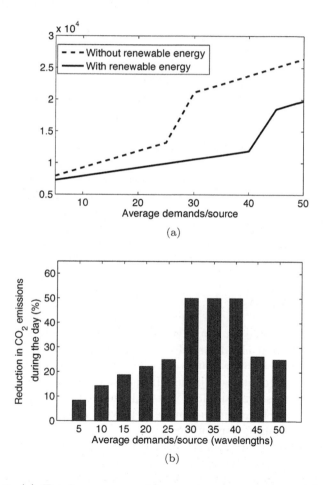

Figure 8.6: (a) Total non-renewable energy consumption with and without solar powered IT servers. (b) Reduction in CO_2 emissions during a day when solar powered IT servers are used.

the scheme presented in [27] where renewable energy sources are not adopted. Comparing these two schemes it is observed that for a broad range of demands the scheme with the solar powered IT servers consumes significantly lower non-green energy per operating period. Furthermore, the average non-renewable energy consumption for the architecture with solar powered IT servers increases almost linearly with the number of demands for up 40wv/source, in contrast to the non-renewable design where a stepwise increase of the power consumption is observed above 20wv/source.

For traffic demands above 20wv/source, the multi-period VI design scheme with renewable energy sources consumes significantly lower non-green energy for serving the same amount of demands compared to the non multi-period VI design scheme in the order of 50%: in the former approach fewer IT servers are activated during the night time to serve the same amount of demands. The benefit in terms of reduction in CO_2 emissions is depicted in Fig. 8.6b where the total CO_2 emissions during a day period has been reduced from 8% up to 50% for low traffic demands and high traffic demands, respectively. Given that the power consumption required for the operation of the IT servers is dominant in this type of network, by appropriately scheduling demands and switching-off the unused IT resources significant reduction of the energy consumption is achieved.

The above observation is also confirmed in Fig. 8.7 where the average utilization of optical network resources as well as the average number of non-green active IT servers for the two approaches is plotted as a function of the average demands/source. It is seen that when the VI is designed employing renewable energy sources more IT servers are activated to cover the same amount of demands since their impact on CO_2 emissions in negligible. Therefore, less optical network resources are employed since data travel shorter distances to arrive at their destination compared to the non-renewable energy scheme.

8.5   Conclusions

This chapter concentrated on converged optical network and IT infrastructures suitable to support cloud services. In order to maximize the utilization and efficiency of these infrastructures the concept of Virtual Infrastructures, over one or more interconnected Physical Infrastructures comprising both network and IT resources, was applied. Through the adoption of VI solutions, optical network and IT resources can be deployed and managed as logical services, rather than physical resources.

Taking into account the energy consumption levels associated with the ICT today and the expansion of the Internet energy-efficient infrastructures with reduced CO_2 emissions become critical. To address this, a hybrid energy power supply system for the high energy consuming IT resources is adopted. In this system conventional and renewable energy sources are cooperating to produce the necessary power for the IT equipment to operate and support the

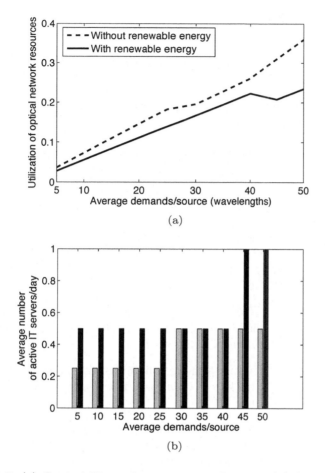

Figure 8.7: (a) Optical Network resources utilization. (b) Active Number of IT servers during a day.

required services.

The reduction in CO_2 emissions is further increased by applying energy aware planning of VIs over the converged PI incorporating both network and IT resources. To quantify the benefits of the proposed approach a Mixed Integer Linear Programming model suitable for planning VIs is proposed and developed. This model takes into account multi-period and multi-service considerations over an integrated hybrid-solar powered IT and optical network infrastructure and aims at minimizing the CO_2 emissions of the planned VIs. Our modelling results indicate significant reduction of the overall CO_2 emissions that varies between 10-50% for different volumes of demand requests.

Acknowledgments

This work was carried out with the support of the GEYSERS (FP7-ICT-248657) project funded by the European Commission through the 7th ICT Framework Program.

Bibliography

[1] M. A. Rappa, "The utility business model and the future of computing systems," *IBM Systems Journal*, vol. 43, no. 1, pp. 32-42, 2004.

[2] A. Buyya, J. Broberg, and A.Gociski, *Cloud Computing Principles and Paradigms*, John Wiley & Sons, 2011.

[3] M. Pickavet, W. Vereecken, S. Demeyer, P. Audenaert, C. Develder, D. Colle, B. Dhoert, and P. Demeester, "Worldwide energy needs of ICT: the rise of power-aware networking," *Proceedings IEEE ANTS 2008*, Mumbai (India), pp. 15-17, Dec. 2008.

[4] A.Tzanakaki et al., "Dimensioning the future Pan-European optical network with energy efficiency considerations," *Journal of Optical Communications and Networking* 3, 272-280, 2011.

[5] G. Shen, and R. S. Tucker, "Energy-minimized design for IP over WDM networks," *Journal of Optical Communications and Networking*, vol. 1, no. 1, pp. 176-186, June 2009.

[6] A. Stavdas, C. (T) Politi, T. Orphanoudakis, and A. Drakos, "Optical packet routers: how they can efficiently and cost-effectively scale to petabits per second," *Journal of Optical Networking*, vol. 7, no. 10, pp. 876-894, October 2008.

[7] E. Yetginer, and G. Rouskas, "Power efficient traffic grooming in optical WDM networks," in *Proc. of IEEE GLOBECOM 2009*, pp. 1-6, Nov.-Dec. 2009.

[8] S. Aleksic, "Analysis of power consumption in future high-capacity network nodes," *Journal of Optical Communications and Networking*, vol. 1, no. 3, pp. 245-258, August 2009.

[9] L. Chiaraviglio, M. Mellia, and F. Neri, "Reducing power consumption in backbone networks," in *Proc. of IEEE ICC 2009*, pp. 1-6, Jun. 2009.

[10] IEEE 802.3 Energy Efficient Ethernet Study Group, [Online]. Available: http://grouper.ieee.org/groups/802/3/eee_study/index.html.

[11] Y. Zhang, Y. Wang, and X. Wang, "GreenWare: greening cloud-scale data centers to maximize the use of renewable energy," the 12th ACM/IFIP/USENIX International Middleware Conference (Middleware 2011), Lisbon, Portugal, December 2011.

[12] X. Dong, I. El-Gorashi, and J.M.H. Elmirghani, "Green IP over WDM networks: solar and wind renewable sources and data centres," *Global Telecommunications Conference (GLOBECOM 2011)*, 2011 IEEE, pp. 1-6, 5-9 Dec. 2011.

[13] A. Tzanakaki et al.,"Energy efficiency in integrated IT and optical network infrastructures: the GEYSERS approach," in *Proc. of IEEE IN-FOCOM 2011, Workshop on Green Communications and Networking* (2011).

[14] M.P. Anastasopoulos, A. Tzanakaki, and K. Georgakilas, "Stochastic virtual infrastructure planning in elastic cloud deploying optical networking," in *Proc. of OFC 2012*, Pres. number: OW1A.3.

[15] A. Tzanakaki, M.P. Anastasopoulos, K. Georgakilas, et al., "Energy efficiency considerations in integrated IT and optical network resilient infrastructures," in *Proc. of ICTON 2011*.

[16] A. Tzanakaki, M.P. Anastasopoulos, K. Georgakilas, D. Simeonidou, "Energy aware planning of multiple virtual infrastructures over converged optical network and IT physical resources," in *Proc. of ECOC 2011*.

[17] A. Tzanakaki et al., "Power considerations towards a sustainable pan-European network," JWA061, OFC2011.

[18] S. Aleksic, "Analysis of power consumption in future high-capacity network nodes," *Journal of Optical Communications and Networking*, vol. 1, no. 3, p. 245, 2009.

[19] K. M. Katrinis, and A. Tzanakaki, "On the dimensioning of WDM optical networks with impairment-aware regeneration," *IEEE/ACM Transactions on Networking*, vol. 19, no. 3, pp. 735-746.

[20] J. Baliga, R. Ayre, K. Hinton, W. V. Sorin, and R. S. Tucker, "Energy consumption in optical IP networks," in *IEEE/OSA Journal of Lightwave Technology*, vol. 27, no. 13, pp. 2391-2403, 2009.

[21] A. Tzanakaki, "Energy efficient VI planning over converged optical networks and IT resources," *Green Communications*, CRC Press, edited by Jinsong Wu, Sundeep Rangan, Honggang Zhang, 2012.

[22] Z. Davis, "Power consumption and cooling in the data center: a survey," http://www.greenbiz.com/sites/default/files/document/Custom016C45F77410.pdf

[23] V. Valancius et al., "Greening the Internet with nano data centers," *Proceedings of the 5th International Conference on Emerging Networking Experiments and Technologies (CoNEXT '09)*, ACM, New York, NY, USA, 37-48, 2009.

[24] Oracle Data Sheet, "Sun Oracle DataBaseMachine," Online: http://www.oracle.com/us/products/database/database-machine-069034.html

[25] NASA, Solar radiation and the earth system, Online: http://www.nasa.gov/centers/goddard/education/index.html

[26] http://www.pveducation.org/pvcdrom/properties-of-sunlight/declination-angle

[27] M.P. Anastasopoulos, A. Tzanakaki and K. Georgakilas, "Virtual infrastructure planning in elastic cloud deploying optical networking," in *Proc. of 3rd IEEE CloudCom 2011*, pp. 685-689, Nov. 29 - Dec. 1, 2011.

[28] E. Kubilinskas, P. Nilsson, and M. Pioro, "Design models for robust multi-layer next generation Internet core networks carrying elastic traffic," in *Proc. of DRCN 2003*, 61-68, 2003.

[29] E. Kubilinskas, F. Aslam, M. Dzida and M. Pioro, "Recovery, routing and load balancing strategy for an IP/MPLS network," *Managing Traffic Performance in Converged Networks, Lecture Notes in Computer Science*, vol. 4516/2007, 65-76, 2008.

[30] Standard Performance Evaluation Corporation (SPEC), Online: www.spec.org

[31] P. Batchelor et al., "Study on the implementation of optical transparent transport networks in the European environment-Results of the research project COST 239," *Photonic Network Communications*, vol. 2, no. 1, pp. 15-32, 2000.

[32] Amazon Elastic Compute Cloud (Amazon EC2), http://aws.amazon.com/ec2/

[33] R.M. Swanson,"Approaching the 29% limit efficiency of silicon solar cells," Photovoltaic Specialists Conference, 2005. *Conference Record of the Thirty-First IEEE*, pp. 889- 894, 3-7 Jan. 2005.

Chapter 9

Low Power Dynamic Scheduling for Computing Systems

Michael J. Neely
University of Southern California, mjneely@usc.edu

This chapter considers energy-aware control for a computing system with two states: *active* and *idle*. In the active state, the controller chooses to perform a single task using one of multiple task processing modes. The controller then saves energy by choosing an amount of time for the system to be idle. These decisions affect processing time, energy expenditure, and an abstract *attribute vector* that can be used to model other criteria of interest (such as processing quality or distortion). The goal is to optimize time average system performance. Applications of this model include a smart phone that makes energy-efficient computation and transmission decisions, a computer that processes tasks subject to rate, quality and power constraints, and a smart grid energy manager that allocates resources in reaction to a time varying energy price.

The solution methodology of this chapter uses the theory of *optimization for renewal systems* developed in [25] [29]. Section 9.1 focuses on a computer system that seeks to minimize average power subject to processing rate constraints for different classes of tasks. Section 9.2 generalizes to treat optimization for a larger class of systems. Section 9.3 extends the model to allow control actions to react to a random event observed at the beginning of each active period, such as a vector of current channel conditions or energy prices.

9.1   Task Scheduling with Processing Rate Constraints

To illustrate the method, this section considers a particular system. Consider a computer system that repeatedly processes tasks. There are N classes of tasks, where N is a positive integer. For simplicity, assume each class always has a new task ready to be performed (this is extended to randomly arriving

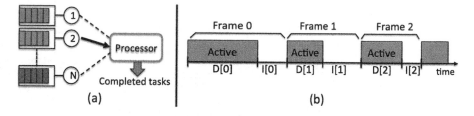

Figure 9.1: (a) A processor that chooses from one of N classes on each frame k. (b) A timeline illustrating the active and idle periods for each frame.

tasks in Section 9.2.3). The system operates over time intervals called *frames*. Each frame $k \in \{0, 1, 2, \ldots\}$ begins with an *active period* of size $D[k]$ and ends with an *idle period* of size $I[k]$ (see Fig. 9.1). At the beginning of each active period k, the controller selects a new task of class $c[k] \in \{1, \ldots, N\}$. It also chooses a *processing mode* $m[k]$ from a finite set $\mathcal{M}$ of possible processing options. These control decisions affect the duration $D[k]$ of the active period and the energy $e[k]$ that is incurred. The controller then selects an amount of time $I[k]$ to be idle, where $I[k]$ is chosen such that $0 \le I[k] \le I_{max}$ for some positive number I_{max}. Choosing $I[k] = 0$ effectively skips the idle period, so there can be back-to-back active periods. For simplicity, this section assumes that no energy is expended in the idle state.

Assume that $D[k]$ and $e[k]$ are *random functions* of the class and mode decisions for frame k. Specifically, assume $D[k]$ and $e[k]$ are conditionally independent of the past, given the current $(c[k], m[k])$ that is used, with mean values given by functions $\hat{D}(c, m)$ and $\hat{e}(c, m)$ defined over $(c, m) \in \{1, \ldots, N\} \times \mathcal{M}$:

$$\hat{D}(c, m) \triangleq \mathbb{E}\left[D[k] | (c[k], m[k]) = (c, m)\right] \quad , \quad \hat{e}(c, m) \triangleq \mathbb{E}\left[e[k] | (c[k], m[k]) = (c, m)\right]$$

where the notation "$a \triangleq b$" represents "a is defined to be equal to b." This chapter uses $\hat{D}(c[k], m[k])$ and $\hat{e}(c[k], m[k])$ to denote expectations given a particular decision $(c[k], m[k])$ for frame k:

$$\hat{D}(c[k], m[k]) = \mathbb{E}\left[D[k] | c[k], m[k]\right] \quad , \quad \hat{e}(c[k], m[k]) = \mathbb{E}\left[e[k] | c[k], m[k]\right]$$

It is assumed there is a positive value D_{min} such that $D[k] \ge D_{min}$ for all frames k, regardless of the $(c[k], m[k])$ decisions. Thus, all frame sizes are at least D_{min} units of time. Further, for technical reasons, it is assumed that second moments of $D[k]$ and $e[k]$ are bounded by a finite constant σ^2, so that:

$$\mathbb{E}\left[D[k]^2\right] \le \sigma^2 \quad , \quad \mathbb{E}\left[e[k]^2\right] \le \sigma^2 \tag{9.1}$$

where (9.1) holds regardless of the policy for selecting $(c[k], m[k])$. The conditional joint distribution of $(D[k], e[k])$, given $(c[k], m[k])$, is otherwise arbitrary, and only the mean values $\hat{D}(c, m)$ and $\hat{e}(c, m)$ are known for each

$c \in \{1, \ldots, N\}$ and $m \in \mathcal{M}$. In the special case of a deterministic system, the functions $\hat{D}(c, m)$ and $\hat{e}(c, m)$ can be viewed as deterministic mappings from a given control action $(c[k], m[k]) = (c, m)$ to the actual delay $D[k] = \hat{D}(c, m)$ and energy $e[k] = \hat{e}(c, m)$ experienced on frame k, rather than as expectations of these values.

9.1.1 Examples of Energy-Aware Processing

Consider an example where a computer system performs computation for different tasks. Suppose the system uses a chip multi-processor that can select between one of multiple processing modes for each task. For example, this might be done using voltage/frequency scaling, or by using a choice of different processing cores [9] [2]. Let $\mathcal{M}$ represent the set of processing modes. For each mode $m \in \mathcal{M}$, define:

$$
\begin{aligned}
T_{setup}(m) &\triangleq \text{Setup time for mode } m. \\
e_{setup}(m) &\triangleq \text{Setup energy for mode } m. \\
DPI(m) &\triangleq \text{Average delay-per-instruction for mode } m. \\
EPI(m) &\triangleq \text{Average energy-per-instruction for mode } m.
\end{aligned}
$$

Further suppose that tasks of class $c \in \{1, \ldots, N\}$ have an average number of instructions equal to $\overline{S}_c$. For simplicity, suppose the number of instructions in a task is independent of the energy and delay of each individual instruction. Then the average energy and delay functions $\hat{e}(c, m)$ and $\hat{D}(c, m)$ are:

$$
\begin{aligned}
\hat{e}(c, m) &= e_{setup}(m) + \overline{S}_c EPI(m) \\
\hat{D}(c, m) &= T_{setup}(m) + \overline{S}_c DPI(m)
\end{aligned}
$$

In cases when the system cannot be modeled using $DPI(m)$ and $EPI(m)$ functions, the expectations $\hat{e}(c, m)$ and $\hat{D}(c, m)$ can be estimated as empirical averages of energy and delay observed when processing type c tasks with mode m. While this example assumes all classes have the same set of processing mode options $\mathcal{M}$, this can easily be extended to restrict each class c to its own subset of options $\mathcal{M}_c$.

As another example, consider the problem of *wireless data transmission*. Here, each task represents a packet of data that must be transmitted. Let $\mathcal{M}$ represent the set of wireless transmission options (such as modulation and coding strategies). For each $m \in \mathcal{M}$, define $\mu(m)$ as the transmission rate (in bits per unit time) under option m, and let $power(m)$ be the power used. For simplicity, assume there are no transmission errors. Let $\overline{B}_c$ represent the average packet size for class $c \in \{1, \ldots, N\}$, in units of bits. Thus:

$$
\begin{aligned}
\hat{e}(c, m) &= power(m)\overline{B}_c/\mu(m) \\
\hat{D}(c, m) &= \overline{B}_c/\mu(m)
\end{aligned}
$$

In the case when each transmission mode $m \in \mathcal{M}$ has a known error probability, the functions $\hat{e}(c, m)$ and $\hat{D}(c, m)$ can be redefined to account for retransmissions. An important alternative scenario is when channel states are time-varying but can be measured at the beginning of each frame. This can be treated using the extended theory in Section 9.3.

9.1.2 Time Averages as Ratios of Frame Averages

The goal is to design a control policy that makes decisions over frames to minimize time average power subject to processing each class $n \in \{1, \ldots, N\}$ with rate at least λ_n, for some desired processing rates $(\lambda_1, \ldots, \lambda_N)$ that are given. Before formalizing this as a mathematical optimization, this subsection shows how to write time averages in terms of frame averages. Suppose there is a control policy that yields a sequence of energies $\{e[0], e[1], e[2], \ldots\}$ and corresponding frame sizes $\{D[0] + I[0], D[1] + I[1], D[2] + I[2], \ldots\}$ for each frame $k \in \{0, 1, 2, \ldots\}$. The *frame averages* $\overline{e}, \overline{D}, \overline{I}$ are defined:

$$\overline{e} \triangleq \lim_{K \to \infty} \frac{1}{K} \sum_{k=0}^{K-1} e[k] \quad , \quad \overline{D} \triangleq \lim_{K \to \infty} \frac{1}{K} \sum_{k=0}^{K-1} D[k] \quad , \quad \overline{I} \triangleq \lim_{K \to \infty} \frac{1}{K} \sum_{k=0}^{K-1} I[k] \quad (9.2)$$

where, for simplicity, it is assumed the limits converge to constants with probability 1. Note that $\overline{e}$ does *not* represent the time average power used by the system, because it does not consider the amount of time spent in each frame. The time average power considers the accumulated energy used divided by the total time, and is written as follows:

$$\lim_{K \to \infty} \frac{\sum_{k=0}^{K-1} e[k]}{\sum_{k=0}^{K-1} (D[k] + I[k])} = \lim_{K \to \infty} \frac{\frac{1}{K} \sum_{k=0}^{K-1} e[k]}{\frac{1}{K} \sum_{k=0}^{K-1} (D[k] + I[k])} = \frac{\overline{e}}{\overline{D} + \overline{I}}$$

Therefore, the time average power is equal to the average energy per frame divided by the average frame size. This simple observation is often used in *renewal-reward theory* [6] [35].

For each class $n \in \{1, \ldots, N\}$ and each frame k, define an *indicator variable* $1_n[k]$ that is 1 if the controller chooses to process a class n task on frame k, and 0 else:

$$1_n[k] \triangleq \begin{cases} 1 & \text{if } c[k] = n \\ 0 & \text{if } c[k] \neq n \end{cases}$$

Then $\overline{1}_n \triangleq \lim_{K \to \infty} \frac{1}{K} \sum_{k=0}^{K-1} 1_n[k]$ is the fraction of frames that choose class n, and the ratio $\overline{1}_n / (\overline{D} + \overline{I})$ is the time average rate of processing class n tasks, in tasks per unit time.

The problem of minimizing time average power subject to processing each class n at a rate of at least λ_n tasks per unit time is then mathematically written as follows:

$$\text{Minimize:} \quad \frac{\overline{e}}{\overline{D} + \overline{I}} \tag{9.3}$$

$$\text{Subject to:} \quad \frac{\overline{I_n}}{\overline{D} + \overline{I}} \geq \lambda_n \quad \forall n \in \{1, \ldots, N\} \tag{9.4}$$

$$(c[k], m[k]) \in \{1, \ldots, N\} \times \mathcal{M} \quad \forall k \in \{0, 1, 2, \ldots\} \tag{9.5}$$

$$0 \leq I[k] \leq I_{max} \quad \forall k \in \{0, 1, 2, \ldots\} \tag{9.6}$$

where the objective (9.3) is average power, the constraint (9.4) ensures the processing rate of each class n is at least λ_n, and constraints (9.5)-(9.6) ensure that $c[k] \in \{1, \ldots, N\}$, $m[k] \in \mathcal{M}$, and $0 \leq I[k] \leq I_{max}$ for each frame k.

9.1.3   Relation to Frame Average Expectations

The problem (9.3)-(9.6) is defined by frame averages. This subsection shows that frame averages are related to frame average *expectations*, and hence can be related to the expectation functions $\hat{D}(c, m)$, $\hat{e}(c, m)$. Consider any (possibly randomized) control algorithm for selecting $(c[k], m[k])$ over frames, and assume this gives rise to well-defined expectations $\mathbb{E}[e[k]]$ for each frame k. By the law of iterated expectations, it follows that for any given frame $k \in \{0, 1, 2, \ldots\}$:

$$\mathbb{E}[e[k]] = \mathbb{E}[\mathbb{E}[e[k]|c[k], m[k]]] = \mathbb{E}[\hat{e}(c[k], m[k])] \tag{9.7}$$

Furthermore, because second moments are bounded by a constant σ^2 on each frame k, the *bounded moment convergence theorem* (given in Appendix A) ensures that if the frame average energy converges to a constant $\overline{e}$ with probability 1, as defined in (9.2), then $\overline{e}$ is the same as the frame average expectation:

$$\lim_{K \to \infty} \frac{1}{K} \sum_{k=0}^{K-1} e[k] = \overline{e} = \lim_{K \to \infty} \frac{1}{K} \sum_{k=0}^{K-1} \mathbb{E}[\hat{e}(c[k], m[k])]$$

The same goes for the quantities $\overline{D}, \overline{I}, \overline{I_n}$. Hence, one can interpret the problem (9.3)-(9.6) using frame average expectations, rather than pure frame averages.

9.1.4   An Example with One Task Class

Consider a simple example with only one type of task. The system processes a new task of this type at the beginning of every busy period. The energy and delay functions can then be written purely in terms of the processing mode $m \in \mathcal{M}$, so that we have $\hat{e}(m)$ and $\hat{D}(m)$. Suppose there are only two processing mode options, so that $\mathcal{M} = \{1, 2\}$, and that each option leads to a deterministic energy and delay, as given below:

$$m[k] = 1 \implies (e[k], D[k]) = (\hat{e}(1), \hat{D}(1)) = (1, 7) \tag{9.8}$$
$$m[k] = 2 \implies (e[k], D[k]) = (\hat{e}(2), \hat{D}(2)) = (3, 4) \tag{9.9}$$

Option $m[k] = 1$ requires 1 unit of energy but 7 units of processing time. Option $m[k] = 2$ is more energy-expensive (requiring 3 units of energy) but is faster (taking only 4 time units). The idle time $I[k]$ is chosen every frame in the interval $[0, 10]$, so that $I_{max} = 10$.

9.1.4.1 No Constraints

For this system, suppose we seek to minimize average power $\overline{e}/(\overline{D} + \overline{I})$, with no processing rate constraint. Consider three possible algorithms:

1. Always use $m[k] = 1$ and $I[k] = I$ for all frames k, for some constant $I \in [0, 10]$.

2. Always use $m[k] = 2$ and $I[k] = I$ for all frames k, for some constant $I \in [0, 10]$.

3. Always use $I[k] = I$ for all frames k, for some constant $I \in [0, 10]$. However, for each frame k, independently choose $m[k] = 1$ with probability p, and $m[k] = 2$ with probability $1 - p$ (for some p that satisfies $0 \le p \le 1$).

Clearly the third algorithm contains the first two for $p = 1$ and $p = 0$, respectively. The time average power under each algorithm is:

$$m[k] = 1 \text{ always} \implies \frac{\overline{e}}{\overline{D} + \overline{I}} = \frac{\hat{e}(1)}{\hat{D}(1) + I} = \frac{1}{7 + I}$$

$$m[k] = 2 \text{ always} \implies \frac{\overline{e}}{\overline{D} + \overline{I}} = \frac{\hat{e}(2)}{\hat{D}(2) + I} = \frac{3}{4 + I}$$

$$\text{probabilistic rule} \implies \frac{\overline{e}}{\overline{D} + \overline{I}} = \frac{p\hat{e}(1) + (1 - p)\hat{e}(2)}{p\hat{D}(1) + (1 - p)\hat{D}(2) + I}$$
$$= \frac{1p + 3(1 - p)}{7p + 4(1 - p) + I}$$

It is clear that in all three cases, we should choose $I = 10$ to minimize average power. Further, it is clear that always choosing $m[k] = 1$ is better than always choosing $m[k] = 2$. However, it is not immediately obvious if a randomized mode selection rule can do even better. The answer is no: In this case, power is minimized by choosing $m[k] = 1$ and $I[k] = 10$ for all frames k, yielding average power $1/17$.

This fact holds more generally: Let $a(\alpha)$ and $b(\alpha)$ be any deterministic real valued functions defined over general *actions* α that are chosen in an abstract *action space* $\mathcal{A}$. Assume the functions are bounded, and that there is a value $b_{min} > 0$ such that $b(\alpha) \geq b_{min}$ for all $\alpha \in \mathcal{A}$. Consider designing a randomized choice of α that minimizes $\mathbb{E}\left[a(\alpha)\right]/\mathbb{E}\left[b(\alpha)\right]$, where the expectations are with respect to the randomness in the α selection. The next lemma shows this is done by *deterministically* choosing an action $\alpha \in \mathcal{A}$ that minimizes $a(\alpha)/b(\alpha)$.

Lemma 1. *Under the assumptions of the preceding paragraph, for any randomized choice of $\alpha \in \mathcal{A}$ we have:*

$$\frac{\mathbb{E}\left[a(\alpha)\right]}{\mathbb{E}\left[b(\alpha)\right]} \geq \inf_{\alpha \in \mathcal{A}} \left[\frac{a(\alpha)}{b(\alpha)}\right]$$

and so the infimum of the ratio over the class of deterministic decisions yields a value that is less than or equal to that of any randomized selection.

Proof. Consider any randomized policy that yields expectations $\mathbb{E}\left[a(\alpha)\right]$ and $\mathbb{E}\left[b(\alpha)\right]$. Without loss of generality, assume these expectations are achieved by a policy that randomizes over a finite set of M actions $\alpha_1, \ldots, \alpha_M$ in $\mathcal{A}$ with some probabilities $p_1, \ldots, p_M$:[1]

$$\mathbb{E}\left[a(\alpha)\right] = \sum_{m=1}^{M} p_m a(\alpha_m) \ , \ \mathbb{E}\left[b(\alpha)\right] = \sum_{m=1}^{M} p_m b(\alpha_m)$$

Then, because $b(\alpha_m) > 0$ for all α_m, we have:

$$
\begin{aligned}
\frac{\mathbb{E}\left[a(\alpha)\right]}{\mathbb{E}\left[b(\alpha)\right]} &= \frac{\sum_{m=1}^{M} p_m a(\alpha_m)}{\sum_{m=1}^{M} p_m b(\alpha_m)} \\
&= \frac{\sum_{m=1}^{M} p_m b(\alpha_m)[a(\alpha_m)/b(\alpha_m)]}{\sum_{m=1}^{M} p_m b(\alpha_m)} \\
&\geq \frac{\sum_{m=1}^{M} p_m b(\alpha_m) \inf_{\alpha \in \mathcal{A}}[a(\alpha)/b(\alpha)]}{\sum_{m=1}^{M} p_m b(\alpha_m)} \\
&= \inf_{\alpha \in \mathcal{A}} [a(\alpha)/b(\alpha)]
\end{aligned}
$$

$\square$

9.1.4.2   One Constraint

The preceding subsection shows that unconstrained problems can be solved by deterministic actions. This is not true for constrained problems. This subsection shows that adding just a single constraint often *necessitates* the use of randomized actions. Consider the same problem as before, with two choices

---

[1]Indeed, because the set $\mathcal{S} = \{(a(\alpha), b(\alpha))$ such that $\alpha \in \mathcal{A}\}$ is bounded, the expectation $(\mathbb{E}\left[a(\alpha)\right], \mathbb{E}\left[b(\alpha)\right])$ is finite and is contained in the convex hull of $\mathcal{S}$. Thus, $(\mathbb{E}\left[a(\alpha)\right], \mathbb{E}\left[b(\alpha)\right])$ is a convex combination of a finite number of points in $\mathcal{S}$.

for $m[k]$, and with the same $\hat{e}(m)$ and $\hat{D}(m)$ values given in (9.8)-(9.9). We want to minimize $\overline{e}/(\overline{D} + \overline{I})$ subject to $1/(\overline{D} + \overline{I}) \geq 1/5$, where $1/(\overline{D} + \overline{I})$ is the rate of processing jobs. The constraint is equivalent to $\overline{D} + \overline{I} \leq 5$.

Assume the algorithm chooses $I[k]$ over frames to yield an average $\overline{I}$ that is somewhere in the interval $0 \leq \overline{I} \leq 10$. Now consider different algorithms for selecting $m[k]$. If we choose $m[k] = 1$ always (so that $(\hat{e}(1), \hat{D}(1)) = (1, 7)$), then:

$$m[k] = 1 \text{ always} \implies \frac{\overline{e}}{\overline{D} + \overline{I}} = \frac{1}{7 + \overline{I}} \quad , \quad \overline{D} + \overline{I} = 7 + \overline{I}$$

and so it is impossible to meet the constraint $\overline{D} + \overline{I} \leq 5$, because $7 + \overline{I} > 5$. If we choose $m[k] = 2$ always (so that $(\hat{e}(2), \hat{D}(2)) = (3, 4)$), then:

$$m[k] = 2 \text{ always} \implies \frac{\overline{e}}{\overline{D} + \overline{I}} = \frac{3}{4 + \overline{I}} \quad , \quad \overline{D} + \overline{I} = 4 + \overline{I}$$

It is clear that we can meet the constraint by choosing $\overline{I}$ so that $0 \leq \overline{I} \leq 1$, and power is minimized in this setting by using $\overline{I} = 1$. This can be achieved, for example, by using $I[k] = 1$ for all frames k. This meets the processing rate constraint with equality: $\overline{D} + \overline{I} = 4 + 1 = 5$. Further, it yields average power $\overline{e}/(\overline{D} + \overline{I}) = 3/5 = 0.6$.

However, it is possible to reduce average power while also meeting the constraint with equality by using the following randomized policy (which can be shown to be optimal): Choose $I[k] = 0$ for all frames k, so that $\overline{I} = 0$. Then every frame k, independently choose $m[k] = 1$ with probability $1/3$, and $m[k] = 2$ with probability $2/3$. We then have:

$$\overline{D} + \overline{I} = (1/3)7 + (2/3)4 + 0 = 5$$

and so the processing rate constraint is met with equality. However, average power is:

$$\frac{\overline{e}}{\overline{D} + \overline{I}} = \frac{(1/3)1 + (2/3)3}{(1/3)7 + (2/3)4 + 0} = 7/15 \approx 0.466667$$

This is a significant savings over the average power of 0.6 from the deterministic policy.

9.1.5 The Linear Fractional Program for Task Scheduling

Now consider the general problem (9.3)-(9.6) for minimizing average power subject to average processing rate constraints for each of the N classes. Assume the problem is *feasible*, so that it is possible to choose actions $(c[k], m[k], I[k])$ over frames to meet the desired constraints (9.4)-(9.6). It can be shown that an optimal solution can be achieved over the class of *stationary and randomized policies* with the following structure: Every frame, independently choose vector $(c[k], m[k])$ with some probabilities $p(c, m) = Pr[(c[k], m[k]) = (c, m)]$. Further, use a constant idle time $I[k] = I$ for all frames k, for some constant I that satisfies $0 \leq I \leq I_{max}$. Thus, the problem can be written as the following *linear fractional program* with unknowns $p(c, m)$ and I and known constants $\hat{e}(c, m)$, $\hat{D}(c, m)$, λ_n, and I_{max}:

$$\text{Minimize:} \quad \frac{\sum_{c=1}^{N} \sum_{m \in \mathcal{M}} p(c, m)\hat{e}(c, m)}{I + \sum_{c=1}^{N} \sum_{m \in \mathcal{M}} p(c, m)\hat{D}(c, m)} \tag{9.10}$$

$$\text{Subject to:} \quad \frac{\sum_{m \in \mathcal{M}} p(n, m)}{I + \sum_{c=1}^{N} \sum_{m \in \mathcal{M}} p(c, m)\hat{D}(c, m)} \geq \lambda_n \quad \forall n \in \{1, \ldots, N\} \tag{9.11}$$

$$0 \leq I \leq I_{max} \tag{9.12}$$

$$p(c, m) \geq 0 \quad \forall c \in \{1, \ldots, N\}, \ \forall m \in \mathcal{M} \tag{9.13}$$

$$\sum_{c=1}^{N} \sum_{m \in \mathcal{M}} p(c, m) = 1 \tag{9.14}$$

where the numerator and denominator in (9.10) are equal to $\overline{e}$ and $\overline{D} + \overline{I}$, respectively, under this randomized algorithm, the numerator in the left-hand-side of (9.11) is equal to $\overline{I}_n$, and the constraints (9.13)-(9.14) specify that $p(c, m)$ must be a valid probability mass function.

Linear fractional programs can be solved in several ways. One method uses a nonlinear change of variables to map the problem to a convex program [3]. However, this method does not admit an *online* implementation, because time averages are not preserved through the nonlinear change of variables. Below, an online algorithm is presented that makes decisions every frame k. The algorithm is *not* a stationary and randomized algorithm as described above. However, it yields time averages that satisfy the desired constraints of the problem (9.3)-(9.6), with a time average power expenditure that can be pushed arbitrarily close to the optimal value. A significant advantage of this approach is that it extends to treat cases with random task arrivals, without requiring knowledge of the $(\lambda_1, \ldots, \lambda_N)$ arrival rates, and to treat other problems with observed random events, without requiring knowledge of the probability distribution for these events. These extensions are shown in later sections.

For later analysis, it is useful to write (9.10)-(9.14) in a simpler form. Let $power^{opt}$ be the optimal time average power for the above linear fractional program, achieved by some probability distribution $p^*(c, m)$ and idle time I^* that satisfies $0 \leq I^* \leq I_{max}$. Let $(c^*[k], m^*[k], I^*[k])$ represent the frame k

decisions under this stationary and randomized policy. Thus:

$$\frac{\mathbb{E}\left[\hat{e}(c^*[k], m^*[k])\right]}{\mathbb{E}\left[I^*[k] + \hat{D}(c^*[k], m^*[k])\right]} = power^{opt} \tag{9.15}$$

$$\frac{\mathbb{E}\left[1_n^*[k]\right]}{\mathbb{E}\left[I^*[k] + \hat{D}(c^*[k], m^*[k])\right]} \geq \lambda_n \quad \forall n \in \{1, \ldots, N\} \tag{9.16}$$

where $1_n^*[k]$ is an indicator function that is 1 if $c^*[k] = n$, and 0 else. The numerator and denominator of (9.15) correspond to those of (9.10). Likewise, the constraint (9.16) corresponds to (9.11).

9.1.6 Virtual Queues

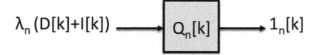

Figure 9.2: An illustration of the virtual queue $Q_n[k]$ from equation (9.18).

To solve the problem (9.3)-(9.6), we first consider the constraints (9.4), which are equivalent to the constraints:

$$\lambda_n(\overline{D} + \overline{I}) \leq \overline{I}_n \quad \forall n \in \{1, \ldots, N\} \tag{9.17}$$

For each constraint $n \in \{1, \ldots, N\}$, define a *virtual queue* $Q_n[k]$ that is updated on frames $k \in \{0, 1, 2, \ldots\}$ by:

$$Q_n[k+1] = \max[Q_n[k] + \lambda_n(D[k] + I[k]) - 1_n[k], 0] \tag{9.18}$$

The initial condition $Q_n[0]$ can be any non-negative value. For simplicity, it is assumed throughout that $Q_n[0] = 0$ for all $n \in \{1, \ldots, N\}$. The update (9.18) can be viewed as a discrete time queueing equation, where $Q_n[k]$ is the backlog on frame k, $\lambda_n(D[k] + I[k])$ is an effective amount of "new arrivals," and $1_n[k]$ is the amount of "offered service" (see Fig. 9.2). The intuition is that if all virtual queues $Q_n[k]$ are *stable*, then the average "arrival rate" $\lambda_n(\overline{D}+\overline{I})$ must be less than or equal to the average "service rate" $\overline{I}_n$, which ensures the desired constraint (9.17). This is made precise in the following lemma.

Lemma 2. *(Virtual Queues) Suppose $Q_n[k]$ has update equation given by (9.18), with any non-negative initial condition.*
  a) For all $K \in \{1, 2, 3, \ldots\}$ we have:

$$\frac{1}{K}\sum_{k=0}^{K-1}[\lambda_n(D[k] + I[k]) - 1_n[k]] \leq \frac{Q_n[K] - Q_n[0]}{K} \tag{9.19}$$

b) If $\lim_{K\to\infty} Q_n[K]/K = 0$ *with probability 1, then:*

$$\limsup_{K\to\infty} \frac{1}{K} \sum_{k=0}^{K-1} [\lambda_n(D[k] + I[k]) - 1_n[k]] \le 0 \quad \text{with probability 1} \qquad (9.20)$$

c) If $\lim_{K\to\infty} \mathbb{E}[Q_n[K]]/K = 0$, *then:*

$$\limsup_{K\to\infty} [\lambda_n(\overline{D}[K] + \overline{I}[K]) - \overline{1}_n[K]] \le 0 \qquad (9.21)$$

where $\overline{D}[K]$, $\overline{I}[K]$, $\overline{1}_n[K]$ *are defined:*

$$\overline{D}[K] \triangleq \frac{1}{K} \sum_{k=0}^{K-1} \mathbb{E}[D[k]] \ , \ \overline{I}[K] \triangleq \frac{1}{K} \sum_{k=0}^{K-1} \mathbb{E}[I[k]]$$

$$\overline{1}_n[K] \triangleq \frac{1}{K} \sum_{k=0}^{K-1} \mathbb{E}[1_n[k]]$$

Proof. From (9.18) we have for all $k \in \{0, 1, 2, \ldots\}$:

$$Q_n[k+1] \ge Q_n[k] + \lambda_n(D[k] + I[k]) - 1_n[k]$$

Fixing a positive integer K and summing the above over $k \in \{0, \ldots, K-1\}$ yields:

$$Q_n[K] - Q_n[0] \ge \sum_{k=0}^{K-1} [\lambda_n(D[k] + I[k]) - 1_n[k]]$$

Dividing the above by K proves part (a). Part (b) follows from (9.19) by taking a lim sup. Part (c) follows by first taking expectations of (9.19) and then taking a lim sup. □

Inequality (9.19) shows that the value $Q_n[K]/K$ bounds the amount by which the desired constraint for class n is violated by the time averages achieved over the first K frames. Suppose that $D[k]$, $I[k]$, and $1_n[k]$ have frame averages that converge to constants $\overline{D}$, $\overline{I}$, $\overline{1}_n$ with probability 1. Part (c) of Lemma 2 indicates that if $\lim_{K\to\infty} \mathbb{E}[Q_n[K]]/K = 0$ for all $n \in \{1, \ldots, N\}$, then $\lambda_n(\overline{D} + \overline{I}) \le \overline{1}_n$ for all $n \in \{1, \ldots, N\}$.

In the language of queueing theory, a discrete time queue $Q[k]$ is said to be *rate stable* if $\lim_{k\to\infty} Q[k]/k = 0$ with probability 1, and is *mean rate stable* if $\lim_{k\to\infty} \mathbb{E}[Q[k]]/k = 0$ [25]. With this terminology, the above lemma shows that if $Q_n[k]$ is rate stable then (9.20) holds, and if $Q_n[k]$ is mean rate stable then (9.21) holds.

9.1.7 The Drift-Plus-Penalty Ratio

To stabilize the queues while minimizing time average power, we use *Lyapunov optimization theory*, which gives rise to the *drift-plus-penalty ratio algorithm*

[25]. First define $L[k]$ as the sum of the squares of all queues on frame k (divided by 2 for convenience later):

$$L[k] \triangleq \frac{1}{2} \sum_{n=1}^{N} Q_n[k]^2$$

$L[k]$ is often called a *Lyapunov function*, and acts as a scalar measure of the size of the queues. Intuitively, keeping $L[k]$ small leads to stable queues, and we should take actions that tend to shrink $L[k]$ from one frame to the next. Define $\Delta[k]$ as the *Lyapunov drift*, being the difference in the Lyapunov function from one frame to the next:

$$\Delta[k] \triangleq L[k+1] - L[k]$$

Taking actions to minimize $\Delta[k]$ every frame can be shown to ensure the desired constraints are satisfied whenever it is possible to satisfy them, but does not incorporate power minimization. To incorporate this, every frame k we observe the current queue vector $\boldsymbol{Q}[k] = (Q_1[k], \ldots, Q_N[k])$ and choose control actions $(c[k], m[k], I[k])$ to minimize a bound on the following *drift-plus-penalty ratio*:

$$\frac{\mathbb{E}\left[\Delta[k] + V e[k] | \boldsymbol{Q}[k]\right]}{\mathbb{E}\left[D[k] + I[k] | \boldsymbol{Q}[k]\right]}$$

where V is a non-negative parameter that weights the extent to which power minimization is emphasized. The intuition is that the numerator incorporates both drift and energy. The denominator "normalizes" this by the expected frame size, with the understanding that average power must include both energy and frame size. We soon show that this intuition is correct, in that all desired time average constraints are satisfied, and the time average power is within $O(1/V)$ of the optimal value *power^{opt}*. Hence, average power can be pushed arbitrarily close to optimal by using a sufficiently large value of V. The tradeoff is that average queue sizes grow with V, which impacts the convergence time required to satisfy the desired constraints.

The drift-plus-penalty ratio method was first developed for the context of restless bandit systems in [18] [19]. The method was used for optimization of renewal systems in [25] [29], which treat problems similar to those considered in this chapter. In the special case when all frame sizes are fixed and equal to one unit of time (a *time slot*), and when $V = 0$, the method reduces to observing queues $\boldsymbol{Q}[k]$ every slot k and taking actions to minimize a bound on $\mathbb{E}\left[\Delta[k] | \boldsymbol{Q}[k]\right]$. This is the rule that generates the classic max-weight scheduling algorithms for queue stability (without performance optimization), developed by Tassiulas and Ephremides in [38] [39]. For systems with unit time slots but with $V > 0$, the *drift-plus-penalty ratio* technique reduces to the *drift-plus-penalty* technique of [24] [8] [28], which treats joint queue stability and penalty minimization in systems with unit size slots.

9.1.7.1 Bounding the Drift-Plus-Penalty Ratio

To construct an explicit algorithm, we first bound the drift-plus-penalty ratio.

Lemma 3. *For all frames $k \in \{0, 1, 2, \ldots\}$, all possible $\boldsymbol{Q}[k]$, and under any decisions for $(c[k], m[k], I[k])$, we have:*

$$\frac{\mathbb{E}\left[\Delta[k] + Ve[k]|\boldsymbol{Q}[k]\right]}{\mathbb{E}\left[D[k] + I[k]|\boldsymbol{Q}[k]\right]} \leq \frac{B}{\mathbb{E}\left[D[k] + I[k]|\boldsymbol{Q}[k]\right]} + \frac{\mathbb{E}\left[V\hat{e}(c[k], m[k])|\boldsymbol{Q}[k]\right]}{\mathbb{E}\left[\hat{D}(c[k], m[k]) + I[k]|\boldsymbol{Q}[k]\right]}$$

$$+ \frac{\sum_{n=1}^{N} Q_n[k]\mathbb{E}\left[\lambda_n(\hat{D}(c[k], m[k]) + I[k]) - 1_n[k]|\boldsymbol{Q}[k]\right]}{\mathbb{E}\left[\hat{D}(c[k], m[k]) + I[k]|\boldsymbol{Q}[k]\right]} \quad (9.22)$$

where B is a constant that satisfies the following for all possible $\boldsymbol{Q}[k]$ and all policies:

$$B \geq \tfrac{1}{2} \sum_{n=1}^{N} \mathbb{E}\left[(\lambda_n(D[k] + I[k]) - 1_n[k])^2|\boldsymbol{Q}[k]\right]$$

Such a constant B exists by the second moment boundedness assumptions (9.1).

Proof. Note by iterated expectations that (similar to (9.7)):[2]

$$\mathbb{E}\left[D[k]|\boldsymbol{Q}[k]\right] = \mathbb{E}\left[\hat{D}(c[k], m[k])|\boldsymbol{Q}[k]\right] \quad , \quad \mathbb{E}\left[e[k]|\boldsymbol{Q}[k]\right] = \mathbb{E}\left[\hat{e}(c[k], m[k])|\boldsymbol{Q}[k]\right]$$

Thus, the denominator is common for all terms of inequality (9.22), and it suffices to prove:

$$\mathbb{E}\left[\Delta[k]|\boldsymbol{Q}[k]\right] \leq B + \sum_{n=1}^{N} Q_n[k]\mathbb{E}\left[\lambda_n(\hat{D}(c[k], m[k]) + I[k]) - 1_n[k]|\boldsymbol{Q}[k]\right] \quad (9.23)$$

To this end, by squaring (9.18) and noting that $\max[x, 0]^2 \leq x^2$, we have for each n:

$$\frac{1}{2}Q_n[k+1]^2 \leq \frac{1}{2}(Q_n[k] + \lambda_n(D[k] + I[k]) - 1_n[k])^2$$

$$= \frac{1}{2}Q_n[k]^2 + \frac{1}{2}(\lambda_n(D[k] + I[k]) - 1_n[k])^2$$

$$+ Q_n[k](\lambda_n(D[k] + I[k]) - 1_n[k])$$

Summing the above over $n \in \{1, \ldots, N\}$ and using the definition of $\Delta[k]$ gives:

$$\Delta[k] \leq \frac{1}{2}\sum_{n=1}^{N}(\lambda_n(D[k] + I[k]) - 1_n[k])^2 + \sum_{n=1}^{N} Q_n[k](\lambda_n(D[k] + I[k]) - 1_n[k])$$

Taking conditional expectations given $\boldsymbol{Q}[k]$ and using the bound B proves (9.23). $\qquad\square$

---

[2] Indeed, by iterated expectations we have $\mathbb{E}\left[D[k]|\boldsymbol{Q}[k]\right] = \mathbb{E}[\mathbb{E}[D[k]|(c[k], m[k]), \boldsymbol{Q}[k]]|\boldsymbol{Q}[k]]$, and $\mathbb{E}\left[D[k]|(c[k], m[k]), \boldsymbol{Q}[k]\right] = \mathbb{E}\left[D[k]|(c[k], m[k])\right]$ because $D[k]$ is conditionally independent of the past given the current $(c[k], m[k])$ used.

9.1.7.2 The Task Scheduling Algorithm

Our algorithm takes actions every frame to minimize the last two terms on the right-hand-side of the drift-plus-penalty ratio bound (9.22). The only part of these terms that we have control over on frame k (given the observed $\boldsymbol{Q}[k]$) is given below:

$$\frac{\mathbb{E}\left[V\hat{e}(c[k],m[k]) - \sum_{n=1}^{N} Q_n[k]1_n[k]|\boldsymbol{Q}[k]\right]}{\mathbb{E}\left[\hat{D}(c[k],m[k]) + I[k]|\boldsymbol{Q}[k]\right]}$$

Recall from Lemma 1 that minimizing the above ratio of expectations is accomplished over a deterministic choice of $(c[k],m[k],I[k])$. Thus, every frame k we perform the following:

- Observe queues $\boldsymbol{Q}[k] = (Q_1[k],\ldots,Q_N[k])$. Then choose $c[k] \in \{1,\ldots,N\}$, $m[k] \in \mathcal{M}$, and $I[k]$ such that $0 \le I[k] \le I_{max}$ to minimize:

$$\frac{V\hat{e}(c[k],m[k]) - Q_{c[k]}[k]}{\hat{D}(c[k],m[k]) + I[k]} \qquad (9.24)$$

- Update queues $Q_n[k]$ for each $n \in \{1,\ldots,N\}$ via (9.18), using the $D[k]$, $I[k]$, and $1_n[k]$ values that result from the decisions $c[k]$, $m[k]$, $I[k]$ that minimized (9.24)

9.1.7.3 Steps to Minimize (9.24)

Here we elaborate on how to perform the minimization in (9.24) for each frame k. For each $c \in \{1,\ldots,N\}$ and $m \in \mathcal{M}$, define $idle(c,m)$ as the value of $I[k]$ that minimizes (9.24), given that we have $(c[k],m[k]) = (c,m)$. It is easy to see that:

$$idle(c,m) = \begin{cases} 0 & \text{if } V\hat{e}(c,m) - Q_c[k] \le 0 \\ I_{max} & \text{otherwise} \end{cases}$$

Now define $val(c,m)$ by:

$$val(c,m) = \frac{V\hat{e}(c,m) - Q_c[k]}{\hat{D}(c,m) + idle(c,m)}$$

Then we choose $(c[k],m[k])$ as the minimizer of $val(c,m)$ over $c \in \{1,\ldots,N\}$ and $m \in \mathcal{M}$, breaking ties arbitrarily, and choose $I[k] = idle(c[k],m[k])$. Note that this algorithm chooses $I[k] = 0$ or $I[k] = I_{max}$ on every frame k. Nevertheless, it results in a *frame average* $\bar{I}$ that approaches optimality for large V.

9.1.8   Performance of the Task Scheduling Algorithm

For simplicity, the performance theorem is presented in terms of zero initial conditions. It is assumed throughout that the problem (9.3)-(9.6) is feasible, so that it is possible to satisfy the constraints.

Theorem 1. *Suppose $Q_n[0] = 0$ for all $n \in \{1, \ldots, N\}$, and that the problem (9.3)-(9.6) is feasible. Then under the above task scheduling algorithm:*
  a) For all frames $K \in \{1, 2, 3, \ldots\}$ we have:[3]

$$\frac{\overline{e}[K]}{\overline{D}[K] + \overline{I}[K]} \leq power^{opt} + \frac{B}{V(\overline{D}[K] + \overline{I}[K])} \tag{9.25}$$

where B is defined in Lemma 3, $power^{opt}$ is the minimum power solution for the problem (9.3)-(9.6), and $\overline{e}[K]$, $\overline{D}[K]$, $\overline{I}[K]$ are defined by:

$$\overline{e}[K] \triangleq \tfrac{1}{K} \sum_{k=0}^{K-1} \mathbb{E}\left[e[k]\right] \;,\;\; \overline{D}[K] \triangleq \tfrac{1}{K} \sum_{k=0}^{K-1} \mathbb{E}\left[D[k]\right] \;,\;\; \overline{I}[K] \triangleq \tfrac{1}{K} \sum_{k=0}^{K-1} \mathbb{E}\left[I[k]\right]$$

  b) The desired constraints (9.20) and (9.21) are satisfied for all $n \in \{1, \ldots, N\}$. Further, we have for each frame $K \in \{1, 2, 3, \ldots\}$:

$$\frac{\mathbb{E}\left[\|\boldsymbol{Q}[k]\|\right]}{K} \leq \sqrt{\frac{2(B + V\beta)}{K}} \tag{9.26}$$

where $\|\boldsymbol{Q}[k]\| \triangleq \sqrt{\sum_{n=1}^{N} Q_n[k]^2}$ is the norm of the queue vector (being at least as large as each component $Q_n[k]$), and β is a constant that satisfies the following for all frames k:

$$\beta \geq \mathbb{E}\left[power^{opt}(D[k] + I[k]) - e[k]\right]$$

Such a constant β exists because the second moments (and hence first moments) are bounded.

In the special case of a deterministic system, all expectations of the above theorem can be removed, and the results hold deterministically for all frames K. Theorem 1 indicates that average power can be pushed arbitrarily close to $power^{opt}$, using the V parameter that affects an $O(1/V)$ performance gap given in (9.25). The tradeoff is that V increases the expected size of $\mathbb{E}\left[Q_n[K]\right]/K$ as shown in (9.26), which bounds the expected deviation from the nth constraint during the first K frames (recall (9.19) from Lemma 2). Under a mild additional "Slater-type" assumption that ensures all constraints can be satisfied with "ϵ-slackness," a stronger result on the virtual queues can be shown, namely, that the same algorithm yields queues with average size $O(V)$ [25]. This typically ensures a tighter constraint tradeoff than that given in (9.26). A related improved tradeoff is explored in more detail in Section 9.2.3.

---

[3]The right-hand-side in (9.25) can be simplified to $power^{opt} + B/(VD_{min})$, since all frames are at least D_{min} in size.

Proof. (Theorem 1 part (a)) Given $\boldsymbol{Q}[k]$ for frame k, our control decisions minimize the last two terms in the right-hand-side of the drift-plus-penalty ratio bound (9.22), and hence:

$$
\frac{\mathbb{E}\left[\Delta[k] + Ve[k] | \boldsymbol{Q}[k]\right]}{\mathbb{E}\left[D[k] + I[k] | \boldsymbol{Q}[k]\right]} \leq \frac{B}{\mathbb{E}\left[D[k] + I[k] | \boldsymbol{Q}[k]\right]}
$$

$$
+ \frac{\mathbb{E}\left[V\hat{e}(c^*[k], m^*[k]) | \boldsymbol{Q}[k]\right]}{\mathbb{E}\left[\hat{D}(c^*[k], m^*[k]) + I^*[k] | \boldsymbol{Q}[k]\right]}
$$

$$
+ \frac{\sum_{n=1}^{N} Q_n[k] \mathbb{E}\left[\lambda_n(\hat{D}(c^*[k], m^*[k]) + I^*[k]) - 1_n^*[k] | \boldsymbol{Q}[k]\right]}{\mathbb{E}\left[\hat{D}(c^*[k], m^*[k]) + I^*[k] | \boldsymbol{Q}[k]\right]}
$$

$$(9.27)$$

where $c^*[k]$, $m^*[k]$, $I^*[k]$, $1_n^*[k]$ are from any alternative (possibly randomized) decisions that can be made on frame k. Now recall the existence of stationary and randomized decisions that yield (9.15)-(9.16). In particular, these decisions are independent of queue backlog $\boldsymbol{Q}[k]$ and thus yield (from (9.15)):

$$
\frac{\mathbb{E}\left[\hat{e}(c^*[k], m^*[k]) | \boldsymbol{Q}[k]\right]}{\mathbb{E}\left[\hat{D}(c^*[k], m^*[k]) + I^*[k] | \boldsymbol{Q}[k]\right]} = \frac{\mathbb{E}\left[\hat{e}(c^*[k], m^*[k])\right]}{\mathbb{E}\left[\hat{D}(c^*[k], m^*[k]) + I^*[k]\right]} = power^{opt}
$$

and for all $n \in \{1, \ldots, N\}$ we have (from (9.16)):

$$
\mathbb{E}\left[\lambda_n(\hat{D}(c^*[k], m^*[k]) + I^*[k]) - 1_n^*[k] | \boldsymbol{Q}[k]\right]
$$

$$
= \mathbb{E}\left[\lambda_n(\hat{D}(c^*[k], m^*[k]) + I^*[k]) - 1_n^*[k]\right] \leq 0
$$

Plugging the above into the right-hand-side of (9.27) yields:

$$
\frac{\mathbb{E}\left[\Delta[k] + Ve[k] | \boldsymbol{Q}[k]\right]}{\mathbb{E}\left[D[k] + I[k] | \boldsymbol{Q}[k]\right]} \leq \frac{B}{\mathbb{E}\left[D[k] + I[k] | \boldsymbol{Q}[k]\right]} + Vpower^{opt}
$$

Rearranging terms gives:

$$
\mathbb{E}\left[\Delta[k] + Ve[k] | \boldsymbol{Q}[k]\right] \leq B + Vpower^{opt}\mathbb{E}\left[D[k] + I[k] | \boldsymbol{Q}[k]\right]
$$

Taking expectations of the above (with respect to the random $\boldsymbol{Q}[k]$) and using the law of iterated expectations gives:

$$
\mathbb{E}\left[\Delta[k] + Ve[k]\right] \leq B + Vpower^{opt}\mathbb{E}\left[D[k] + I[k]\right] \tag{9.28}
$$

The above holds for all $k \in \{0, 1, 2, \ldots\}$. Fixing a positive integer K and summing (9.28) over $k \in \{0, 1, \ldots, K-1\}$ yields, by the definition $\Delta[k] = L[k+1] - L[k]$:

$$
\mathbb{E}\left[L[K] - L[0]\right] + V \sum_{k=0}^{K-1} \mathbb{E}\left[e[k]\right] \leq BK + Vpower^{opt} \sum_{k=0}^{K-1} \mathbb{E}\left[D[k] + I[k]\right]
$$

Noting that $L[0] = 0$ and $L[K] \geq 0$ and using the definitions of $\overline{e}[K]$, $\overline{D}[K]$, $\overline{I}[K]$ yields:

$$VK\overline{e}[K] \leq BK + VKpower^{opt}(\overline{D}[K] + \overline{I}[K])$$

Rearranging terms yields the result of part (a).   □

Proof. (Theorem 1 part (b)) To prove part (b), note from (9.28) we have:

$$\mathbb{E}\left[\Delta[k]\right] \leq B + V\mathbb{E}\left[power^{opt}(D[k] + I[k]) - e[k]\right] \leq B + V\beta$$

Summing the above over $k \in \{0, 1, \ldots, K - 1\}$ gives:

$$\mathbb{E}\left[L[K]\right] - \mathbb{E}\left[L[0]\right] \leq (B + V\beta)K$$

Using the definition of $L[K]$ and noting that $L[0] = 0$ gives:

$$\sum_{l=1}^{K} \mathbb{E}\left[Q_n[K]^2\right] \leq 2(B + V\beta)K$$

Thus, we have $\mathbb{E}\left[\|\boldsymbol{Q}[K]\|^2\right] \leq 2(B + V\beta)K$. Jensen's inequality for $f(x) = x^2$ ensures $\mathbb{E}\left[\|\boldsymbol{Q}[k]\|\right]^2 \leq \mathbb{E}\left[\|\boldsymbol{Q}[k]\|^2\right]$, and so for all positive integers K:

$$\mathbb{E}\left[\|\boldsymbol{Q}[k]\|\right]^2 \leq 2(B + V\beta)K \qquad (9.29)$$

Taking a square root of both sides of (9.29) and dividing by K proves (9.26). From (9.26) we have for each $n \in \{1, \ldots, N\}$:

$$\lim_{K \to \infty} \frac{\mathbb{E}\left[Q_n[k]\right]}{K} \leq \lim_{K \to \infty} \frac{\mathbb{E}\left[\|\boldsymbol{Q}[K]\|\right]}{K} \leq \lim_{K \to \infty} \frac{\sqrt{2(B + V\beta)}}{\sqrt{K}} = 0$$

and hence by Lemma 2 we know constraint (9.21) holds. Further, in [30] it is shown that (9.29) together with the fact that second moments of queue changes are bounded implies $\lim_{k \to \infty} Q_n[k]/k = 0$ with probability 1. Thus, (9.20) holds.   □

9.1.9   Simulation

We first simulate the task scheduling algorithm for the simple deterministic system with one class and one constraint, as described in Section 9.1.4. The $\hat{e}(m)$ and $\hat{D}(m)$ functions are defined in (9.8)-(9.9), and the goal is to minimize average power subject to a processing rate constraint $1/(\overline{D} + \overline{I}) \geq 0.2$. We already know the optimal power is $power^{opt} = 7/15 \approx 0.466667$. We expect the algorithm to approach this optimal power as V is increased, and to approach the desired behavior of using $I[k] = 0$ for all k, meeting the constraint with equality, and using $m[k] = 1$ for $1/3$ of the frames. This is indeed what happens, although in this simple case the algorithm seems insensitive to the

V parameter and locks into a desirable periodic schedule even for very low (but positive) V values. Using $V = 1$ and one million frames, the algorithm gets average power 0.466661, uses $m[k] = 1$ a fraction of time 0.333340, has average idle time $\bar{I} = 0.000010$ and yields a processing rate 0.199999 (almost exactly equal to the desired constraint of 0.2). Increasing V yields similar performance. The constraint is still satisfied when we decrease the value of V, but average power degrades (being 0.526316 for $V = 0$).

We next consider a system with 10 classes of tasks and two processing modes. The energy and delay characteristics for each class $i \in \{1, \dots, 10\}$ and mode $m \in \{1, 2\}$ are:

$$\text{Mode 1: } (\hat{e}(i, 1), \hat{D}(i, 1)) = (1i, 5i) \tag{9.30}$$

$$\text{Mode 2: } (\hat{e}(i, 2), \hat{D}(i, 2)) = (2i, 3i) \tag{9.31}$$

so that mode 1 uses less energy but takes longer than mode 2, and the computational requirements for each class increase with $i \in \{1, \dots, 10\}$. We assume desired rates are given by $\lambda_i = \rho/(30i)$ for $i \in \{1, \dots, 10\}$, for some positive value ρ. The problem is feasible whenever $\rho \le 1$. We use $\rho = 0.8$ and run the simulation for 10 million frames. Fig. 9.3 shows the resulting average power as V is varied between 0 and 3, which converges to near optimal after $V = 0.3$. All 10 processing rate constraints are met within 5 decimal points of accuracy after the 10 million frames. An illustration of how convergence time is affected by the V parameter is shown in Fig. 9.4, which illustrates the average processing rate $\bar{I}_{10}/\bar{T}$ for class 10, as compared to the desired constraint $\lambda_{10} = \rho/300$. It is seen, for example, that convergence is faster for $V = 0.05$ than for $V = 1$. Convergence times can be improved using non-zero initial queue backlog and the theory of *place holder backlog* in [25], although we omit this topic for brevity.

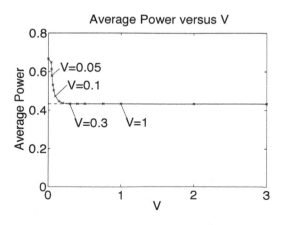

Figure 9.3: Average power versus V.

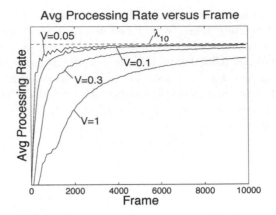

Figure 9.4: Processing rate $\overline{1}_{10}/\overline{T}$ versus frame index.

9.2 Optimization with General Attributes

This section generalizes the problem to allow time average optimization for abstract *attributes*. Consider again a frame-based system with frame index $k \in \{0, 1, 2, \ldots\}$. Every frame k, the controller makes a *control action* $\alpha[k]$, chosen within an abstract set $\mathcal{A}$ of allowable actions. The action $\alpha[k]$ affects the *frame size* $T[k]$ and an *attribute vector* $\boldsymbol{y}[k] = (y_0[k], y_1[k], \ldots, y_L[k])$. Specifically, assume these are random functions that are conditionally independent of the past given the current $\alpha[k]$ decision, with mean values given by functions $\hat{T}(\alpha)$ and $\hat{y}_l(\alpha)$ for all $\alpha \in \mathcal{A}$:

$$\hat{T}(\alpha) = \mathbb{E}\left[T[k]|\alpha[k] = \alpha\right] \ , \ \hat{y}_l(\alpha) = \mathbb{E}\left[y_l[k]|\alpha[k] = \alpha\right]$$

Similar to the previous section, it is assumed there is a minimum frame size $T_{min} > 0$ such that $T[k] \geq T_{min}$ for all k, and that second moments are bounded by a constant σ^2, regardless of the policy $\alpha[k]$. The joint distribution of $(T[k], y_0[k], y_1[k], \ldots, y_L[k])$ is otherwise arbitrary.

Define frame averages $\overline{T}$ and $\overline{y}_l$ by:

$$\overline{T} = \lim_{K \to \infty} \frac{1}{K} \sum_{k=0}^{K-1} T[k] \ , \ \overline{y}_l = \lim_{K \to \infty} \frac{1}{K} \sum_{k=0}^{K-1} y_l[k]$$

As discussed in Section 9.1.2, the value $\overline{y}_l/\overline{T}$ represents the *time average* associated with attribute $y_l[k]$. The general problem is then:

$$\text{Minimize:} \quad \overline{y}_0/\overline{T} \tag{9.32}$$

$$\text{Subject to:} \quad \overline{y}_l/\overline{T} \leq c_l \ \ \forall l \in \{1, \ldots, L\} \tag{9.33}$$

$$\alpha[k] \in \mathcal{A} \ \ \forall k \in \{0, 1, 2, \ldots\} \tag{9.34}$$

where $c_1, \ldots, c_L$ are given constants that specify the desired time average constraints.

9.2.1 Mapping to the Task Scheduling Problem

To illustrate the generality of this framework, this subsection uses the new notation to exactly represent the task scheduling problem from Section 9.1. For that problem, one can define the control action $\alpha[k]$ to have the form $\alpha[k] = (c[k], m[k], I[k])$, and the action space $\mathcal{A}$ is then the set of all (c, m, I) such that $c \in \{1, \ldots, N\}$, $m \in \mathcal{M}$, and $0 \leq I \leq I_{max}$.

The frame size is $T[k] = D[k] + I[k]$, and $\hat{T}(\alpha[k])$ is given by:

$$\hat{T}(\alpha[k]) = \hat{D}(c[k], m[k]) + I[k]$$

We then define $y_0[k]$ as the energy expended in frame k, so that $y_0[k] = e[k]$ and $\hat{y}_0(\alpha[k]) = \hat{e}(c[k], m[k])$. There are N constraints, so define $L = N$. To express the desired constraints $\overline{1}_n/\overline{T} \geq \lambda_n$ in the form $\overline{y}_n/\overline{T} \leq c_n$, one can define $y_n[k] = -1_n[k]$ and $c_n = -\lambda_n$ for each $n \in \{1, \ldots, N\}$, and $\hat{y}_n(\alpha[k]) = -1_n[k]$. Alternatively, one could define $y_n[k] = \lambda_n T[k] - 1_n[k]$ and enforce the constraint $\overline{y}_n \leq 0$ for all $n \in \{1, \ldots, N\}$.

This general setup provides more flexibility. For example, suppose the idle state does not use 0 energy, but operates at a low power p_{idle} and expends total energy $p_{idle}I[k]$ on frame k. Then total energy for frame k can be defined as $y_0[k] = e[k] + p_{idle}I[k]$, where $e[k]$ is the energy spent in the busy period. The setup can also handle systems with multiple idle mode options, each providing a different energy savings but incurring a different wakeup time.

9.2.2 The General Algorithm

The algorithm for solving the general problem (9.32)-(9.34) is described below. Each constraint $\overline{y}_l \leq c_l \overline{T}$ in (9.33) is treated using a virtual queue $Q_l[k]$ with update:

$$Q_l[k+1] = \max[Q_l[k] + y_l[k] - c_l T[k], 0] \quad \forall l \in \{1, \ldots, L\} \qquad (9.35)$$

Defining $L[k]$ and $\Delta[k]$ as before (in Section 9.1.7) leads to the following bound on the drift-plus-penalty ratio, which can be proven in a manner similar to Lemma 3:

$$\frac{\mathbb{E}\left[\Delta[k] + V y_0[k] | \boldsymbol{Q}[k]\right]}{\mathbb{E}\left[T[k] | \boldsymbol{Q}[k]\right]} \leq \frac{B}{\mathbb{E}\left[T[k] | \boldsymbol{Q}[k]\right]}$$

$$+ \frac{\mathbb{E}\left[V \hat{y}_0(\alpha[k]) + \sum_{l=1}^{L} Q_l[k] \hat{y}_l(\alpha[k]) | \boldsymbol{Q}[k]\right]}{\mathbb{E}\left[\hat{T}(\alpha[k]) | \boldsymbol{Q}[k]\right]}$$

$$(9.36)$$

where B is a constant that satisfies the following for all $\boldsymbol{Q}[k]$ and all possible actions $\alpha[k]$:

$$B \geq \frac{1}{2} \sum_{l=1}^{L} \mathbb{E}\left[(y_l[k] - c_l T[k])^2 | \boldsymbol{Q}[k]\right]$$

Every frame, the controller observes queues $Q[k]$ and takes an action $\alpha[k] \in \mathcal{A}$ that minimizes the second term on the right-hand-side of (9.36). We know from Lemma 1 that minimizing the ratio of expectations is accomplished by a deterministic selection of $\alpha[k] \in \mathcal{A}$. The resulting algorithm is:

- Observe $Q[k]$ and choose $\alpha[k] \in \mathcal{A}$ to minimize (breaking ties arbitrarily):

$$\frac{V\hat{y}_0(\alpha[k]) + \sum_{l=1}^{L} Q_l[k]\hat{y}_l(\alpha[k])}{\hat{T}(\alpha[k])} \qquad (9.37)$$

- Update virtual queues $Q_l[k]$ for each $l \in \{1, \ldots, L\}$ via (9.35).

One subtlety is that the expression (9.37) may not have an achievable minimum over the general (possibly infinite) set $\mathcal{A}$ (for example, the infimum of the function $f(x) = x$ over the open interval $0 < x < 1$ is not achievable over that interval). This is no problem: Our algorithm in fact works for any *approximate* minimum that is an additive constant C away from the exact infimum every frame k (for any arbitrarily large constant $C \geq 0$). This effectively changes the "B" constant in our performance bounds to a new constant "$B+C$" [25]. Let $ratio^{opt}$ represent the optimal ratio of $\overline{y}_0/\overline{T}$ for the problem (9.32)-(9.34). As before, it can be shown that if the problem is feasible (so that there exists an algorithm that can achieve the constraints (9.33)-(9.34)), then any C-additive approximation of the above algorithm satisfies all desired constraints and yields $\overline{y}_0/\overline{T} \leq ratio^{opt} + O(1/V)$, which can be pushed arbitrarily close to $ratio^{opt}$ as V is increased, with the same tradeoff in the queue sizes (and hence convergence times) with V. The proof of this is similar to that of Theorem 1, and is omitted for brevity (see [25] [29] for the full proof).

9.2.3   Random Task Arrivals and Flow Control

Again consider a system with N classes of tasks, as in Section 9.1. Each frame k again has a busy period of duration $D[k]$ and an idle period of duration $I[k]$ as in Fig. 9.1. However, rather than always having tasks available for processing, this subsection assumes tasks arrive *randomly* with rates $(\lambda_1, \ldots, \lambda_N)$, where λ_n is the rate of task arrivals per unit time (see Fig. 9.5). At the beginning of each busy period, the controller chooses a variable $c[k]$ that specifies which type of task is performed. However, $c[k]$ can now take values in the set $\{0, 1, \ldots, N\}$, where $c[k] = 0$ is a *null* choice that selects no task on frame k. If $c[k] = 0$, the busy period has some positive size D_0 and may spend a small amount of energy to power the electronics, but does not process any task. The mode selection variable $m[k]$ takes values in the same set $\mathcal{M}$ as before. The idle time variable $I[k]$ is again chosen in the interval $[0, I_{max}]$.

Further, for each $n \in \{1, 2, \ldots, N\}$ we introduce *flow control variables* $\gamma_n[k]$, chosen in the interval $0 \leq \gamma_n[k] \leq 1$. The variable $\gamma_n[k]$ represents the probability of admitting each new randomly arriving task of class n on frame

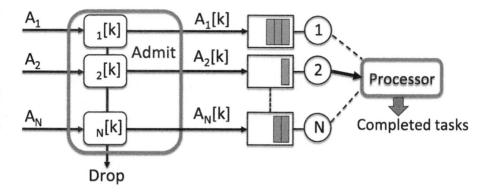

Figure 9.5: A task processing system with random arrivals and flow control.

k (see Fig. 9.5). This enables the system to drop tasks if the raw arrival rates $(\lambda_1, \ldots, \lambda_N)$ cannot be supported. Let $\boldsymbol{\gamma}[k] = (\gamma_1[k], \ldots, \gamma_N[k])$ be the vector of these variables.

We thus have $\alpha[k] = (c[k], m[k], I[k], \boldsymbol{\gamma}[k])$, with action space $\mathcal{A}$ being the set of all $(c, m, I, \boldsymbol{\gamma})$ such that $c \in \{0, 1, \ldots, N\}$, $m \in \mathcal{M}$, $0 \leq I \leq I_{max}$, and $0 \leq \gamma_n \leq 1$ for $n \in \{1, \ldots, N\}$. Define $e[k]$ and $D[k]$ as the energy and busy period duration for frame k. Assume $e[k]$ depends only on $(c[k], m[k], I[k])$, and $D[k]$ depends only on $(c[k], m[k])$, with averages given by functions $\hat{e}(c[k], m[k], I[k])$ and $\hat{D}(c[k], m[k])$:

$$\hat{e}(c[k], m[k], I[k]) = \mathbb{E}\left[e[k]|c[k], m[k], I[k]\right] \ , \ \ \hat{D}(c[k], m[k]) = \mathbb{E}\left[D[k]|c[k], m[k]\right]$$

Finally, for each $n \in \{1, \ldots, N\}$, define $A_n[k]$ as the random number of new arrivals admitted on frame k, which depends on the total frame size $D[k] + I[k]$ and the admission probability $\gamma_n[k]$. Formally, assume the arrival vector $(A_1[k], \ldots, A_N[k])$ is conditionally independent of the past given the current $\alpha[k]$ used, with expectations:

$$\mathbb{E}\left[A_n[k]|\alpha[k]\right] = \lambda_n \gamma_n[k][\hat{D}(c[k], m[k]) + I[k]] \tag{9.38}$$

The assumption on independence of the past holds whenever arrivals are independent and Poisson, or when all frame sizes are an integer number of fixed size slots, and arrivals are independent and identically distributed (i.i.d.) over slots with some general distribution.

We seek to maximize a weighted sum of admission rates subject to supporting all of the admitted tasks, and to maintaining average power to within a given positive constant P_{av}:

$$\text{Maximize:} \quad \frac{\sum_{n=1}^{N} w_n \overline{A}_n}{\overline{D} + \overline{I}} \qquad (9.39)$$

$$\text{Subject to:} \quad \overline{A}_n/(\overline{D} + \overline{I}) \leq \overline{1}_n/(\overline{D} + \overline{I}) \quad \forall n \in \{1, \dots, N\} \qquad (9.40)$$

$$\frac{\overline{e}}{\overline{D} + \overline{I}} \leq P_{av} \qquad (9.41)$$

$$\alpha[k] \in \mathcal{A} \quad \forall k \in \{0, 1, 2, \dots\} \qquad (9.42)$$

where $(w_1, \dots, w_N)$ are a collection of positive weights that prioritize the different classes in the optimization objective.

We thus have $L = N + 1$ constraints. To treat this problem, define $T[k] = D[k] + I[k]$, so that $\hat{T}(\alpha[k]) = \hat{D}(c[k], m[k]) + I[k]$. Further define $y_0[k]$, $y_n[k]$ for $n \in \{1, \dots, N\}$, and $x[k]$ as:

$$y_0[k] = -\sum_{n=1}^{N} w_n A_n[k] \implies \hat{y}_0(\alpha[k])$$

$$= -\sum_{n=1}^{N} w_n \lambda_n \gamma_n[k][\hat{D}(c[k], m[k]) + I[k]]$$

$$y_n[k] = A_n[k] - 1_n[k] \implies \hat{y}_n(\alpha[k])$$
$$= \lambda_n \gamma_n[k][\hat{D}(c[k], m[k]) + I[k]] - 1_n[k]$$

$$x[k] = e[k] - [D[k] + I[k]]P_{av} \implies \hat{x}(\alpha[k]) = \hat{e}(c[k], m[k], I[k])$$
$$- [\hat{D}(c[k], m[k]) + I[k]]P_{av}$$

Then:

- Minimizing $\overline{y}_0/\overline{T}$ is equivalent to (9.39).

- The constraints $\overline{y}_n \leq 0$ for $n \in \{1, \dots, N\}$ are equivalent to (9.40).

- The constraint $\overline{x} \leq 0$ is equivalent to (9.41).

Note that the above problem does not specify any explicit queueing for the randomly arriving tasks. The algorithm will in fact construct explicit queues (so that the virtual queues can be viewed as actual queues). Note also that the constraint $\alpha[k] \in \mathcal{A}$ does not allow restrictions on actions based on the queue state, such as when the queue is empty or not. Thus, in principle, we allow the possibility of "processing" a task of class n even when there is no such task available. In this case, we assume this processing is still costly, in that it incurs time equal to $\hat{D}(c[k], m[k])$ and energy equal to $\hat{e}(c[k], m[k], I[k])$. Our algorithm will naturally learn to avoid the inefficiencies associated with such actions.

9.2.3.1   The Dynamic Algorithm for Random Task Arrivals

To enforce the constraints $\bar{y}_n \leq 0$ for each $n \in \{1, \ldots, N\}$, define queue $Q_n[k]$ with update:

$$Q_n[k+1] = \max[Q_n[k] + A_n[k] - 1_n[k], 0] \tag{9.43}$$

To enforce $\bar{x} \leq 0$, define a virtual queue $Z[k]$ with update:

$$Z[k+1] = \max[Z[k] + e[k] - [D[k] + I[k]]P_{av}, 0] \tag{9.44}$$

It can be seen that the queue update (9.43) is the same as that of an *actual queue* for class n tasks, with random task arrivals $A_n[k]$ and task service $1_n[k]$. The minimization of (9.37) then becomes the following: Every frame k, observe queues $\boldsymbol{Q}[k]$ and $Z[k]$. Then choose $c[k] \in \{0, 1, \ldots, N\}$, $m[k] \in \mathcal{M}$, $I[k] \in [0, I_{max}]$, and $\gamma_n[k] \in [0, 1]$ for all $n \in \{1, \ldots, N\}$ to minimize:

$$\frac{-V \sum_{n=1}^{N} w_n \lambda_n \gamma_n[k][\hat{D}(c[k], m[k]) + I[k]]}{\hat{D}(c[k], m[k]) + I[k]}$$

$$+ \frac{\sum_{n=1}^{N} Q_n[k](\lambda_n \gamma_n[k][\hat{D}(c[k], m[k]) + I[k]] - 1_n[k])}{\hat{D}(c[k], m[k]) + I[k]}$$

$$+ \frac{Z[k](\hat{e}(c[k], m[k], I[k]) - [\hat{D}(c[k], m[k]) + I[k]]P_{av})}{\hat{D}(c[k], m[k]) + I[k]} \tag{9.45}$$

After a simplifying cancellation of terms, it is easy to see that the $\gamma_n[k]$ decisions can be separated from all other decisions (see Exercise 2). The resulting algorithm then observes queues $\boldsymbol{Q}[k]$ and $Z[k]$ every frame k and performs the following:

- (Flow Control) For each $n \in \{1, \ldots, N\}$, choose $\gamma_n[k]$ as:

$$\gamma_n[k] = \begin{cases} 1 & \text{if } Q_n[k] \leq V w_n \\ 0 & \text{otherwise} \end{cases} \tag{9.46}$$

- (Task Scheduling) Choose $c[k] \in \{0, 1, \ldots, N\}$, $m[k] \in \mathcal{M}$, $I[k] \in [0, I_{max}]$ to minimize:

$$\frac{Z[k]\hat{e}(c[k], m[k], I[k]) - \sum_{n=1}^{N} Q_n[k]1_n[k]}{\hat{D}(c[k], m[k]) + I[k]} \tag{9.47}$$

- (Queue Update) Update $Q_n[k]$ for each $n \in \{1, \ldots, N\}$ by (9.43), and update $Z[k]$ by (9.44).

The minimization problem (9.47) is similar to (9.24), and can be carried out in the same manner as discussed in Section 9.1.7. A key observation about the above algorithm is that *it does not require knowledge of the arrival rates* $(\lambda_1, \ldots, \lambda_N)$. Indeed, the λ_n terms cancel out of the minimization, so that the flow control variables $\gamma_n[k]$ in (9.46) make "bang-bang" decisions that admit all newly arriving tasks of class n on frame k if $Q_n[k] \leq V w_n$, and admit none otherwise. This property makes the algorithm naturally adaptive to situations when the arrival rates change, as shown in the simulations of Section 9.2.4.

Note that if $\hat{e}(0, m, I) < \hat{e}(c, m, I)$ for all $c \in \{1, \ldots, N\}$, $m \in \mathcal{M}$, $I \in [0, I_{max}]$, so that the energy associated with processing *no task* is less than the energy of processing any class $c \neq 0$, then the minimization in (9.47) will never select a class c such that $Q_c[k] = 0$. That is, the algorithm naturally will never select a class for which no task is available.

9.2.3.2 Deterministic Queue Bounds and Constraint Violation Bounds

In addition to satisfying the desired constraints and achieving a weighted sum of admitted rates that is within $O(1/V)$ of optimality, the flow control structure of the task scheduling algorithm admits *deterministic queue bounds*. Specifically, assume all frame sizes are bounded by a constant T_{max}, and that the raw number of class n arrivals per frame (before admission control) is at most $A_{n,max}$. By the flow control policy (9.46), new arrivals of class n are only admitted if $Q_n[k] \leq V w_n$. Thus, assuming that $Q_n[0] \leq V w_n + A_{n,max}$, we must have $Q_n[k] \leq V w_n + A_{n,max}$ for all frames $k \in \{0, 1, 2, \ldots\}$. This specifies a worst-case queue backlog that is $O(V)$, which establishes an explicit $[O(1/V), O(V)]$ performance-backlog tradeoff that is superior to that given in (9.26).

With mild additional structure on the $\hat{e}(c, m, I)$ function, the deterministic bounds on queues $Q_n[k]$ lead to a deterministic bound on $Z[k]$, so that one can compute a value Z_{max}, of size $O(V)$, such that $Z[k] \leq Z_{max}$ for all k. This is explored in Exercise 3 (see also [28]).

9.2.4 Simulations and Adaptiveness of Random Task Scheduling

Here we simulate the dynamic task scheduling and flow control algorithm (9.46)-(9.47), using the 10-class system model defined in Section 9.1.9 with $\hat{e}(c, m)$ functions given in (9.30)-(9.31). For consistency with that model, we remove the $c[k] = 0$ option, so that the decision (9.31) chooses $c[k] = 1, m[k] = 1$, incurring one unit of energy, in case no tasks are available. Arrivals are from independent Bernoulli processes with rates $\lambda_i = \rho/(30i)$ for each class $i \in \{1, \ldots, 10\}$, with $\rho = 0.8$. We use weights $w_n = 1$ for all n, so that the objective is to maximize total throughput, and $P_{av} = 0.5$, which we know is feasible from results in Fig. 9.3 of Section 9.1.9. Thus, we expect the algorithm to learn to admit everything, so that the admission rate approaches the total arrival rate as V is increased. We simulate for 10 million frames, using V in the interval from 0 to 200. Results are shown in Figs. 9.6 and 9.7. Fig 9.6 shows the algorithm learns to admit everything for large V (100 or above), and Fig.

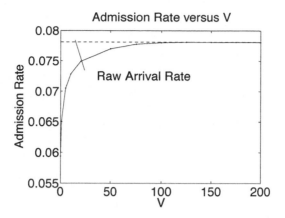

Figure 9.6: Total admission rate versus V.

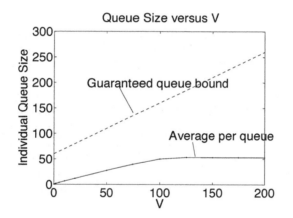

Figure 9.7: Average queue size per queue, and the deterministic bound from Section 9.2.3.2.

9.7 plots the resulting average queue size (in number of tasks) per queue, together with the deterministic bound $Q_n[k] \leq V + 60$ (where $A_{n,max} = 60$ in this case because there is at most one arrival per slot, and the largest possible frame is $D_{max} + I_{max} = 50 + 10 = 60$). The average power constraint was satisfied (with slackness) for all cases. The average queue size in Fig. 9.7 grows linearly with V until $V = 100$, when it saturates by admitting everything. The saturation value is the average queue size associated with admitting the raw arrival rates directly.

We now illustrate that the algorithm is robust to abrupt rate changes. We consider the same system as before, run over 10 million frames with $V = 100$. However, we break the simulation timeline into three equal size phases.

Low Power Dynamic Scheduling for Computing Systems

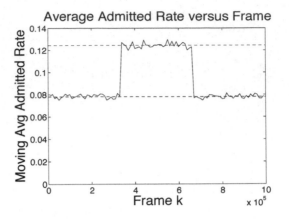

Figure 9.8: Moving average admission rate versus frame index k for the system with abrupt rate changes.

During the first and third phase, we use arrival rates $\lambda_i = \rho_1/(30i)$ for $i \in \{1,\dots,10\}$, where $\rho_1 = 0.8$ as in the previous experiment. Then, during the second (middle) phase, we double the rates to $\lambda_i = \rho_2/(30i)$, where $\rho_2 = 1.6$. Because $\rho_2 > 1$, these rates are *infeasible* and the algorithm must learn to optimally drop tasks so as to maximize the admission rate subject to the power constraint. Recall that the algorithm is unaware of the arrival rates and must adapt to the existing system conditions. The results are shown in Figs. 9.8 and 9.9. Fig. 9.8 shows a moving average admission rate versus time. During the first and third phases of the simulation, we have $\rho_1 = 0.8$ and the admitted rates are close to the raw arrival rate (shown as the lower dashed

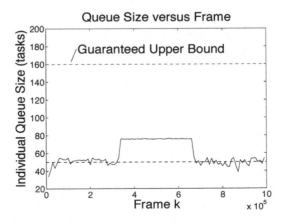

Figure 9.9: Individual queue size (average and guaranteed bound) versus frame index k for the system with abrupt rate changes.

horizontal line). During the middle interval (with $\rho_2 = 1.6$), the algorithm quickly adapts to the increased arrivals and yields admitted rates that are close to those that should be used in a system with loading $\rho_2 = 1.6$ always (shown as the higher dashed horizontal line). Fig. 9.9 plots the corresponding moving average queue backlog per queue. The lower dashed horizontal line indicates the value of average backlog that would be achieved in a system with loading $\rho_1 = 0.8$ always. Also shown is the deterministic queue bound $V + 60 = 160$, which holds for all frames, regardless of the raw arrival rates.

9.2.5 Task Scheduling: Extensions and Further Reading

This section considered minimizing time averages subject to time average constraints. Extended techniques for minimizing convex functions of time average vectors (or maximizing concave functions) subject to similar constraints are treated in the renewal optimization theory of [25] [29]. This is useful for extending the flow control problem (9.39)-(9.42) to optimize a sum of concave utility functions $\phi_n(x)$ of the time average admission rates for each class:

$$\sum_{n=1}^{N} \phi_n \left(\overline{A}_n / (\overline{D} + \overline{I}) \right)$$

Using logarithmic functions $\phi_n(x)$ leads to *proportional fairness* [15], while other fairness properties are achieved with other concave functions [23] [37] [41]. Related utility optimization for particular classes of wireless systems with fixed size frames of T slots, and with probabilistic reception every slot, is treated using a different technique in [11] [10]. Related work on utility optimal scheduling for single-slot data networks is treated using convex optimization in [16] [22] [43] [21], stochastic "primal-dual" algorithms in [17] [1] [36] [4] [20] and stochastic "dual" algorithms in [24] [32] [5] [8] [34].

For the problem with random arrivals, queueing delay can often be improved by changing the constraint $\overline{A}_n \le \overline{I}_n$ of (9.40) to $\overline{A}_n + \epsilon \le \overline{I}_n$, for some $\epsilon > 0$. This specifies that the output processing rate should be ϵ larger than the arrival rate. However, unlike the case $\epsilon = 0$, such a constraint yields a queue $Q_n[k]$ that contains some "fake" tasks, and can lead to decisions that serve the queue when no actual tasks are available. This can be overcome using the theory of ϵ-*persistent service queues* in [33] [26].

Note that our flow control algorithm rewards admission of new data on each frame. This fits the general framework of this section, where attributes $y_l[k]$ are (possibly random) functions of the control action $\alpha[k] \in \mathcal{A}$. One might attempt to solve the problem by defining a reward upon *completion* of service. This does not fit our framework: That is because the algorithm might "complete" a service in a queue that is empty, simply to accumulate the "fake" reward. One could eliminate fake rewards by an augmented model that allows rewards $\hat{y}_l(\alpha[k], \boldsymbol{Q}[k])$ and/or action spaces $\mathcal{A}(\boldsymbol{Q}[k])$ that depend on the current backlog, although this creates a much more complex Markov decision model that is easily avoided by defining rewards at admission. However, there are some problems that require such a reward structure, such as problems of stock market trading where prior ownership of a particular stock is required to reap the benefits of selling at an observed desirable price. These problems can be treated with a modified Lyapunov function of the form $L[k] = \sum_{n=1}^{N}(Q_n[k] - \theta)^2$, which pushes backlog toward a non-empty state θ. This approach is used for stock trading [27], inventory control [31], energy harvesting [13] and smart-grid energy management [40]. The first algorithms for optimal energy harvesting used a related technique [7], and related problems in processing networks that assemble components by combining different sub-components are treated in [14] [12].

9.2.6   Exercises for Section 9.2

Exercise 1. *Consider a system with N classes, each of which always has new tasks available. Every frame k we choose a class $c[k] \in \{1, \ldots, N\}$, and select a single task of that class for processing. We can process using mode $m[k] \in \mathcal{M}$. The result yields frame size $\hat{T}(c[k], m[k])$, energy $\hat{e}(c[k], m[k])$ and processing quality $\hat{q}(c[k], m[k])$. Design an algorithm that selects $c[k]$ and $m[k]$ every frame k to maximize $\overline{q}/\overline{T}$ subject to an average power constraint $\overline{e}/\overline{T} \le P_{av}$ and to a processing rate constraint of at least λ_n for each class n.*

Exercise 2. *Show that minimization of (9.45) leads to the separable flow control and task scheduling algorithm of (9.46)-(9.47).*

Exercise 3. *Consider the flow control and task scheduling algorithm (9.46)-(9.47), and recall that $Q_n[k] \le Q_{max}$ for all $n \in \{1, \ldots, N\}$ and all frames k, where $Q_{max} \stackrel{\Delta}{=} \max_{n \in \{1, \ldots, N\}}[Vw_n + A_{n,max}]$. Suppose there is a constant $e_{min} > 0$ such that $\hat{e}(c, m, I) \ge e_{min}$ for all $c \ne 0$, and that $\hat{e}(0, m, I) = 0$ for all $m \in \mathcal{M}$, $I \in [0, I_{max}]$.*

a) Show that the minimization of (9.47) chooses $c[k] = 0$ whenever $Z[k] > Q_{max}/e_{min}$.

b) Suppose there is a constant e_{max} such that $e[k] \le e_{max}$ for all k. Conclude that $Z[k] \le Z_{max} \stackrel{\Delta}{=} Q_{max}/e_{min} + e_{max}$ for all frames k.

c) Use the queueing equation (9.44) to show that over any sequence of K frames $\{j, j+1, \ldots, j+K-1\}$ (starting at some frame j), the total energy usage satisfies:

$$\sum_{k=j}^{j+K-1} e[k] \le P_{av} \sum_{k=j}^{j+K-1} T[k] + Z_{max}$$

where $T[k]$ is the size of frame k. Hint: Make an argument similar to the proof of Lemma 2.

Exercise 4. *Consider the same general attributes $y_l[k]$ of Section 9.2, with $\hat{y}_l(\alpha[k])$ for $\alpha[k] \in \mathcal{A}$. State the algorithm for solving the problem of minimizing $\overline{y}_0$ subject to $\overline{y}_l/\overline{T} \le c_l$. Hint: Define appropriate attributes $x_l[k]$ on a system with an "effective" frame size of 1 every frame, and minimize $\overline{x}_0/1$ subject to $\overline{x}_l/1 \le 0$ for $l \in \{1, \ldots, L\}$.*

Exercise 5. *Modify the task scheduling algorithm of Section 9.1.7 to allow the controller to serve more than one task per frame. Specifically, every frame k the controller chooses a* service action $s[k]$ *from a set $\mathcal{S}$ of possible actions. Each service action $s \in \mathcal{S}$ determines a* clearance vector $\mathbf{c}(s) = (c_1(s), \ldots, c_N(s))$, *where $c_i(s)$ is the number of tasks of type i served if action s is used on a given frame. It also incurs a delay $\hat{D}(s)$ and energy $\hat{e}(s)$.*

Exercise 6. *Consider the linear fractional problem of finding a vector $(x_1, \ldots, x_M)$ to minimize $(a_0 + \sum_{i=1}^{M} a_i x_i)/(b_0 + \sum_{i=1}^{M} b_i x_i)$ subject to $\sum_{i=1}^{M} c_{il} x_i \leq d_l$ for $l \in \{1, \ldots, L\}$ and $0 \leq x_i \leq 1$ for all $i \in \{1, \ldots, M\}$. Assume constants a_i, b_i, c_{il}, d_l are given, that $b_0 > 0$, and that $b_i \geq 0$ for all $i \in \{1, \ldots, M\}$. Treat this as a time average problem with action $\alpha[k] = (x_1[k], \ldots, x_M[k])$, action space $\mathcal{A} = \{(x_1, \ldots, x_M) | 0 \leq x_i \leq 1 \; \forall i \in \{1, \ldots, M\}\}$, frame size $T[k] = b_0 + \sum_{i=1}^{M} b_i x_i[k]$, and where we seek to minimize $(a_0 + \sum_{i=1}^{M} a_i \overline{x}_i)/(b_0 + \sum_{i=1}^{M} b_i \overline{x}_i)$ subject to $\sum_{i=1}^{M} c_{il} \overline{x}_i \leq d_l$ for all $l \in \{1, \ldots, L\}$.*

a) State the drift-plus-penalty ratio algorithm (9.37) for this, and conclude that:

$$\lim_{K \to \infty} \frac{a_0 + \sum_{i=1}^{M} a_i \overline{x}_i[K]}{b_0 + \sum_{i=1}^{M} b_i \overline{x}_i[K]} \leq ratio^{opt} + B/V,$$

$$\lim_{K \to \infty} \sum_{i=1}^{M} c_{il} \overline{x}_i[K] \leq d_l \quad \forall l \in \{1, \ldots, L\}$$

for some finite constant B, where $ratio^{opt}$ is the optimal value of the objective function, and $\overline{x}_i[K] \triangleq \frac{1}{K} \sum_{k=0}^{K-1} x_i[k]$. Thus, the limiting time average satisfies the constraints and is within B/V from optimality. Your answer should solve for values $\phi_i[k]$ such that on frame k we choose $(x_1[k], \ldots, x_M[k])$ over $\mathcal{A}$ to minimize $\frac{\phi_0[k] + \sum_{i=1}^{M} \phi_i[k] x_i[k]}{b_0 + \sum_{i=1}^{M} b_i x_i[k]}$. Note: It can be shown a solution is: Define $\mathcal{I} \triangleq \{i \in \{1, \ldots, M\} | b_i = 0\}$ and $\mathcal{J} \triangleq \{j \in \{1, \ldots, M\} | b_j > 0\}$. For all $i \in \mathcal{I}$, choose $x_i[k] = 0$ if $\phi_i[k] \geq 0$, and $x_i[k] = 1$ if $\phi_i[k] < 0$. Next, temporarily select $x_j[k] = 0$ for all $j \in \mathcal{J}$. Then rank order the indices $j \in \mathcal{J}$ from smallest to largest value of $\phi_j[k]/b_j$, and, using this order, greedily change $x_j[k]$ from 0 to 1 if it improves the solution.

b) Note that the case $b_0 = 1$ and $b_i = 0$ for $i \neq 0$ is a linear program. Give an explicit decision rule for each $x_i[k]$ in this case (the solution should be separable for each $i \in \{1, \ldots, M\}$).

9.3 Reacting to Randomly Observed Events

Consider a problem with general attributes $y_0[k], y_1[k], \ldots, y_L[k]$ and frame size $T[k]$ for each frame k, as in Section 9.2. However, now assume the controller observes a random event $\omega[k]$ at the beginning of each frame k, and this can influence attributes and frame sizes. The value of $\omega[k]$ can represent a vector of channel states and/or prices observed for frame k. Assume $\{\omega[k]\}_{k=0}^{\infty}$ is independent and identically distributed (i.i.d.) over frames. The controller chooses an action $\alpha[k] \in \mathcal{A}(\omega[k])$, where the action space $\mathcal{A}(\omega[k])$ possibly depends on the observed $\omega[k]$. Values of $(T[k], y_0[k], \ldots, y_L[k])$ are conditionally independent of the past given the current $\omega[k]$ and $\alpha[k]$, with mean values:

$$\hat{y}_l(\omega[k], \alpha[k]) = \mathbb{E}\left[y_l[k]|\omega[k], \alpha[k]\right] \quad , \quad \hat{T}(\omega[k], \alpha[k]) = \mathbb{E}\left[T[k]|\omega[k], \alpha[k]\right]$$

The goal is to solve the following optimization problem:

$$\text{Minimize:} \quad \overline{y}_0/\overline{T} \tag{9.48}$$
$$\text{Subject to:} \quad \overline{y}_l/\overline{T} \leq c_l \;\; \forall l \in \{1, \ldots, L\} \tag{9.49}$$
$$\alpha[k] \in \mathcal{A}(\omega[k]) \;\; \forall k \in \{0, 1, 2, \ldots\} \tag{9.50}$$

As before, the constraints $\overline{y}_l \leq c_l\overline{T}$ are satisfied via virtual queues $Q_l[k]$ for $l \in \{1, \ldots, L\}$:

$$Q_l[k+1] = \max[Q_l[k] + y_l[k] - c_lT[k], 0] \tag{9.51}$$

The random $\omega[k]$ observations make this problem more complex than those considered in previous sections of this chapter. We present two different algorithms from [25] [29].

9.3.0.1   Algorithm 1

Every frame k, observe $Q[k]$ and $\omega[k]$ and choose $\alpha[k] \in \mathcal{A}(\omega[k])$ to minimize the following ratio of expectations:

$$\frac{\mathbb{E}\left[V\hat{y}_0(\omega[k], \alpha[k]) + \sum_{l=1}^{L} Q_l[k]\hat{y}_l(\omega[k], \alpha[k])|Q[k]\right]}{\mathbb{E}\left[\hat{T}(\omega[k], \alpha[k])|Q[k]\right]} \tag{9.52}$$

Then update the virtual queues via (9.51).

9.3.0.2   Algorithm 2

Define $\theta[0] = 0$, and define $\theta[k]$ for $k \in \{1, 2, 3, \ldots\}$ as a running ratio of averages over past frames:

$$\theta[k] \triangleq \sum_{i=0}^{k-1} y_0[i] / \sum_{i=0}^{k-1} T[i] \tag{9.53}$$

Then every frame k, observe $Q[k]$, $\theta[k]$, and $\omega[k]$, and choose $\alpha[k] \in \mathcal{A}(\omega[k])$ to minimize the following function:

$$V[\hat{y}_0(\omega[k], \alpha[k]) - \theta[k]\hat{T}(\omega[k], \alpha[k])] + \sum_{l=1}^{L} Q_l[k][\hat{y}_l(\omega[k], \alpha[k]) - c_l\hat{T}(\omega[k], \alpha[k])] \tag{9.54}$$

Then update the virtual queues via (9.51).

9.3.0.3 Comparison of Algorithms 1 and 2

Both algorithms are introduced and analyzed in [25] [29], where they are shown to satisfy the desired constraints and yield an optimality gap of $O(1/V)$. Algorithm 1 can be analyzed in a manner similar to the proof of Theorem 1, and has the same tradeoff with V as given in that theorem. However, the ratio of expectations (9.52) is *not necessarily* minimized by observing $w[k]$ and choosing $\alpha[k] \in \mathcal{A}(w[k])$ to minimize the deterministic ratio given $w[k]$. In fact, the minimizing policy depends on the (typically unknown) probability distribution for $w[k]$. A more complex *bisection algorithm* is needed for implementation, as specified in [25] [29].

Algorithm 2 is much simpler and involves a greedy selection of $\alpha[k] \in \mathcal{A}(w[k])$ based only on observation of $w[k]$, without requiring knowledge of the probability distribution for $w[k]$. However, its mathematical analysis does not yield as explicit information regarding convergence time as does Algorithm 1. Further, it requires a running average to be kept starting at frame 0, and hence may not be as adaptive when system statistics change. A more adaptive approximation of Algorithm 2 would define the average $\theta[k]$ over a moving window of some fixed number of frames, or would use an exponentially decaying average.

Both algorithms reduce to the following simplified *drift-plus-penalty* rule in the special case when the frame size $\hat{T}(w[k], \alpha[k])$ is a fixed constant T for all $w[k], \alpha[k]$: Every frame k, observe $w[k]$ and $Q[k]$ and choose $\alpha[k] \in \mathcal{A}(w[k])$ to minimize:

$$V\hat{y}_0(w[k], \alpha[k]) + \sum_{l=1}^{L} Q_l[k]\hat{y}_l(w[k], \alpha[k]) \qquad (9.55)$$

Then update the virtual queues via (9.51). This special case algorithm was developed in [28] to treat systems with fixed size time slots.

A simulation comparison of the algorithms is given in [29]. The next subsection describes an application to energy-aware computation and transmission in a wireless smart phone. Exercise 7 considers opportunistic scheduling where wireless transmissions can be deferred by waiting for more desirable channels. Exercise 8 considers an example of price-aware energy consumption for a network server that can process computational tasks or outsource them to another server.

9.3.1 Efficient Computation and Transmission for a Wireless Smart Device

Consider a wireless smart device (such as a smart phone or sensor) that always has tasks to process. Each task involves a computation operation, followed by a transmission operation over a wireless channel. On each frame k, the device takes a new task and looks at its *meta-data* $\beta[k]$, being information that characterizes the task in terms of its computational and transmission requirements. Let d represent the time required to observe this meta-data. The device then chooses a *computational processing mode* $m[k] \in \mathcal{M}(\beta[k])$, where $\mathcal{M}(\beta[k])$ is the set of all mode options under $\beta[k]$. The mode $m[k]$ and meta-data $\beta[k]$ affect a computation time $D_{comp}[k]$, computation energy $e_{comp}[k]$, computation quality $q[k]$, and generate $A[k]$ bits for transmission over the channel. The expectations of $D_{comp}[k]$, $e_{comp}[k]$, $q[k]$ are:

$$\hat{D}_{comp}(\beta[k], m[k]) = \mathbb{E}\left[D_{comp}[k]|\beta[k], m[k]\right]$$
$$\hat{e}_{comp}(\beta[k], m[k]) = \mathbb{E}\left[e_{comp}[k]|\beta[k], m[k]\right]$$
$$\hat{q}(\beta[k], m[k]) = \mathbb{E}\left[q[k]|\beta[k], m[k]\right]$$

For example, in a wireless sensor, the mode $m[k]$ may represent a particular sensing task, where different tasks can have different qualities and thus incur different energies, times and $A[k]$ bits for transmission. The full conditional distribution of $A[k]$, given $\beta[k]$, $m[k]$, will play a role in the transmission stage (rather than just its conditional expectation).

The $A[k]$ units of data must be transmitted over a wireless channel. Let $S[k]$ be the state of the channel on frame k, and assume $S[k]$ is constant for the duration of the frame. We choose a *transmission mode* $g[k] \in \mathcal{G}$, yielding a transmission time $D_{tran}[k]$ and transmission energy $e_{tran}[k]$ with expectations that depend on $S[k]$, $g[k]$, and $A[k]$. Define random event $\omega[k] = (\beta[k], S[k])$ and action $\alpha[k] = (m[k], g[k])$. We can then define expectation functions $\hat{D}_{tran}(\omega[k], \alpha[k])$ and $\hat{e}_{tran}(\omega[k], \alpha[k])$ by:

$$\hat{D}_{tran}(\omega[k], \alpha[k]) = \mathbb{E}\left[D_{tran}[k]|\omega[k], \alpha[k]\right],$$
$$\hat{e}_{tran}(\omega[k], \alpha[k]) = \mathbb{E}\left[e_{tran}[k]|\omega[k], \alpha[k]\right]$$

where the above expectations are defined via the *conditional distribution* associated with the number of bits $A[k]$ at the computation output, given the meta-data $\beta[k]$ and computation mode $m[k]$ selected in the computation phase (where $\beta[k]$ and $m[k]$ are included in the $\omega[k]$, $\alpha[k]$ information). The total frame size is thus $T[k] = d + D_{comp}[k] + D_{tran}[k]$.

The goal is to maximize frame processing quality per unit time $\overline{q}/\overline{T}$ subject to a processing rate constraint of $1/\overline{T} \geq \lambda$, and subject to an average power constraint $(\overline{e}_{comp} + \overline{e}_{tran})/\overline{T} \leq P_{av}$ (where λ and P_{av} are given constants). This fits the general framework with observed random events $\omega[k] = (\beta[k], S[k])$, control actions $\alpha[k] = (m[k], g[k])$ and action space $\mathcal{A}(\omega[k]) = \mathcal{M}(\beta[k]) \times \mathcal{G}$. We can define $y_0[k] = -q[k]$, $y_1[k] = T[k] - 1/\lambda$ and $y_2[k] = e_{comp}[k] + e_{tran}[k] - P_{av}T[k]$, and solve the problem of minimizing $\overline{y}_0/\overline{T}$ subject to $\overline{y}_1 \leq 0$ and $\overline{y}_2 \leq 0$. To do so, let $Q[k]$ and $Z[k]$ be virtual queues for the two constraints:

$$Q[k+1] = \max[Q[k] + T[k] - 1/\lambda, 0] \tag{9.56}$$
$$Z[k+1] = \max[Z[k] + e_{comp}[k] + e_{tran}[k] - P_{av}T[k], 0] \tag{9.57}$$

Using Algorithm 2, we define $\theta[0] = 0$ and $\theta[k]$ for $k \in \{1, 2, \dots\}$ by (9.53). The Algorithm 2 minimization (9.54) amounts to observing $(\beta[k], S[k])$, $Q[k]$,

$Z[k]$, and $\theta[k]$ on each frame k and choosing $m[k] \in \mathcal{M}(\beta[k])$ and $g[k] \in \mathcal{G}$ to minimize:

$$V[-\hat{q}(\beta[k], m[k]) - \theta[k](d + \hat{D}_{comp}(\beta[k], m[k]) + \hat{D}_{tran}(\omega[k], \alpha[k]))]$$
$$+Q[k][d + \hat{D}_{comp}(\beta[k], m[k]) + \hat{D}_{tran}(S[k], g[k]) - 1/\lambda]$$
$$+Z[k][\hat{e}_{comp}(\beta[k], m[k]) + \hat{e}_{tran}(\omega[k], \alpha[k])$$
$$-P_{av}(d + \hat{D}_{comp}(\beta[k], m[k]) + \hat{D}_{tran}(\omega[k], \alpha[k]))]$$

The computation and transmission operations are coupled and cannot be separated. This yields the following algorithm: Every frame k:

- Observe $\omega[k] = (\beta[k], S[k])$ and values $Q[k]$, $Z[k]$, $\theta[k]$. Then *jointly* choose action $m[k] \in \mathcal{M}(\beta[k])$ and $g[k] \in \mathcal{G}$, for a combined action $\alpha[k] = (m[k], g[k])$, to minimize:

$$-V\hat{q}(\beta[k], m[k]) + \hat{D}_{comp}(\beta[k], m[k])[Q[k] - V\theta[k] - P_{av}Z[k]]$$
$$+Z[k]\hat{e}_{comp}(\beta[k], m[k]) + \hat{D}_{tran}(\omega[k], \alpha[k])[Q[k] - V\theta[k] - P_{av}Z[k]]$$
$$+Z[k]\hat{e}_{tran}(\omega[k], \alpha[k])$$

- (Updates) Update $Q[k]$, $Z[k]$, $\theta[k]$ via (9.56), (9.57), and (9.53).

Exercise 9 shows the algorithm can be implemented without $\theta[k]$ if the goal is changed to maximize $\overline{q}$, rather than $\overline{q}/\overline{T}$. Further, the computation and transmission decisions can be *separated* if the system is modified so that the bits generated from computation are handed to a separate transmission layer for eventual transmission over the channel, rather than requiring transmission on the same frame, similar to the structure used in Exercise 8.

9.3.2 Exercises for Section 9.3

Exercise 7. *(Energy-Efficient Opportunistic Scheduling [28]) Consider a wireless device that operates over fixed size time slots $k \in \{0, 1, 2, \ldots\}$. Every slot k, new data of size $A[k]$ bits arrives and is added to a queue. The data must eventually be transmitted over a time-varying channel. At the beginning of every slot k, a controller observes the channel state $S[k]$ and allocates power $p[k]$ for transmission, enabling transmission of $\mu[k]$ bits, where $\mu[k] = \hat{\mu}(S[k], p[k])$ for some given function $\hat{\mu}(S, p)$. Assume $p[k]$ is chosen so that $0 \le p[k] \le P_{max}$ for some constant $P_{max} > 0$. We want to minimize average power $\overline{p}$ subject to supporting all data, so that $\overline{A} \le \overline{\mu}$. Treat this as a problem with all frames equal to 1 slot, observed random events $\omega[k] = (A[k], S[k])$, and actions $\alpha[k] = p[k] \in [0, P_{max}]$. Design an appropriate queue update and power allocation algorithm, using the policy structure (9.55).*

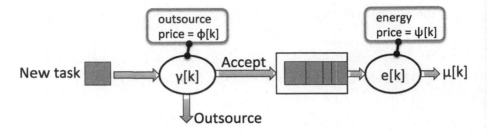

Figure 9.10: The client/server system of Exercise 8, with decision variables $(\gamma[k], e[k])$ and observed prices $(\phi[k], \psi[k])$.

Exercise 8. *(Energy Prices and Network Outsourcing) Consider a computer server that operates over fixed length time slots $k \in \{0, 1, 2, \ldots\}$. Every slot k, a new task arrives and has size $S[k]$ (if no task arrives then $S[k] = 0$). The server decides to either accept the task, or outsource it to another server (see Fig. 9.10). Let $\gamma[k]$ be a binary decision variable that is 1 if the server accepts the task on slot k, and zero else. Define $A[k] = \gamma[k]S[k]$ as the total workload admitted on slot k, which is added to the queue of work to be done. Let $\phi[k]$ be the (possibly time-varying) cost per unit size for outsourcing, so the outsourcing cost is $c_{out}[k] = \phi[k](1 - \gamma[k])S[k]$. Every slot k, the server additionally decides to process some of its backlog by purchasing an amount of energy $e[k]$ at a per unit energy price $\psi[k]$, serving $\mu[k] = \hat{\mu}(e[k])$ units of backlog with cost $c_{energy}[k] = \psi[k]e[k]$, where $\hat{\mu}(e)$ is some given function. Assume $e[k]$ is chosen in some interval $0 \leq e[k] \leq e_{max}$. The goal is to minimize time average cost $\overline{c}_{out} + \overline{c}_{energy}$ subject to supporting all tasks, so that $\overline{A} \leq \overline{\mu}$. Treat this as a problem with all frames equal to 1 slot, observed random events $\omega[k] = (S[k], \phi[k], \psi[k])$, actions $\alpha[k] = (\gamma[k], e[k])$, and action space $\mathcal{A}(\omega[k])$ being the set of all (γ, e) such that $\gamma \in \{0, 1\}$ and $0 \leq e \leq e_{max}$. Design an appropriate queue and state the dynamic algorithm (9.55) for this problem. Show that it can be implemented without knowledge of the probability distribution for $(S[k], \phi[k], \psi[k])$, and that the $\gamma[k]$ and $e[k]$ decisions are separable.*

Exercise 9. *Consider the same problem of Section 9.3.1, with the modified objective of maximizing $\overline{q}$ subject to $1/\overline{T} \geq \lambda$ and $(\overline{e}_{comp} + \overline{e}_{tran})/\overline{T} \leq P_{av}$. Use the observation from Exercise 4 that this can be viewed as a problem with an "effective" fixed frame size equal to 1, and give an algorithm from the policy structure (9.55).*

9.4 Conclusions

This chapter presents a methodology for optimizing time averages in systems with variable length frames. Applications include energy and quality aware task scheduling in smart phones, cost effective energy management at computer servers, and more. The resulting algorithms are dynamic and often do not require knowledge of the probabilities that affect system events. While the performance theorem of this chapter was stated under simple i.i.d. assumptions, the same algorithms are often provably robust to non-i.i.d. situations, including situations where the events are non-ergodic [25]. Simulations in Section 9.2.4 show examples of how such algorithms adapt to changes in the event probabilities. Exercises in this chapter were included to help readers learn to design dynamic algorithms for their own optimization problems.

The solution technique of this chapter uses the theory of optimization for renewal systems from [25] [29], which applies to general problems. Performance for individual problems can often be improved using enhanced techniques, such as using place-holder backlog, exponential Lyapunov functions, LIFO scheduling and ϵ-persistent service queues, as discussed in [25] and references therein. However, the drift-plus-penalty methodology described in this chapter provides much of the insight needed for these more advanced techniques. It is also simple to work with and typically yields desirable solution quality and convergence time.

Appendix A — Bounded Moment Convergence Theorem

This appendix provides a *bounded moment convergence theorem* that is often more convenient than the standard Lebesgue dominated convergence theorem (see, for example, [42] for the standard Lebesgue dominated convergence theorem). We are unaware of a statement and proof in the literature, so we give one here for completeness. Let $X(t)$ be a random process defined either over non-negative real numbers $t \geq 0$, or discrete time $t \in \{0, 1, 2, \ldots\}$. Recall that $X(t)$ converges *in probability* to a constant x if for all $\epsilon > 0$ we have:

$$\lim_{t \to \infty} Pr[|X(t) - x| > \epsilon] = 0$$

Theorem 2. *Suppose there is a real number x such that $X(t)$ converges to x in probability. Further suppose there are finite constants $C > 0$, $\delta > 0$ such that for all t we have $\mathbb{E}\left[|X(t)|^{1+\delta}\right] \leq C$. Then $\lim_{t \to \infty} \mathbb{E}[X(t)] = x$.*

Proof. Without loss of generality, assume $x = 0$ (else, we can define $Y(t) = X(t) - x$). Fix $\epsilon > 0$. By definition of $X(t)$ converging to 0 in probability, we have:

$$\lim_{t \to \infty} Pr[|X(t)| > \epsilon] = 0 \tag{9.58}$$

Further, for all t we have $\mathbb{E}[X(t)] \leq \mathbb{E}[|X(t)|]$, and so:

$$\mathbb{E}[X(t)] \leq \epsilon + \mathbb{E}[|X(t)| \mid |X(t)| > \epsilon] Pr[|X(t)| > \epsilon] \tag{9.59}$$

We want to show the final term in the right-hand-side above converges to 0 when $t \to \infty$. To this end, note that for all t we have:

$$
\begin{aligned}
C &\geq \mathbb{E}\left[|X(t)|^{1+\delta}\right] \\
&\geq \mathbb{E}\left[|X(t)|^{1+\delta} \mid |X(t)| > \epsilon\right] Pr[|X(t)| > \epsilon] \\
&\geq \mathbb{E}\left[|X(t)| \mid |X(t)| > \epsilon\right]^{1+\delta} Pr[|X(t)| > \epsilon]
\end{aligned}
\tag{9.60}
$$

where (9.60) follows by Jensen's inequality applied to the conditional expectation of the function $f(|X(t)|)$, where $f(x) = x^{1+\delta}$ is convex over $x \geq 0$. Multiplying inequality (9.60) by $Pr[|X(t)| > \epsilon]^{\delta}$ yields:

$$
CPr[|X(t)| > \epsilon]^{\delta} \geq \left(\mathbb{E}\left[|X(t)| \mid |X(t)| > \epsilon\right] Pr[|X(t)| > \epsilon]\right)^{1+\delta} \geq 0
$$

Taking a limit of the above as $t \to \infty$ and using (9.58) yields:

$$
0 \geq \lim_{t \to \infty} \left(\mathbb{E}\left[|X(t)| \mid |X(t)| > \epsilon\right] Pr[|X(t)| > \epsilon]\right)^{1+\delta} \geq 0
$$

It follows that:

$$
\lim_{t \to \infty} \mathbb{E}\left[|X(t)| \mid |X(t)| > \epsilon\right] Pr[|X(t)| > \epsilon] = 0
$$

Using this equality and taking a lim sup of (9.59) yields $\limsup_{t \to \infty} \mathbb{E}[X(t)] \leq \epsilon$. This holds for all $\epsilon > 0$, and so $\limsup_{t \to \infty} \mathbb{E}[X(t)] \leq 0$. Similarly, it can be shown that $\liminf_{t \to \infty} \mathbb{E}[X(t)] \geq 0$. Thus, $\lim_{t \to \infty} \mathbb{E}[X(t)] = 0$. □

Recall that convergence with probability 1 is stronger than convergence in probability, and so the above result also holds if $\lim_{t \to \infty} X(t) = x$ with probability 1. Theorem 2 can be applied to the case when $\lim_{K \to \infty} \frac{1}{K} \sum_{k=0}^{K-1} e[k] = \overline{e}$ with probability 1, for some finite constant $\overline{e}$, and when there is a constant C such that $\mathbb{E}\left[e[k]^2\right] \leq C$ for all k. Indeed, one can define $X(K) = \frac{1}{K} \sum_{k=0}^{K-1} e[k]$ for $K \in \{1, 2, 3, \ldots\}$ and use the Cauchy-Schwartz inequality to show $\mathbb{E}\left[X(K)^2\right] \leq C$ for all $K \in \{1, 2, 3, \ldots\}$, and so $\lim_{K \to \infty} \frac{1}{K} \sum_{k=0}^{K-1} \mathbb{E}[e[k]] = \overline{e}$.

Bibliography

[1] R. Agrawal and V. Subramanian. Optimality of certain channel aware scheduling policies. *Proc. 40th Annual Allerton Conference on Communication, Control, and Computing, Monticello, IL*, Oct. 2002.

[2] M. Annavaram, E. Grochowski, and J. Shen. Mitigating amdahl's law through epi throttling. *Proc. 32nd International Symposium on Computer Architecture (ISCA)*, pp. 298-309, June 2005.

[3] S. Boyd and L. Vandenberghe. *Convex Optimization*. Cambridge University Press, 2004.

[4] A. Eryilmaz and R. Srikant. Joint congestion control, routing, and mac for stability and fairness in wireless networks. *IEEE Journal on Selected Areas in Communications, Special Issue on Nonlinear Optimization of Communication Systems*, vol. 14, pp. 1514-1524, Aug. 2006.

[5] A. Eryilmaz and R. Srikant. Fair resource allocation in wireless networks using queue-length-based scheduling and congestion control. *Proc. IEEE INFOCOM*, March 2005.

[6] R. Gallager. *Discrete Stochastic Processes*. Kluwer Academic Publishers, Boston, 1996.

[7] M. Gatzianas, L. Georgiadis, and L. Tassiulas. Control of wireless networks with rechargeable batteries. *IEEE Transactions on Wireless Communications*, vol. 9, no. 2, pp. 581-593, Feb. 2010.

[8] L. Georgiadis, M. J. Neely, and L. Tassiulas. Resource allocation and cross-layer control in wireless networks. *Foundations and Trends in Networking*, vol. 1, no. 1, pp. 1-149, 2006.

[9] E. Grochowski, R. Ronen, J. Shen, and H. Wang. Best of both latency and throughput. *Proc. IEEE Conf. on Computer Design (ICCD)*, pp. 236-243, Oct. 2004.

[10] I. Hou, V. Borkar, and P. R. Kumar. A theory of qos for wireless. *Proc. IEEE INFOCOM*, April 2009.

[11] I. Hou and P. R. Kumar. Utility maximization for delay constrained qos in wireless. *Proc. IEEE INFOCOM*, March 2010.

[12] L. Huang and M. J. Neely. Utility optimal scheduling in processing networks. *Proc. IFIP, Performance*, 2011.

[13] L. Huang and M. J. Neely. Utility optimal scheduling in energy harvesting networks. *Proc. Mobihoc*, May 2011.

[14] L. Jiang and J. Walrand. *Scheduling and Congestion Control for Wireless and Processing Networks*. Morgan & Claypool, 2010.

[15] F. Kelly. Charging and rate control for elastic traffic. *European Transactions on Telecommunications*, vol. 8, no. 1 pp. 33-37, Jan.-Feb. 1997.

[16] F.P. Kelly, A. Maulloo, and D. Tan. Rate control for communication networks: Shadow prices, proportional fairness, and stability. *Journ. of the Operational Res. Society*, vol. 49, no. 3, pp. 237-252, March 1998.

[17] H. Kushner and P. Whiting. Asymptotic properties of proportional-fair sharing algorithms. *Proc. of 40th Annual Allerton Conf. on Communication, Control, and Computing*, 2002.

[18] C. Li and M. J. Neely. Network utility maximization over partially observable markovian channels. *Arxiv Technical Report: arXiv:1008.3421v1*, Aug. 2010.

[19] C. Li and M. J. Neely. Network utility maximization over partially observable markovian channels. *Proc. Intl. Symposium on Modeling and Optimization in Mobile, Ad Hoc, and Wireless Networks (WiOpt)*, May 2011.

[20] Q. Li and R. Negi. Scheduling in wireless networks under uncertainties: A greedy primal-dual approach. *Arxiv Technical Report: arXiv:1001:2050v2*, June 2010.

[21] X. Lin and N. B. Shroff. Joint rate control and scheduling in multihop wireless networks. *Proc. of 43rd IEEE Conf. on Decision and Control, Paradise Island, Bahamas*, Dec. 2004.

[22] S. H. Low. A duality model of TCP and queue management algorithms. *IEEE Trans. on Networking*, vol. 11, no. 4, pp. 525-536, August 2003.

[23] J. Mo and J. Walrand. Fair end-to-end window-based congestion control. *IEEE/ACM Transactions on Networking*, vol. 8, no. 5, Oct. 2000.

[24] M. J. Neely. *Dynamic Power Allocation and Routing for Satellite and Wireless Networks with Time Varying Channels*. PhD thesis, Massachusetts Institute of Technology, LIDS, 2003.

[25] M. J. Neely. *Stochastic Network Optimization with Application to Communication and Queueing Systems*. Morgan & Claypool, 2010.

[26] M. J. Neely. Opportunistic scheduling with worst case delay guarantees in single and multi-hop networks. *Proc. IEEE INFOCOM*, 2011.

[27] M. J. Neely. Stock market trading via stochastic network optimization. *Proc. IEEE Conference on Decision and Control (CDC)*, Atlanta, GA, Dec. 2010.

[28] M. J. Neely. Energy optimal control for time varying wireless networks. *IEEE Transactions on Information Theory*, vol. 52, no. 7, pp. 2915-2934, July 2006.

[29] M. J. Neely. Dynamic optimization and learning for renewal systems. *Proc. Asilomar Conf. on Signals, Systems, and Computers*, Nov. 2010.

[30] M. J. Neely. Queue stability and probability 1 convergence via lyapunov optimization. *Arxiv Technical Report, arXiv:1008.3519v2*, Oct. 2010.

[31] M. J. Neely and L. Huang. Dynamic product assembly and inventory control for maximum profit. *Proc. IEEE Conf. on Decision and Control (CDC)*, Atlanta, GA, Dec. 2010.

[32] M. J. Neely, E. Modiano, and C. Li. Fairness and optimal stochastic control for heterogeneous networks. *Proc. IEEE INFOCOM*, March 2005.

[33] M. J. Neely, A. S. Tehrani, and A. G. Dimakis. Efficient algorithms for renewable energy allocation to delay tolerant consumers. *1st IEEE International Conference on Smart Grid Communications*, Oct. 2010.

[34] A. Ribeiro and G. B. Giannakis. Separation principles in wireless networking. *IEEE Transactions on Information Theory*, vol. 56, no. 9, pp. 4488-4505, Sept. 2010.

[35] S. Ross. *Introduction to Probability Models*. Academic Press, 8th edition, Dec. 2002.

[36] A. Stolyar. Maximizing queueing network utility subject to stability: Greedy primal-dual algorithm. *Queueing Systems*, vol. 50, no. 4, pp. 401-457, 2005.

[37] A. Tang, J. Wang, and S. Low. Is fair allocation always inefficient. *Proc. IEEE INFOCOM*, March 2004.

[38] L. Tassiulas and A. Ephremides. Stability properties of constrained queueing systems and scheduling policies for maximum throughput in multihop radio networks. *IEEE Transacations on Automatic Control*, vol. 37, no. 12, pp. 1936-1948, Dec. 1992.

[39] L. Tassiulas and A. Ephremides. Dynamic server allocation to parallel queues with randomly varying connectivity. *IEEE Transactions on Information Theory*, vol. 39, no. 2, pp. 466-478, March 1993.

[40] R. Urgaonkar, B. Urgaonkar, M. J. Neely, and A. Sivasubramaniam. Optimal power cost management using stored energy in data centers. *Proc. SIGMETRICS*, June 2011.

[41] W.-H. Wang, M. Palaniswami, and S. H. Low. Application-oriented flow control: Fundamentals, algorithms, and fairness. *IEEE/ACM Transactions on Networking*, vol. 14, no. 6, Dec. 2006.

[42] D. Williams. *Probability with Martingales*. Cambridge Mathematical Textbooks, Cambridge University Press, 1991.

[43] L. Xiao, M. Johansson, and S. P. Boyd. Simultaneous routing and resource allocation via dual decomposition. *IEEE Transactions on Communications*, vol. 52, no. 7, pp. 1136-1144, July 2004.

Part III

Smart Grid Communications and Networking

Chapter 10

Smart Grid Communication Network and Its Applications

Zhao Li, Dmitry Ishchenko, Fang Yang and *Yanzhu Ye
ABB US Corp. Research Center, USA, {Zhao.Li, Dmitry.Ishchenko, Fang.Yang}@us.abb.com
*Energy Management Department, NEC Laboratories America, Inc. Cupertino, CA yanzhuye@sv.nec-labs.com

Abstract

This chapter briefly reviews the recent development of utility communication networks, including the advanced metering infrastructure (AMI) and the supervisory control and data acquisition (SCADA). The standardizations of communication protocols in both AMI and SCADA systems are the major focus of this chapter. In addition, some potential grid management applications that are facilitated by the real-time communication mechanism and enable efficient grid operations are also discussed.

10.1 Introduction

A promising solution for challenges presented by rising global energy demand, aging infrastructures, increasing fuel costs and renewable portfolio standards in light of climate change is the smart grid [1]. Technically the smart grid is the convergence of monitoring and control technology applied to the electric grid, the primary objective of which is to optimize consumer energy usage based on environmental and/or price preferences. The smart grid allows sustainable options to customers and improves security, reliability and efficiency for utilities [2], [3].

Since one of primary objectives of the smart grid is to improve energy efficiency, it is often referred to as the "green grid." Two green features of the smart grid are that it maximizes the utilization of renewable resources (e.g., solar and wind) and improves the efficiency of power consumption through dynamic pricing. However, because of the intermittence of renewable resources and dynamic price driven power consumption, both power production and consumption are dynamic and unpredictable, complicating grid management.

To better manage the smart grid, utility control centers must be able to monitor and control the grid in real-time. Hence, building a large scale real-time communication system becomes one of the important pre-requests for constructing smart grid systems [4].

Considered as an extension of the Internet, the smart grid is the so-called "Internet of things" [5], [6], in which connected machines and/or devices communicate with each other across society's infrastructure. The smart grid is not only an information transportation system but also a closed loop online monitoring and control system. Its duty is to balance the power production and power consumption of a grid constrained by Kirchhoff's law under environments in which power production and consumption are dynamic. Hence, the smart grid consists of two types of information flows: monitoring flow and control flow. Through the monitoring flow, utility control centers collect measurements from end devices (e.g., residential smart meters and sensors); Through the control flow, a center sends control commands to individual end devices. As transported messages always carry important information, particularly when the grid is affected by a big disturbance, the utility network should ensure the instantaneous deliverability of important messages. If a control center fails to deliver or allows the delay of such a message, it loses the opportunity to govern such a disturbance and potentially renders the grid system unstable and unreliable.

Nowadays, utility communication networks roughly fall into two categories: supervisory control and data acquisition (SCADA) [7], [8], a real-time communication network (e.g., SCADA) and advanced metering infrastructure (AMI) [9] and advanced metering reading (AMR) network, a non-real-time communication network. In the past several decades, SCADA has played a key role in the online monitoring and control of the grid system. However, in practice, SCADA has been constrained by high construction costs, so it has been deployed in only a small part of the grid system (e.g., the power transmission system) for the real-time monitoring and control of important devices. By contrast, because of its relatively low construction cost, AMI has been widely deployed in the grid system, reaching the feeder and residential customer levels. Rather than functioning as a monitoring and control network, AMI primarily collects energy usage automatically from residential smart meters and transports it back to the control center on a monthly basis. Therefore, it improves the efficiency of the data collection process and the quality of collected measurements, further improving the quality of customer service.

As discussed, building a large-scale and real-time communication network that reaches both the feeder and residential levels is one of the important objectives of the smart grid. To reach this goal, researchers have taken the following two directions in their studies of the smart grid communication: extending the scope of SCADA and/or enhancing the real-time capability of AMI. However, with the scope of the monitored network reaching the feeder and residential level, the number of monitored data points is significantly increased, reaching millions. Therefore, handling such a large amount of data has posed considerable challenges for both strategies.

In this chapter, we review the efforts to build real-time and near real-time smart grid communication networks in the past few years. Overall we primarily focus on efforts on extending the scope of the existing SCADA [11] system and enhancing the real-time capability of the AMI [12]. The rest of this chapter is structured as follows: focusing on extending AMI infrastructure, the second section introduces components of AMI systems, discusses the standardization of the AMI protocol and reviews some potential distribution management applications [10] triggered by real-time and near real-time AMI infrastructures. The third section discusses SCADA, the legacy real-time utility communication system, and its standardization process in the context of the smart grid, based on which the application of distribution automation is discussed. The fourth section concludes the chapter.

10.2 AMI and Its Applications

10.2.1 Background

The AMI (advanced metering infrastructure) [12] consists of metering, communication, and data management functions, offering the two-way transportation of customer energy usage data and meter control signals between customers and utility control centers. The AMI was originally developed from advanced meter reading (AMR) [14], [15], [16], [17], and [18], a one-way communication infrastructure that implements automatically collection of meter measurements from residential smart meters to utility control centers for calculating monthly bills and fulfilling other related activities. Partially as the next generation of "AMR," AMI not only enhances the traditional data collection functionality (i.e., improving monthly meter data collection to real time or near real time meter data collection) but also develops the remote control capability from the control center to smart meters.

In the past few years, motivated by the economic stimulus plan of the U.S. government, most U.S. states have begun the process of deploying smart meters within AMI infrastructures. At the beginning of 2009, for example, Texas initiated a project of deploying six million smart meters and expected to complete it by 2012; and California plans to install 10 million smart meters by the end of 2012. The deployment of smart meters is taking place not only in the United States but throughout the world. Based on current estimates, by 2015, smart meter installations are expected to reach 250 million worldwide [19].

10.2.2 The AMI Infrastructure

This section briefly reviews the components of the AMI infrastructure, including the meter system, the communication network and the meter data management system.

(a) Electromechanical meter

(b) Solid-State Meter

Figure 10.1: Electromechanical meter and solid-state meter [20].

10.2.2.1 The AMI Metering System

As the end device of the AMI infrastructure, the AMI metering system refers to all electricity meters that have been installed at customer sites, performing both measuring and communication functions. AMI metering systems fall into two categories: electromechanical meters and digital solid-state electricity meters (Fig. 10.1). The electromechanical meter is antiquated technology now being replaced by digital solid-state electricity meters.

A solid-state electricity meter, a meter constructed by digital signal processing technologies, is essentially a computer system that utilizes a microprocessor to convert analog signals to digital signals and further processes these digital signals into user-friendly results. Adding a new function to solid-state meters is as easy as installing a new application in a general computer. Hence, its functionalities can be easily expanded to adapt to various application scenarios. For example, beyond the traditional kWh consumption measurement, a solid-state meter provides demand interval information, time-of-use (TOU), load profile recording, voltage monitoring, reverse flow and tamper detection, power outage notification, a service control switch and other applications.

To communicate with other smart meters or utility control centers, a smart meter is generally equipped with a communication module. Popular communication modules in the current market are: low-power radios, the global system for mobile communications (GSM), general packet radio services (GPRS), Bluetooth and others (Table 10.1). In general, each AMI vendor develops its own proprietary communication modules that, in most cases, are not interoperable with the communication modules produced by other vendors.

10.2.2.2 AMI Communication Network

The AMI communication network is a two-way data transportation channel that transports meter measurements and meter control signals back and forth between individual meters and utility control centers. Generally, a classic AMI communication network can be classified into three layers: the home area net-

Table 10.1: Communication Modules Supported by AMI Vendors

AMI Vendors	Communication Modules
Landis + Gyr	Unlicensed RF, PLC
Itron	Zigbee, unlicensed RF, public carrier network (OpenWay)
Elster	Unlicensed RF, public carrier network
Echelon	PLC, RF, Ethernet
GE	PLC, public carrier network, RF
Sensus	Licensed RF (FlexNet)
Eka	Unlicensed RF (EkaNet)
SmartSynch	Public carrier network
Tantalus	RF(TUNet)
Triliant	ZigBee, public wireless network

work (HAN), the meter local area network (LAN) and the wide area network (WAN) [6].

WAN (Wide Area Network)

As the highest-level of aggregation in an AMI network, the WAN handles connectivity between a high-level meter data collector and a utility control center or among high-level meter data collectors. The WAN is the backbone of the AMI communication network, through which numerous AMI measurements and control signals are transported.

Meter LAN (Local Area Network)

This distribution network handles connectivity between data concentrators or some distribution automation devices (e.g., monitors, re-closers, switches, capacitor controllers) and high-level data collectors. Compared with the WAN, the meter LAN has larger geographical coverage, but less data transportation.

HAN (Home Area Network)

For utilities, the HAN has been defined or viewed as a grouping of home appliances and consumer electronic devices that allow for remote interface, analysis, control and maintenance. The electric meter acts as the gateway of the HAN: collecting measurements (e.g., electricity, water and gas) and sending them to the utility control center while executing control commands received from the utility control center.

The WAN, the LAN and the HAN are generally constructed by wired and wireless network technologies. In the current AMI communication network, while widely applied wired communication technologies include communication via telephone systems, Ethernet, power line carriers, and broadband over power lines, widely applied wireless technologies include communication via mobile systems, cellular networks, and wireless mesh networks. Table 10.2 demonstrates the features of above network technologies in the current market.

The various layers of the AMI network require different performance, cov-

Table 10.2: Features of Some Classic Communication Technologies

	Name	Data Rate	Range
Wired	PLC	100K bps	Same with power network
	BPL	< 200M bps	Same with power network
	Fiber Optic	10~40Gbps	30~50 miles with repeaters
Wireless	WiMAX	<70M bps	Up to 30 miles
	WiFi	11~54Mbps	<100m
	Zigbee (802.15.4)	20k~250kbps	< 1mile

Key: PLC Power Line Communication,
BPL Broadband over PowerLine [20]

erage, and security, so they are constructed by various wired or wireless communication technologies. For the HAN, which requires self-healing, plug-in-play, low power and low cost, ZigBee is the preferred technology. For the LAN, which requires good coverage and relatively low performance, PLC, unlicensed spectrum radio and WiFi are likely choices. For the WAN, which requires both high performance and good coverage, BPL, WiMAX, and the licensed/unlicensed spectrum radio are preferable.

10.2.2.3 The Meter Data Management System

The meter data management system (MDMS) of the AMI provides a set of advanced software tools that manage large volumes of meter data: It collects, validates, and stores meter data in a central data repository and allows utilities to take full advantage of AMI information in the following activities: network monitoring, load research, operational analyses and decision making. In addition, it also enables meter data to be shared with end customers, who can access the data whenever they wish to make decisions about how and when they use energy.

Since MDMS collects meter readings from millions of meters at a certain time interval (i.e., 15 minutes), the volume of meter data will increase exponentially. Therefore, the challenge for MDMS is to store and manage the huge amount of data and then extract valuable information from them to support various utility applications, tasks that cannot be controlled using traditional database technologies. However, a well-defined solution that provides sufficient scalability to manage such an enormous meter data set is still in the development phase in both theory and practice.

10.2.3 The Standardization of the AMI Infrastructure

In the current market, smart meters from different vendors are generally non-interoperable. However, for most utilities, deploying millions of smart meters is a long-term investment indeed, once a utility adopts smart meters from an AMI vendor, it must follow up with related products from the same vendor for the sake of compatibility. However, because utilities are reluctant to commit to a certain meter vendor, particularly in the early stages of smart grid development, enabling interoperability between AMI products from different vendors is an effective way to protect utilities' investment. Therefore, most important standard committees in the world (e.g., AEIC, ANSI, EPRI and NIST) are currently attempting to meet the needs of utilities for interoperability. Table 10.3 lists popular standard communication protocols and meter information models in the current market, separately defined by ANSI, IEC and IEEE. Most of the above standards have recently been revised (e.g., C12.18 and C12.19) or newly defined (e.g., C12.22) to support new requirements (i.e., demands responses) from the smart grid [13].

Generally, the standardization of the AMI infrastructure includes the standardization of both AMI communication protocols and AMI information models.

10.2.3.1   The Standardization of AMI Communication Protocols

In the past two years, the focus of the standardization of AMI communication protocols has gradually shifted from the physical level (e.g., ANSI C12.18 [21]) and the device level (e.g., ANSI C12.21 [23]) to the application level (e.g., ANSI C12.22 [24]) because application level communication protocols effectively isolate the details of underlying physical network configurations and implementation.

In the following section, we introduce two application level communication protocols that are popular in both the United States (i.e., C12.19 [22] and C12.22) and European markets (i.e., IEC 62056-53 and IEC 62056-62).

ANSI C12.22

Historically, after a set of standard table contents and formats were defined in ANSI C12.19 (the details for the C12.19 standard are discussed in the next section), a point-to-point standard protocol (ANSI C12.18) that transported the table data over an optical connection was developed. Afterwards, the "Protocol Specification for Telephone Modem Communication" (ANSI C12.21), which allowed devices to transport tables over telephone modems was developed. The C12.22 standard, expanding on the concepts of both the ANSI C12.18, and C12.21 standards, allows the transport of table data over any reliable networking communications system.

The goal of the ANSI C12.22 standard is to define a meshed network infrastructure customized for AMI applications. The standard contains the following functionalities:

To define a Datagram that may convey ANSI C12.19 data tables through any network, which must include the AMI network and optionally includes the Internet

Table 10.3: Popular Standard Information Model and Communication Protocols

	Name	Time to Market	Category	Content	Domain
ANSI	C12.19 [22]	2005	Information model	Utility industry end device data tables	Gas, Water, and Electricity
	C12.22 [24]	2008	Communication protocol	Protocol specification for interfacing to data communication networks	Gas, Water, and Electricity
	C12.18 [21]	2005	Communication protocol	Protocol specification for ANSI Type 2 optical port	Gas, Water, and Electricity
	C12.21 [24]	2005	Communication protocol	Protocol for telephone modern communication	Gas, Water, and Electricity
IEC	61968-9 [25]	2009	Information model	Meter data model in power distribution system	Electricity only
	62065-53 [27]	2007	Communication protocol	COSEM application layer protocol	Gas, Water, and Electricity
	62056-61 [28]	2007	Information protocol	Meter object identification system	Gas, Water, and Electricity
	62056-62 [29]	2007	Information protocol	Interface for data exchanging	Gas, Water, and Electricity
IEEE	1701 [30]	2011	Communication protocol	Optical Port Communication Protocol (compatible with C12.18)	Gas, Water, and Electricity
	1702 [31]	2011	Communication protocol	Telephone modern communication protocol (compatible with C12.21)	Gas, Water, and Electricity

Table 10.3: *Continued.*

P1377/D9 [32]	2011	Information model	End device data tables (compatible with C12.19)	Gas, Water, and Electricity
P1703/D8 [33]	2011	Communication protocol	LAN/WAN Node communication protocol (compatible with C12.22)	Gas, Water, and Electricity

Key:
>   **ANSI** – American National Standards Institute
>   **IEC** – International Electrotechnical Commission
>   **IEEE** – Institute of Electrical and Electronics Engineers

To provide a seven-layer communication infrastructure for interfacing a C12.22 device with a C12.22 communication module

To provide an infrastructure for point-to-point communication that will be used over local ports such as optical ports, or modems

To provide an infrastructure for efficient one-way messaging

The ANSI C12.22 mesh network consists of the C12.22 nodes and network. A C12.22 node, a point on the network that attaches to a C12.22 network, is a combination of both a C12.22 device and communication module. The C12.22 communication module is a hardware module that attaches a C12.22 device to a C12.22 network. The C12.22 device contains meter data in the forms of tables defined in the C12.19. The interface between the communication module and the device is completely defined by the C12.22 standard.

The C12.22 network defines an AMI specific mesh communication infrastructure that consists of one or more C12.22 network segments (a sub-network) or a C12.22 LAN (Fig. 10.2). Within a network segment, a collection of C12.22 nodes communicates without forwarding messages through either a C12.22 relay or a C12.22 gateway. The C12.22 network segments can be connected into a C12.22 WAN through C12.22 relays and gateways, where meters from various network segments can communicate with one another.

Similar to the open system interconnection (OSI) model, the C12.22 communication protocol consists of seven layers (Fig. 10.3): an application layer (layer 7), a presentation layer (layer 6), a session layer (layer 5), a transport layer (layer 4), a network layer (layer 3), a data link layer (layer 2), and a physical layer (layer 1). Unlike OSI, C12.22 is customized for meter data transportation. For example, the application layer of C12.22 supports only ANSI C12.19 tables, EPSEM and ACSE (EPSEM and ACSE are languages that encapsulate C12.19 meter data). The standard services provided by the

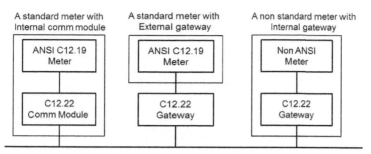

A standard meter with Internal comm module

A standard meter with External gateway

A non standard meter with Internal gateway

| ANSI C12.19 Meter | ANSI C12.19 Meter | Non ANSI Meter |
| C12.22 Comm Module | C12.22 Gateway | C12.22 Gateway |

C12.22 Network Segment

Figure 10.2: C12.22 mesh communication network.

layer seven of C12.22 include an identification service, a read service, a write service, a security service, a trace service, and others; layers 1 through 6 support various physical network connections in the meter industry as well as the standard Internet connection.

IEC62056 [26]

IEC62056, which defines the meter interface classes for the Companion Specification for the Energy Metering (COSEM) model, includes a series of standards on data exchange for meter reading, tariffs and load control. Similar to ANSI C12.22, IEC62056-53, the application layer communication protocol in the COSEM model, is defined based on the several other IEC62056 [26] series protocols, including IEC62056-21, 42, 46, and 47. Except for the IEC62056-21, which is used in hand-held devices for locally exchanging data with meters, the remaining protocols define the various layers of the communication network that support application level communication: the physical layer (IEC62056-42), the data link layer (IEC62056-46), and the transport layer (IEC62056-47). Similar to ANSI C12.22, the meter data carried by IEC62056-53 are defined by IEC62056-61 and IEC62056-62, which are dedicated meter data models in the IEC62056 series (Fig. 10.4). As an application layer communication protocol, IEC62056-53 primarily provides three services to application-level semantics: the GET service (.request, .confirm), the SET service (.request, .confirm), and the ACTION service (.request, .confirm).

Although both IEC62056 and ANSI C12.22 provide a way of constructing the advanced mesh AMI network, each has a unique market focus: IEC62056 primarily focuses on the European market, while ANSI C12.22 focuses on the North American market. In the current North American market, most AMI vendors support C12.18 and C12.21, but few support C12.22 since it has only recently been defined. Itron [35], Elstor [36], and Trilliant [37] were the pioneers supporting the C12 communication protocols. Because of the advantages of C12.22, we predict that in the near future, most major meter vendors will support C12.22 standard communication protocols in the North American market.

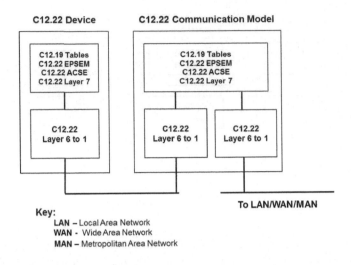

Figure 10.3: The interconnection model of the C12.22 communication protocol.

10.2.3.2   The Standardized AMI Information Model

An information model is a representation of concepts, relationships, constraints, rules, and operations that specify data semantics for a chosen domain of discourse [38]. In the AMI communication infrastructure, it is necessary that an information model, in which all communication participants can semantically reach a certain level of understanding, be maintained.

In this section, we briefly discuss major standard information models in today's market: ANSI C12.19 and IEC62056-62. The former is widely used in the North American market and the latter in the European market.

ANSI C12.19-2008

ANSI C12.19 resulted from comprehensive cooperative effort among utilities, meter manufacturers, automated meter reading service companies, the ANSI, Measurement Canada (for Industry Canada), NEMA, the IEEE, Utilimetrics and other interested parties. Currently, it has two versions: ANSI C12.19-1997 and ANSI C12.19-2008. As the latter is intended to accommodate the concepts of the most recently identified advanced metering infrastructure, it is primarily the focus of this chapter.

The heart of ANSI C12.19 is a set of defined standard tables and procedures: The former are methods of storing the collected meter data and controlling parameters, and the latter are methods of invoking certain actions against the above data and parameters [22]. The standard tables in C12.19 are typically classified into sections, referred to as "decades." Each decade pertains to a particular feature set and a related function. Transferring data from or to an end device that adheres to the C12.19 standard entails reading

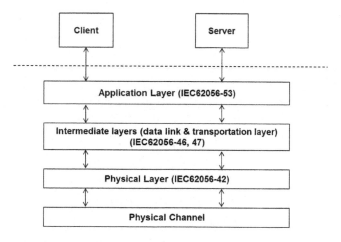

Figure 10.4: The layer based communication model of IEC62056.

or writing a particular table or a portion of a table. Even though the C12.19 standard covers a broader range of tables and procedures, it is highly unlikely that any smart meter will be able to embed all of the tables or even a majority of those defined in ANSI C12.19. Hence, implementers are encouraged to choose an appropriate subset that suits their needs.

The C12.19 standard is a general meter information model that serves various domains, including electricity, water, and gas. As an example, Fig. 10.5 illustrates the electricity information abstracted from the tables defined in Decade 1 of the C12.19 standard. In addition, the tables in C12.19 can be customized through several standard operations.

IEC 62056-62

Unlike ANSI C12-19, which uses tables to package meter measurements, IEC 62056-62 models meter information through a series of interface classes. As the information modeled by C12.19 and IEC 62056-62 are nearly identical, we do not duplicate our efforts to further introduce the content of IEC 62056-62. Similar to ANSI C12.19, as a general meter data model, IEC 62056 supports not only electricity meters but also gas and water meters.

For AMI vendors, the preference to support certain standards reveals a strong geographical bias. For example, most smart meter vendors in the North American market are more likely to choose ANSI series standards (i.e., C12.19 and C12.22) while those in the European market are more likely to select IEC standards. As of today, only Elster completely supports IEC62056 series standards, including IEC62056-42, 46, 53, 61, and 62. Other vendors, such as Itron support only a portion of the IEC 62056 standards, and some such as GE and Sensus do not support the series protocols at all.

Triggered by the rapid development of the smart grid, beyond supporting proprietary communication protocols, most AMI vendors have begun to support the standard communication protocols and meter data models. As of today, most AMI vendors have accepted C12.19 and C12.22 (Table 10.4), standards for the future AMI network.

In summary, AMI is a two-way communication network ranging from residential houses to control centers. As an information provider, it is complementary, to some extent, to DMS providing real-time or near real-time system state information, and as a command executor, conducting control commands sent from the utility control centers to residential smart meters. As real or near real-time system state information can significantly improve the quality of DMS applications, the integration of the AMI with the DMS may represent a feasible, efficient solution for improving the quality of DMS applications.

10.2.4　Advanced DMS Applications

Having evolved from a traditional passive system to a more proactive smart grid, the AMI is one of the enabler technologies for the distribution system. In particular, the AMI system can provide a large amount of measurements and two-way communication systems that facilitate the enhancement of existing distribution management applications and the development of new smart applications. Some examples of new enhanced applications enabled by AMI include distribution system state estimation, advanced outage management and demand response. These advanced applications are essential to the enhancement of distribution system monitoring, analysis and automation control techniques. This section discusses the functionalities of these advanced applications and their requirements for the AMI system.

10.2.4.1　Distribution System State Estimation

Distribution system state estimation (DSSE) refers to the procedure in which the best estimation of a system state by statistical methods given the system three-phase detailed model and a set of redundant system measurements. Generally, a DSSE solution attempts to provide information that (1) determines whether the system state can be computed from existing measurement data, (2) detects, identifies, and rejects bad measurement data, (3) filters out measurement errors and computes the system state with minimal error and (4) evaluates the quality of the state estimation results.

Historically, constrained by economic and other factors, measurement instruments such as RTUs and other sensors could not be installed in every location that required measurements. As a result, distribution systems usually do not have enough measurements to execute DSSE. Under this circumstance, the distribution management system (DMS) is used to apply load allocation and power flow functions as an alternative solution for estimating the operating state of a distribution grid. The load allocation assumes that the loading level on each service transformer is proportional to its nominal rating and distributes the total power demand measured at the feeder head to every service transformer accordingly. The assumption behind load allocation may be less than accurate in practice. Therefore, this alternative solution may produce results that do not best represent the system operation status.

In recent years, motivated by the trend toward the smart grid, many utilities have been installing and/or upgrading advanced metering infrastructures to improve their communication mechanisms. The AMI system with millions of meters located at customer sites can provide a large number of customer level measurements, enabling the comprehensive observation of the distribution network. Because of the availability of a large amount meter dataset, DSSE can produce the best estimate of system operating states. As of today, although large-scale smart meters can provide redundant measurements for DSSE purposes, because of the performance limitations on communication channels, most meter data cannot be transmitted to the utility control center in real-time or near real-time. In fact, the delay may be as much as 24 hours. Current DSSE research considers meter data only pseudo measurements with low quality compared to the SCADA data and data from other sensors with fast communication capability.

A DSSE application usually executes periodically, such as every 5 to 15 minutes. To take the most advantage of meter data in a DSSE application, the AMI communication system must be able to transmit an enormous quantity of meter measurements, including voltage, power, energy consumption and time stamp in real or near real time. The high performance of the AMI communication system provides the DSSE application with high-quality meter measurements that are crucial to facilitating the timely and accurate monitoring of the distribution system operating conditions. In addition, it builds a solid foundation on which various other smart grid technologies, such as online voltage/var control, load balancing and service restoration can be implemented.

10.2.4.2 Advanced Outage Management

Between 80% and 90% power system faults occur in the distribution system. After a fault occurs but before crew is dispatched to the field for fault repair and service restoration to affected customers (while the faulty circuit is being repaired), the location of the outage root (a protective device or the open conductor) and the outage area boundary must first be located. This process is usually referred to as "outage scoping analysis" (OSA). As a fundamental function in the outage management system (OMS), the performance and accuracy of an OSA determines how efficient and effective the crews are in being dispatched to particular locations for fault reparation and service restoration tasks.

Because of the very limited real-time (SCADA) information available in the distribution network, the conventional source of primary outage information for conducting an OSA are customer trouble calls (i.e., some customers in the outage area who experience the loss of power supply may call the utility to report the power outage). However, such trouble call-based outage scoping typically leads to a prolonged OSA procedure because of customer absences (e.g., customers are at work or in bed). In addition, the accuracy of the outage scoping analysis results are affected by the number of trouble calls received. In practical terms, the length of time it takes the utility to send a crew into the field often relies on the confirmation of an outage, which may take hours.

The OSA function can be greatly enhanced with the installation of the AMI in the distribution network. Smart meter data from the AMI system can transmit additional outage information to the OSA. In particular, the "last gasp" sent by smart meters before they lose power during an outage event not only carries more precise outage information (e.g., notification of the location) but also arrives at the utility control center earlier than customer trouble calls. In other words, the utility does not have to wait for a sufficient number of trouble calls to initiate the OSA. In addition, the two-way communication channels of the AMI system also allow utility control centers to conduct on-demand polling of a meter status for the purpose of outage confirmation.

When a fault results in a large outage area, many meters in the outage area will send out their "last gasp" message simultaneously, which will impose a burst of communication load on the AMI communication system. Therefore, enhancement of the OSA functionality with meter "last gasp" information largely depends on the performance of the AMI communication system. Thus, the AMI communication system must be able to transmit a large quantity of high-priority meter data to the control center within a very short period of time.

10.2.4.3 Demand Response

A utility (or the demand response service provider) can take a semi-emergency preventive action called a "demand response" (DR), which the utility typically executes during a peak load, when its system approaches full utilization capacity. The result of DR is a change in the level of the energy consumption of end customers in response to changes in power system loading and/or energy prices. In such a situation, a utility can either reduce of shift (postpone) energy consumption to a different time to reduce the total demand at peak load. In this way, the utility can avoid having to purchase power from the spot market or impose forced outages on customers.

In this context, demand responsive customers are either reactive (i.e., responding to a signal from the utility) or proactive (i.e., unilaterally taking the initiative as a result of changes in energy prices). The former is the focus of this chapter. In the reactive case, which is referred to as an "incentive-based DR," a set of demand reduction signals (i.e., DR signals) is issued by the utility and communicated to participating customers. These signals could be in the form of either demand reduction requests or mandatory commands. Various types of resources can be utilized under this program, namely, directly controllable loads (mostly at residential customer sites) and loads that can be interrupted or reduced upon receipt of a signal from the utility (mostly at commercial and industrial customer sites).

The successful implementation of a DR requires an efficient two-way communication system between the utility and individual customers. The utility would need to poll meter data from the customers, process it in the DR engine, and generate signals that will be transmitted to the appropriate customer meters accordingly.

10.2.4.4 Dependency on the AMI System

Several typical advanced applications listed above for the smart distribution system have one common feature: Their implementation largely depends on real or near real-time meter information provided by the AMI system.

With regard to DSSE, meter data constitute the majority of the required measurements. Such meter data generally include residential real and reactive power and the voltage magnitude. Ideally, these meter data measurements should be updated in real time for the DSSE application. However, for a practical distribution circuit, this process may involve the communication of millions of meters. Therefore, to ensure efficient and effective outage management, the process requires "last gasp" meter messages that notify the DMS system of the occurrence of an outage event; at the same time, it requires an on-demand polling functionality on a meter that verifies the outage boundary. Both functionalities rely on the AMI system. A large outage can trigger a burst of "last gasp" meter messages sent to the DMS system.

10.3 SCADA – The Utility Monitoring and Control Network and Its Applications

10.3.1 Background

One of the most mature applications of communication technologies by the electrical utilities is telemetry and telecontrol, which are commonly fulfilled by a supervisory control and data acquisition (SCADA) system. Concretely, the primary purpose for a SCADA system is the remote monitoring of the electrical quantities of the electrical grid, transporting the measured data to the network control center and sending the control commands from the controller (or operator) to the remote devices. SCADA has traditionally only been deployed at the transmission level. Stimulated by the smart grid, it has recently been deployed to the distribution level.

In the United States, the smart grid is typically associated with more intelligent and advanced SCADA algorithms as applied primarily to distribution systems (e.g., feeder automation, substation automation) [39]. Similar to AMI, SCADA consists of field/end devices, communication networks, and the SCADA data storage system. More importantly, SCADA mainly focuses on real-time communication.

As SCADA systems for transmission have been in place since the late 1960s and transitioned through several generations of communication technologies and protocols, rationalizing the evolution of the above communication technologies and protocols are the major focus of the following section. In addition, some smart applications (e.g., the feeder automation) based on a modern SCADA system are also discussed.

10.3.2   Components of SCADA

This section provides a review of the field devices or end devices comprising a typical utility SCADA system. Generally, intelligent electronic devices (IEDs) and high precision meters are widely deployed in SCADA systems as end devices. As meters have been introduced in Section 10.2, in this section, we focus on IEDs.

10.3.2.1   Intelligent Electronic Devices (IEDs)

An IED is typically a microprocessor device used for monitoring and control of primary power equipment responsible for taking action when the operating conditions of the power system are functioning abnormally and when not taking a control action would critically damage the costly system components such as high power and high voltage generators or transformers based on certain control logic. Typically IEDs are transmission and distribution protective relays and standard control units for re-closers, switches, voltage regulators capacitor banks and transformer tap changers.

Functionally, the IED can be considered an industrial embedded controller, controlling the primary high voltage/ high power equipment. It can be installed either indoors at the substation or outdoors in the field, (e.g., at the top of the distribution line utility pole). Similar to smart meters, the IED is usually equipped with a standard communication module (e.g., RS-232, RS-485, Modbus, DNP3, and IEC60870-5-101), which makes it easy to integrate with the utility SCADA system.

10.3.2.2   Remote Terminal Units (RTU)

Located between the field devices and the SCADA master, a remote terminal unit (RTU) is a device acting as a master for series-connected field devices [40], the purpose of which is to minimize the number of point-to-point connections from SCADA masters to field devices. Hence, the downstream of RTU is a large number of field devices and the upstream is SCADA masters. The RTU functions as a broker or a data collector, transporting information back and forth between the SCADA masters and the field devices (e.g., the IED and smart meters) primarily using serial communication protocols.

10.3.2.3 Automation Controllers

In recent years, traditional RTUs are being replaced by automation controllers, an industrial computation system equipped with embedded Windows or Linux systems. The automation controllers primarily use Ethernet rather than serial communication protocols to communicate with the IP-based field devices, such as substation IEDs [41]. In addition, wireless communications are also commonly used in conjunction with these controllers to integrate the remote devices, for instance re-closer controls installed on the top of a utility pole, where utility workers have difficulty reaching.

As these automation controllers are often capable of converting data and services among the various utility communication protocols to provide flexibility for integration with SCADA, they are also called "substation gateways," or "protocol converters."

10.3.3 Communication Protocols in SCADA

This section begins by briefly reviewing the development of utility communication protocols, and then it focuses on introducing two of the most commonly used protocols: DNP3, which is employed by a majority of utilities in the United States, and IEC61850, which is currently overtaking IEC60870-5 in Europe and the rest of the world, including North America.

Initially, RS-232 and RS-485 were used as physical interfaces with Modbus, DNP or IEC 60870-5-101 [42] as communication protocols in SCADA. However, since the late 1990s, Ethernet interfaces have become the de facto standard for most of the IEDs, and traditional serial-based protocols have begun to support Ethernet interfaces in their physical layers; for example, Modbus and IEC 60870 evolved into Modbus over TCP and IEC 60870-5-104 [43]. During the above transitions, DNP3 has become the most important utility serial communication protocol in the US and was developed into an IEEE Standard, IEEE 1815, [44] in 2010.

Besides DNP3, IEC61850, a network-based communication protocol, is identified as one of the key standards that lay the foundation of the future Smart Grid framework by the US Federal Energy Regulatory Commission (FERC) in 2010. Released in 2003, IEC 61850 has experienced rapid growth in the past few years and has been widely accepted by utilities, vendors and the research community. Different from traditional SCADA communication protocols, IEC61850 not only defines communication specifications but also delimits an information model that is compatible with the common information model (CIM) [34]. As CIM is a popular information model in the power system domain, applications based on IEC61850 have strong interoperability with most power system applications in state of the art.

The following section briefly reviews DNP3 and IEC61850, two most popular communication protocols in the current market.

294 *Smart Grid Communication Network and Its Applications*

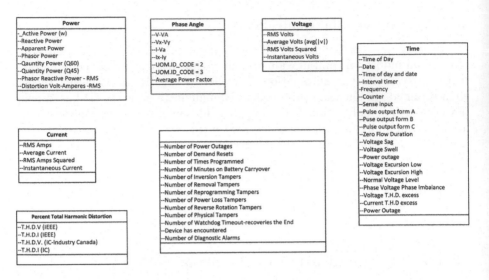

Figure 10.5: Information modeled by ANSI C12.19.

10.3.3.1 Distributed Network Protocol 3 (DNP3)

Developed and maintained by the DNP3 Users Group, an open community for vendors and utilities, DNP3 is recognized as an important utility communication protocol, particularly in the US market. Even though DNP3 supports Ethernet in its physical layer, it still keeps the serial nature in its application layer. Hence, from application's point of view, DNP3 is still a serial communication protocol. Since saving communication bandwidth was one of the most critical goals that the DNP series protocols were trying to achieve, DNP3 is suitable for situations in which communication resources (e.g., the communication bandwidth) are limited. Technically, DNP3 is constructed based on a simplified the seven-layer OSI model, including only three layers: the application layer the data link layer and the physical layer.

Fig. 10.6 demonstrates the master-outstation communication model of DNP3 [44], which includes a master device or information collectors (e.g., an RTU or SCADA master) and one or more outstation devices or end devices in SCADA (e.g., IEDs) connected to each other using serial or Ethernet links. Here, inputs are values collected by outstation and outputs are commands sent by masters. Generally, an outstation contains several arrays of data points in different types, mapping to various internal device parameters in the outstation firmware.

DNP3 defines several Classes to organize the data: Class 0: Static data – the snapshot of the outstation current point values; Classes 1,2,3 – the historical point data from the outstation event buffer. These Classes are assigned to a range of data points in outstations. Additionally, the data points can support different object variations so that besides the actual value of the pa-

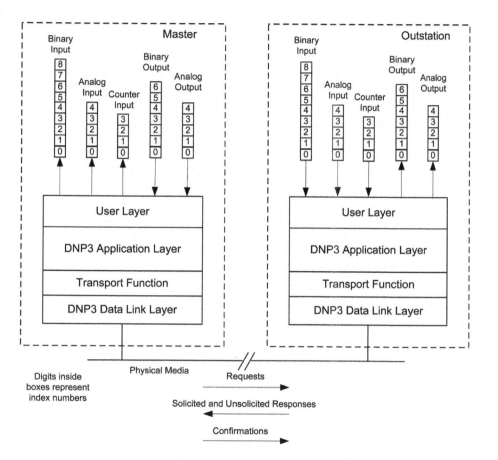

Figure 10.6: DNP3 communication model.

rameter some additional information such as the timestamp for the event is also reported.

To reduce the network traffic, the concept of deadband is also introduced to allow generating events based on monitoring the analog inputs against thresholds (deadbands). An analog input generates an event only when its absolute value changes exceed a predefined threshold (deadband) so that the small fluctuations of the analog value mapped to the analog input are filtered out and will not be reported.

In DNP3, a communication instance is typically initiated by the master sending a request to the outstations sequentially to report the data, which is referred to as polling. Masters can poll the outstation for different Class events. Meanwhile, a mechanism of spontaneous unsolicited reports has also been added to DNP3, through which the outstations can push data to a master. In this case the Outstation initiates the reporting of the event data without the request from the master when it accumulates a pre-determined number of class events. If the original unsolicited reporting attempt failed, the outstation initiates a retry until the application layer confirmation from the master is received or the final confirmation timeout is expired. To save the bandwidth, DNP3 supports the report-by-exception (RBE) function, by which the outstation only reports the data that has changed since the last poll. The RBE function significantly reduces the amount of information that must be transferred, saving the bandwidth.

10.3.3.2   IEC 61850

IEC61850 is maintained by the International Electro-technical Commission Technical Committee 57. Technically, IEC61850 has developed based on utility communications architecture (UCA), an open communications standard for the electric power utilities defined by Electrical Power Research Institute (EPRI) in the early 1990s. Originally, IEC61850 was designed for communications within substations. Afterwards, it expands concepts and principles outlined in the original version beyond the substation fence.

IEC 61850 is a series of specifications, consisting of ten major parts [45], which primarily focus on the basic communication structure (IEC61850-7), specific communication service mapping to MMS (IEC61850-8), specific communication service mapping – sampled values (IEC61850-9) and configuration description language (IEC61850-6).

Based on Extensible Markup Language (XML), the IEC61850 models various field devices of the utility automation system using substation configuration language (SCL) provided by IEC 61850-6, describing substation equipment (e.g., such as settings of substation IEDs), their topology relationships and the communication data exchange procedures using XML schemas and UML diagrams.

Similar to high-level computer programming languages, SCL describes substation objects using the following concepts: Data Attributes (DA), Common

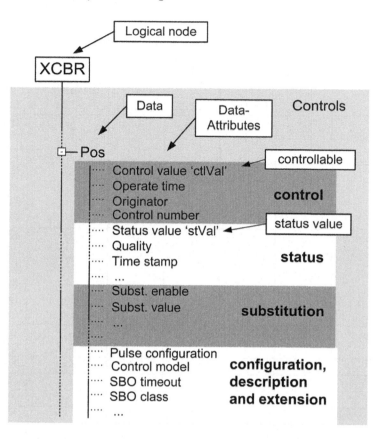

Figure 10.7: Hierarchical structure of IEC61850 object model.

Data Classes (CDC), Logical Nodes (LN), and Logical Devices (LD). Data Attributes define the basic data types (e.g., float, Boolean, integer) in SCL. Constructed by Data Attributes, Common Data Classes define complex data structures that represent objects in the substation automation environment.

A Logical Node, defined in IEC 61850-7-4, typically represents logical functions performed by a physical device (e.g., monitoring, measuring, and control). It is the combination of data objects and their functionalities. In IEC61850, a logical node falls into the following groups: System, Protection, Control and Metering.

A group of Logical Nodes is called a Logical Device. A Physical Device has one or more Logical Devices, which are used to distinguish between various functions performed by the devices, such as protection, control and disturbance monitoring. Fig. 10.7 illustrates the data model for a circuit breaker position defined in IEC61850 [45] using the above concepts.

In IEC61850, the standard object models described in SCL can be accessed through abstract communication service interfaces (ACSI). The ACSI defines

Table 10.4: AMI Vendors Supporting Standard Information Models and Communication Protocols

	C12.18	C12.21	C12.22	C12.19	IEC61968/ CIM	IEC62056/ DLMS/ COSEM
Landis + Gyr	√	√	√	√	√	IEC62056-21 and DLMS
Itron	√	√	√	√	√	IEC62056-21, DLMS/ COSEM
Elster	√	√	√	√	√	IEC62056-42, 46, 53, 61, 62 and DLMS/ COSEM
Echelon	√			√	√	IEC62056-21
GE	√	√	√	√	√	N/A
Sensus	√	√		√		N/A
Eka	√			√		N/A
SmartSynch			√	√		N/A
Tantalus	√	√		√		N/A
Trilliant	√	√	√	√	√	N/A

services on the application layer level as the specific communication service mapping. Some typical basic IEC61850 services are described as follows:

- The SETTING-GROUP-CONTROL BLOCK allows the field device to be switched from one set of preconfigured control parameters to another.

- The REPORT-CONTROL-BLOCK, LOG-CONTROL-BLOCK define the mechanisms for reporting and logging the event data, including immediate real-time reporting, polled reports, integrity scans; buffered and unbuffered reports. The data are reported through the Data Sets that group the data attributes according to the Functional Constraints (i.e., the status or measurements).

- A generic substation event (GSE) and a generic object-oriented substation event (GOOSE) enable fast peer-to-peer communications among the field devices, allowing the implementation of de-centralized control schemes based on local intelligence. A GSE can transmit only binary data, while a GOOSE can also send the analog information arranged in a data set.

- Control class model, defined for performing the control actions, supports both direct control and select-before-operate control with enhanced security.

The protocol profiles of IEC 61850 are defined by the Manufacturing Message Specification (MMS) protocol, the Internet and OSI protocol stacks, which are mainly full OSI seven-layer profiles with associated overhead. In addition, direct usage of the Ethernet (IEEE 802.3x) LAN is also specified for high-speed protection functions, such as GOOSE and the processing of digitized waveforms of high sampling rates (Sampled Values).

10.3.3.3 Discussion

DNP3 is a serial communication protocol, suitable for a low bandwidth. Even after supporting Ethernet on the physical level, DNP3 still maintains the serial manner in its logical level. While IEC61850 is a network-based communication protocol widely used in the substation LAN environment. Compared to IEC61850, DNP3 has low communication overhead and is more suitable for low bandwidth network in practice.

Unlike DNP3, in which data points are grouped according to type, IEC61850 has a well-defined information model that is compatible with CIM, the most popular information model used in the power system domain. However, the last introduced object – the dataset in DNP3 enhances the information model as well. In addition, recent developments in DNP3 enable the creation of an XML-based device profile and the mapping of real-time data to the XML-based information model. To simplify protocol conversion, the DNP3 XML device profile also allows the cross-mapping of the IC61850 objects to DNP3 points.

10.3.4 Distribution Automation

The purpose of distribution automation (DA) is to provide a set of technologies that enable electric utilities to monitor, coordinate and operate distribution components in a real-time mode from remote locations [46]. The DA system consists of individual devices, an intelligent decision system, and communication systems. The communication infrastructure is critical because it enables remote system monitoring and control functions in the automated system. Two examples of distribution automation application are fault detection, isolation, and service restoration (FDIR) and voltage and var control (VVC). This section addresses the functionalities of these two applications and their reliance on the communication system.

10.3.4.1 Fault Detection, Isolation and Service Restoration

As discussed, electric utilities often rely on a trouble call system in which customers can report outages to the utility. Generally, when a fault occurs and customers experience a power outage, customers may call the utility and report the power outage. After receiving a power outage report, the utility may send a crew to the field. The crew investigates the fault location and figures out and implements a switching scheme that first isolates the fault and then restores service to as many of the affected customers as possible while the faulty feeder part is being repaired. This procedure may take several hours to implement.

Nowadays, many utilities have deployed feeder switching devices (e.g., re-closers, circuit breakers and sectionalizers) equipped with IEDs, the intelligent electronic devices that have monitoring, protection, control and communication functionality. The use of IEDs in connection with network communication and distributed control facilitates enables automatic fault detection, isolation and service restoration.

With the automatic FDIR application, the distribution system operating condition can be monitored and controlled by an FDIR controller and multiple IEDs that are associated with switches in substations and along feeders [47]. As IEDs can communicate system information to the controller, when a permanent fault occurs downstream from a recloser after a reclosing sequence, the recloser goes to the lockout status from a normally closed status, and the IED associated with the recloser sends the status change as well as other measurements such as voltage and current magnitudes to the controller in real time. In response, the controller executes FDIR logic, which (1) identifies the location of the fault that causes the lockout of the recloser, (2) determines the switching devices that should be opened so that the faulty feeder section is further isolated and (3) selects appropriate switches that should operate to restore power for the unfaulty out-of-service area. The output of FDIR logic includes a set of switching device status changes, and the commands that switch these devices are then communicated to corresponding IEDs to implement the switch status change accordingly.

By using automatic FDIR technology, we can implement fault detection, isolation and service restoration in a closed loop without human intervention. Throughout the process, FDIR logic automatically alters the topological structure of feeder systems by changing the open/close status of switches with the help of a real or near real-time two-way communication system. Communication between the field IEDs and the controller can be implemented using Modbus, DNP, IEC61850 or any other protocol. In this way, automatic FDIR technology can reduce the duration of power outages resulting from the dispatch of crews to troubleshoot a fault from hours to minutes and significantly improve the reliability of the distribution system reliability level.

10.3.4.2 Voltage and Var Control

Voltage and var control is one of the critical applications in the distribution management system (DMS). Traditionally, its main purpose has been to reduce system energy loss and total demand. The reduction of energy loss is achieved by optimizing the var flow in the distribution network through the control of capacitor banks and other reactive power resources. The reduction of demand is realized by flattening and lowering the overall voltage profile via the control of the substation load tap changer (LTC), line voltage regulators, capacitors and so on.

Voltage and var control is generally autonomous in that each controllable device is equipped with a local controller with sensors that measure the local voltage, the current, the power flow and other electric parameters, and the local controller determines the best course of action based on these local measurements. In this type of control, the control logic and settings are predefined based on operation experience, and the coordination among various device control actions is purely determined by the predefined time delay settings. As a result, VVC does not require a real-time communication mechanism for transmitting measurements as well as the control commands.

Currently, with the increasing penetration level of renewable DERs such as wind and solar resources, the intermittent power output of these types of DERs and the bidirectional power flow brought by the integration of DERs may significantly impact the system voltage and var control. For instance, bidirectional power flow complicates var flow control because of the additional DER control devices as well as the frequent power flow direction changing. In addition, because of the intermittent DER characteristics, the system voltage may exhibit increasing fluctuation and even voltage stability problems, and the voltage control devices tend to operate more often to mitigate these negative effects.

In these new circumstances, the traditional autonomous local control method is no longer capable of satisfying the system operation requirements. Thus, centralized coordinated voltage and var control that can effectively handle the more challenging situation has become essential. Accordingly, the corresponding requirements for communication systems have increased: All information regarding the measurements and controllable device status should be sent to the center controller located in the control center or substations in real time and likewise, the control commands issued from the center controller should be communicated to the controllable devices in real time. This process requires that each control device and any strategic measurement point be equipped with communication modules and the establishment of a real-time communication mechanism between the field devices and measurement points and the control center.

10.4   Conclusion

This chapter briefly reviews the recent development of utility communication networks, including AMI and SCADA.

As a large-scale communication network in the utility, the AMI system has recently received considerable attention. Nowadays, for many utilities, the deployment of the AMI enables real-time or near real-time monitoring and control capabilities in the customer level and accelerates the emergence of advanced grid management applications, such as distribution state estimation, demand response, and outage management, which improve the efficiency of the grid system operation. On the other hand, as the legacy real-time monitoring and control network in utilities, SCADA cannot easily be extended further to the feeder and the customer levels because of its limited bandwidth and the serial-based communication infrastructures which cannot support increased data points caused by such an extension. Complementary to each other, SCADA and AMI systems will coexist for many years to come.

To protect the utility investments and enable the interoperability among devices from various vendors, many new standards of communication protocols have been defined for both AMI and SCADA lately to replace legacy proprietary communication protocols. The new standards outlined not only the standard communication infrastructure (e.g., mesh network infrastructure defined by ANSI C12.12) but also the standard information models (e.g., ANSI C12.19 and IEC61850). Even though the newly defined communication protocols have more advantages than legacy protocols, because of a large number of legacy facilities currently running in grid systems, the legacy communication protocols (e.g., the serial communication protocol) that are supported by these facilities will continue to exist for a relatively long period of time.

Bibliography

[1] Cupp, J. and Beehler, M.E., "Implementing Smart Grid Communications," *TECHBriefs*, no. 4, 2008.

[2] Apel, R., "Energy Automation, Enabling the Intelligent Grid," Next Generation Utilities: Technology Summit, Evian, France, 10/02/2008.

[3] Dugan, R. C. and McGranaghan, M., "Sim City," *IEEE Power & Energy Magazine*, vol. 9, no. 5, Sept./Oct. 2011.

[4] Roy, S., Nordell, D., and Venkata, M. "Lines of Communication," *IEEE Power & Energy Magazine*, vol. 9, no. 5, Sept./Oct. 2011.

[5] Internet of things, http://www.greenbang.com/internet-of-things-is-energys-future_15342.html

[6] Conner, M., "*Sensors empower the "Internet of Things,*" pp. 32–38. ISSN 0012-7515.

[7] Girotti, T.B. Tweed, N.B., and Houser, N.R. "Real-time Var control by SCADA", IEEE Transactions on Power Systems, vol. 5, no. 1, Feb. 1990.

[8] Marihart, D.J. "Communications technology guidelines for EMS/SCADA systems," *IEEE Transactions on Power Delivery*, vol. 16, Apr. 2001.

[9] Bennett, C. and Wicker, S.B. "Decreased time delay and security enhancement recommendations for AMI smart meter networks," *Innovative Smart Grid Technologies* (ISGT), Jan. 2010.

[10] Cassel, W.R., "Distribution management systems: functions and payback," *IEEE Transactions on Power Systems*, vol. 8, issue 3, 1993.

[11] "Supervisory Control and Data Acquisition (SCADA) Systems," Office of the Manager, National Communication System, October 2004.

[12] Yang, F. "Advanced Metering Infrastructure Technology," Technical Report, No. PT-07045, ABB US Corp. Research Center, 2007.

[13] "NIST Framework and Roadmap for Smart Grid Interoperability Standards," Release 1.0 (Draft), Office of the National Coordinator for Smart Grid Interoperability, NIST, 2009.

[14] Fischer, R. A., Laakonen, A. S., and Schulz, N., N., "A Generation Polling Algorithm Using a Wireless AMR System for Restoration Confirmation," *IEEE Transactions on Power Systems.*, vol. 16, no. 2, pp. 312-316, 2001.

[15] Dorey, H., "Advanced Metering in Old and New Worlds," *Power Engineering Journal*, vol. 10, no. 4, pp.146-148, August, 1996.

[16] Jin, Y., and Cox, M. D. "A Pipelined Automatic Meter Reading Scheme," Instrumentation and Measurement Technology Conference, pp. 715-720, 1993.

[17] Mak, S. and Radford, D. "Design Considerations for Implementation of Large Scale Automatic Meter Reading Systems," *IEEE Transactions on Power Delivery*, vol. 10, no. 1, pp. 97-103, 1995.

[18] Clay, M. R. J. and McEntee, A. J. "Advanced Meter Reading Tokenless Prepayment," *Power Engineering Journal*, vol. 10, no. 4, pp. 149-153, August 1996.

[19] "SmartMeter[TM]Installation Progress," PG&E, April 2010, http://www.pge.com/myhome/customerservice/meter/smartmeter/deployment/

[20] Electromechanical meter and solid-state meter http://en.wikipedia.org/wiki/Electric_energy_meter

[21] ANSI C12.18-2006, "American National Standard Protocol Specification for ANSI Type 2 Optical Port," ANSI Inc. 2006.

[22] ANSI C12.19-2008, "American National Standard - Utility Industry End Device Data Tables," ANSI Inc. 2009.

[23] ANSI C12.21-2006, "American National Standard Protocol Specification for Telephone Mode," ANSI Inc. 2006.

[24] ANSI C12.22-2008, "American National Standard - Protocol Specification for Interfacing to Data Communication Networks," ANSI Inc. 2009.

[25] IEC61968-9: "Application integration at electric utilities - System interfaces for distribution management - Part 9: Interfaces for meter reading and control," IEC, 2009.

[26] IEC62056: Electricity metering – Data exchange for meter reading, tariff and load control, IEC62056.

[27] IEC 62056-53: Electricity metering – Data exchange for meter reading, tariff and load control – Part 53: COSEM application layer, IEC 2006.

[28] IEC 62056-61: Electricity metering – Data exchange for meter reading, tariff and load control – Part 61: Object Identification System, IEC 2006.

[29] IEC 62056-62: Electricity metering – Data exchange for meter reading, tariff and load control – Part 62: Interface classes, IEC 2006.

[30] IEEE 1701-2011: IEEE Standard for Optical Port Communication Protocol to Complement the Utility Industry End Device Data Tables, IEEE, 2011.

[31] IEEE 1702-2011: IEEE Standard for Telephone Modem Communication Protocol to Complement the Utility Industry End Device Data Tables, IEEE, 2011.

[32] IEEE P1377/D9, IEEE Draft Standard for Utility Industry Metering Communication Protocol Application Layer (End Device Data Tables), IEEE, 2011.

[33] IEEE P1703/D8, IEEE Draft Standard for Local Area Network/Wide Area Network (LAN/WAN) Node Communication Protocol to complement the Utility Industry End Device Data Tables, IEEE, 2011.

[34] IEC61970, Energy management system application program interface, part 301, common information model (CIM) base, IEC, 2011.

[35] The AMI/AMR solution from Itron Inc. http://www.itron.com/pages/products_category.asp?id=itr_000238.xml

[36] EnergyAxis from Elster Electricity LLC
    http://www.elsterelectricity.com/internet_Content_1.nsf/SResults/
    D72B4A78CC3B0A1B85256DFF006EF2C3

[37] "Trilliant – A Trusted Solution Partner," Solution Brief, Trilliant
    Incorporated, 2009 http://www.trilliantinc.com/4_Rsrcs/_PDFs/TSB_
    TrustedPartner.pdf

[38] Lee, Y. T. "Information modeling from design to implementation" National Institute of Standards and Technology, 1999.

[39] Venkata, S.S., Uluski, R.W. and McGranahan, M. "Critical Elements –
    Distribution Management Systems," *IEEE Power and Energy Magazine*,
    vol. 9, no. 5, September/October 2011.

[40] Clarke, G., and Reynders, D. "Practical Modern SCADA Protocols:
    DNP3, 60870.5 and Related Systems," Elsevier, 2004.

[41] Olovson, H.-E., Werner, T., and Rietman, P. "Next Generation Substations. Impact of the Process Bus," *ABB Review*, Special Report IEC
    61850, 2010.

[42] IEC 60870-5-101: Telecontrol equipment and systems - Part 5-101: Transmission protocols - Companion standard for basic telecontrol tasks, Second edition, IEC, 2003-02.

[43] IEC 60870-5-104: Telecontrol equipment and systems – Part 5-104: Transmission protocols – Network access for IEC 60870-5-101 using standard transport profiles, Second Edition, IEC, 2006-06.

[44] Distributed Network Protocol (DNP3). IEEE Standard 1815-2010, July 2010.

[45] IEC61850, Part 1~10, International Standard, Communication networks and systems for power utility automation, 2003-04.

[46] IEEE PES DA Tutorial, 1998.

[47] Yang, F., Li, Z., Vaibhav, D., Wang, Z., and Stoupis, J. "Graph Theory-Based Feeder Automation Logic for Low-End Controller Application,"
    IEEE PES Generation Meeting, July, 2009.

Chapter 11

Demand and Response in Smart Grid

*Qifen Dong, *Li Yu and +WenZhan Song
*Zhejiang University of Technology, China, qdong@cs.gsu.edu, lyu@zjut.edu.cn
+Georgia State University, USA, wsong@gsu.edu

Smart grid is envisioned as the modernization of the nation's electricity transmission and distribution system to maintain a reliable and secure electricity infrastructure that can meet future demand growth and integrate renewable energy sources. It involves significant new research challenges. Demand and Response (DR), which refers to the dynamic demand mechanisms to manage electricity demand in response to supply conditions, is one of the most important functions of smart grid. DR offers several benefits, including reduction of peak demand, participant financial benefits, integration of renewable resources, and provision of ancillary services. This chapter surveys the ongoing research through elaborating a representative number of DR methods and discusses future directions.

11.1   Demand and Response Overview

Actually, traditional DR mechanisms, such as Real-Time Pricing [1], Critical Peak Pricing [2], Demand Side Bidding [3] and Emergency Demand Response [4], are relatively mature in traditional electricity grids. This section will explain the significance of DR, the traditional DR methods in power grid, and the new requirements of DR in future smart grid are stated.

11.1.1   Significance of Demand Response

Demand Response, defined broadly, is that the users adapt their electricity usages in response to power grid supply conditions, economic signals from a competitive wholesale market or special retail rates [5]. In [6], it is defined more specifically as: *Changes in electric usage by end-use customers from their normal consumption patterns in response to changes in the price of electricity over time, or to incentive payments designed to induce lower electricity use at times of high wholesale market prices or when system reliability is jeopardized.* Raising the temperature of the thermostat in response to short-term high prices, dimming/shutting off lights to match limited available electric energy, and slowing down or stopping production at an industrial operation when the grid system reliability is jeopardized are common examples of DR.

Demand Response has received much attention, because it not only reflects consumers' abilities to reduce electricity consumption when wholesale prices are high or the reliability of the electric grid is threatened, but also improves resource-efficiency of electricity production and social welfare maximization due to closer alignment between customers' electricity charges and the value they place on electricity. Specifically, Demand Response can yield several benefits, which include, but are not limited to:

1. *Peak demand reduction.* The power grid usually needs to provide extra generation, transmission and distribution capacities to cope with the few peak demand hours. For example, in Spain about 4000 MW are required to attend 300 hours of peak consumption per year [7]. It not only increases the operation cost but also causes energy waste. By successful deployment of DR program, electrical load can be shifted from high demand periods to others, thereby flattening the load demand. It improves reliability and operational security of power grid and lowers maintenance cost.

2. *Participant financial benefits.* Participants save on electricity bills by adjusting their demand in response to time-varying electricity rates and

even earn incentive payments because of certain incentive-based programs. In addition, Demand Response contributes to reduced peak demand. Over the longer term, sustained Demand Response lowers aggregate system capacity requirements, allowing electricity utilities and other retail suppliers to purchase or build less new capacity. Eventually these savings may be passed on to most retail customers as bill savings.

3. *Integration of renewable resources.* As various distributed renewable energy resources are increasingly integrated into power system, Demand Response plays an important role in better using these renewable resources. Both demand and renewable energy supplies are dynamic and fluctuating. As pointed out in [8], the objective of a smart grid is not to match the supply to the demand, but in contrast, to match the demand to the available supplies by using DR technology.

4. *Provision of ancillary services.* Using DR to supply ancillary services, i.e., regulation, load following, frequency responsive spinning reserve and supplement reserve, yields several advantages. These benefits include reduced transmission and distribution losses, increased transmission capacity and increased margin to voltage collapse [9].

11.1.2 Demand Response in Traditional Grid

In traditional grid, demand response is classified into two types [10], i.e., price-based demand response and incentive-based demand response. Each contains several typical methods, shown in Fig. 11.1. There are also various DR algorithms developed on these typical methods. Here, we only give a brief introduction of the typical ones. Interested readers can refer to [6, 11, 12] for details.

In price-based demand response programs, customers voluntarily adjust their electricity consumption based on time-varying pricing signals [13], mainly including Time of Use Pricing (TOU), Real Time Pricing (RTP) and Critical Peak Pricing (CPP). Customers can reduce their electricity bills if they adjust the timing of their electricity usage in order to take advantage of lower-priced periods or avoid consuming when prices are higher. The three typical price-based demand response methods are summarized as follows:

1. *Time of Use Pricing.* Electricity prices are fluctuant at different time blocks but fixed in specified periods. The prices are pre-established and known to consumers in advance, allowing them to vary their energy consumption in response to the prices and to manage their energy charges by shifting energy usage to a lower cost period or reducing their consumption. TOU rates reflect the average cost of generating and delivering power during those time periods, and lead customers to change their power consuming mode in order to flatten the load curve. However, the

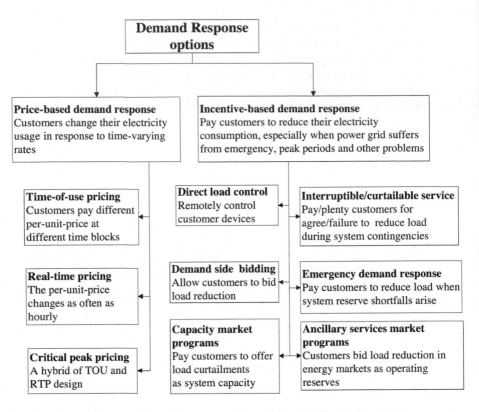

Figure 11.1: Taxonomy of traditional DR methods.

energy prices are mostly based on electricity supply cost of utilities and seldom take the consumer feedbacks to the prices into consideration [14].

2. *Real Time Pricing* [1]. It is an idealized instantaneous dynamic pricing policy. In practice, electricity prices may change as often as hourly (exceptionally more often), reflecting changes in the utility's generation cost and/or the wholesale price of electricity. RTP prices are provided to customers on a day-ahead or hour-ahead basis so that customers can plan energy consumption ahead in response to energy market. However, it increases the cost of installing communication and control equipment.

3. *Critical Peak Pricing* [2]. In power system, critical peak loads occur a few times due to weather or system conditions, but it is a threat to the whole grid system once critical peaks happen. To prevent critical peaks, a CPP method which is a combination of the TOU and RTP design was proposed. In fact, it is an advanced TOU pricing. The difference lies in the improvement that CPP event price under specified trigger conditions (e.g., when system reliability is compromised or supply cost is very high) is much higher than the normal peak price. Customers not

only save money on their electric bills if they can reduce energy demands during these times, but also help to reduce greenhouse gas emissions and defer the construction of additional power plants.

Other incentive-based DR programs reward customers for reducing their electricity consumption relative to some administratively set baselines when the grid operator thinks reliability conditions are compromised. Some of these DR programs even penalize customers who have enrolled but failed to respond or fulfill their contractual commitments when events are declared. These types of DR approaches include [13]:

1. *Direct load control.* It is an approach in which the electricity utilities remotely shut down/start customers' electrical equipment during demand-peak periods [15]. The contracted customers win the incentive payment, usually in the form of credits on their electricity bill. DLC programs are primarily offered to residential or small commercial customers. The controlled electrical devices are those which would not affect customers' normal life a short-term break, e.g., by taking air conditioning, water heaters, pool pumps, clothes dryers.

2. *Demand Side Bidding* [3]. It is a mechanism that encourages customers to participate in the wholesale electricity market. Customers offer a bid which specifies the amount of demand reduction at a given time and the lowest reward that they can accept. Once the bid is accepted, customers are offered financial reward in the form of reduced electricity prices, or via a direct payment for electricity they have "not consumed." DSB has become an important feature of energy markets, and has the potential to grow in importance as its operation becomes increasingly better understood. In contrast to DLC, DSB is mainly offered to large customers.

3. *Interruptible load.* An agreement about retail tariffs with curtailment options is signed between customers and electricity utilities [16]. Customers enjoy a rate discount or bill credit if they do reduce load during system contingencies. However, penalties may be applied due to failure of curtailment. IL programs have traditionally been offered only to the largest industrial (or commercial) customers.

4. *Emergency Demand Response Programs.* EDR provides incentive payments to customers for load reductions during periods when grid reliability issues arise, such as significant transmission constraints, shortages of generation or extremely high demand for electricity. It follows two models [4], i.e., voluntary programs and capacity programs. In voluntary programs, customers have the option to participate. Payments are issued for the amount that customers reduce during a system emergency. In capacity programs, customers are expected to participate and they

are paid for their availability to participate, even if the utility does not call for a reduction in consumption.

5. *Capacity Market Programs.* Customers commit to providing pre-specified load reductions when system contingencies arise. In exchange for being obligated to curtail load when directed, participants receive guaranteed rewards. Customers also face penalties for failure to curtail when called upon to do so. Capacity market programs can be viewed as a form of insurance, i.e., in some years load curtailments will not be called, but participants are still paid in order to be on call at any hour [17].

6. *Ancillary Services Market Programs* [10]. Customers bid a load curtailment in their transmission network operator (TSO) or regional transmission operator (RTO) markets as operating reserves. If their bids are accepted, they get payments for committing to be on stand-by, and if the curtailment is needed, they are called by the TSO/RTO and are paid according to their bid or the market price.

11.1.3   New Requirements in Smart Grid

The traditional DR methods introduced above are relatively mature, but they are difficult to apply to the future smart grid. As pointed out in [18], in ideal DR paradigm, electricity consumption should be treated symmetrically with production and the demand-side customers should be full participants. However, in most of these traditional DR methods, customers are rewarded to reduce their consumption based on an administratively baseline level or cut down their energy usage when informed by a high energy price. In other words, customers just select the energy consumption pattern according to their willingness and receive information. Human selfishness and the lack of interaction among consumers and generators lead to the difficulty of improving energy efficiency and dealing with an emergency of power grid. The characteristic and advantages of smart grid technologies require more participation of customers in the Demand Response. It can be explained from the following two aspects.

Various renewable energy supplies such as solar or wind are being increasingly used in smart grid. In 2008, 11.8% of electricity was generated from renewable energy sources in California U.S. [19]. According to California's Renewable Energy Programs, by 2020, 33% of electricity will be generated from renewable energy resources. In Europe, 8.5% electricity was generated from renewable sources in 2005, and their goal is to achieve 20% by 2020 [20]. Renewable energy resources will spread to different regions and even each household so that energy users become both electricity consumers and producers. This could fundamentally change the power grid from a one-way energy broadcasting network to a two-way energy transmission network to support users to upload their extra energy to the grid and share energy with others.

It is desirable that DR solution is achieved cooperatively by energy producers together with customers.

On the other hand, the advent of smart grid technologies such as digital communication devices and advanced metering infrastructures facilitates a better environment for sharing information and data more readily among energy consumers and producers. Besides, home area networks connected with smart grid [21] allows end-consumers to participate in demand side management. These make DR more intelligent. It has attracted attention and several results have been published. We will elaborate representative DR algorithms proposed in the literature under each following category.

11.2   Representative DR Algorithms in Smart Grid

11.2.1   Classifications

Although there are already various DR results in smart grid, they fall into several categories. Based on the demand management approach, DR algorithms fall into two categories known as centralized and distributed management:

- *Centralized demand management.* In this approach, the electric utilities control the domestic appliances according to a complex centralized algorithm. The large information flow as well as social and legal barriers of centralized solutions hinders their applications in smart grid [7].

- *Distributed demand management.* Under this strategy, the demand decisions are made locally and directly by the end-users. Several theories, such as game theory, consensus methods, and subgradient optimization provide potential solutions for distributed DR [7, 22].

The DR solutions can also be roughly classified in two categories according to the scheduling variables:

- *When to start requested electrical appliances.* In this group, such as [23, 24], it aims at controlling when the devices shall run with the consideration of several factors, e.g., available energy, and pre-defined deadlines. For example, a refrigerator could delay or advance the start time of its cooling cycle within a certain time period. In most literatures of this group, the energy price is dynamic but deterministic, e.g., it is provided by the energy market.

- *How much energy to allocate to users in a time slot.* The goal of this category is estimating the energy demand of consumers in a given time slot, subject to several constraints, e.g., minimum consumption requirements of the consumers, and maximum generation capacity during this time

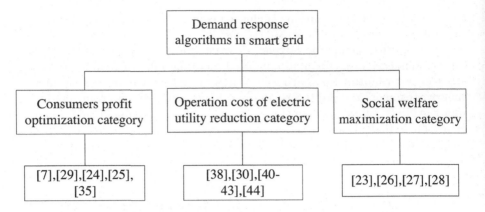

Figure 11.2: Taxonomy of DR algorithms in smart grid

slot. For example, in summer, people will feel much cooler when the air-conditioner is set at $22°C$. However, people can tolerate a temperature under $28°C$. Thus, the demand is adjusted to match available generation in this time slot. Usually, in this case, it computes scheduling variable along with energy price which is related to the energy consumption in the time slot. Particularly, it has been demonstrated in [22, 25–27] that a set of Locational Marginal Prices, which achieve a market equilibrium point, emerge as the Lagrangian multipliers corresponding to power flow balance constraints.

However, regardless of how to deal with the DR problem, the purpose is to increase profit or reduce cost for electric utility or end users. Here, we summarize the representative DR methods into three categories according to optimization objective. The classification is shown in Fig. 11.2.

- *Customer profit optimization category.* The DR methods in this class aim to improve consumer welfare within generation capacity. The welfare can be usage benefit minus electric charge paid to the electric utility [28], or other similar expressions.

- *Operating cost of electric utility reduction category.* The goal of these DR methods is to reduce the operating cost of electric utility while guaranteeing users' minimum demand as much as possible. In some situations, electric utility will pay a rebate to users who are willing to reduce their energy consumption [29].

- *Social welfare maximization category.* Such DR methods aim at optimally matching demand and supply with the objective of maximizing the social welfare [27], which refers to the difference between consumers' efficiencies and producers' operation cost. Currently, in this class, the

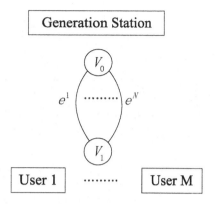

Figure 11.3: Model a power grid system to a directed graph.

energy price is determined dynamically to optimize the profits for both consumers and producers.

11.2.2 Customer Profit Optimization Algorithms

Christian et al. in [7] utilize network congestion game [30] to achieve distributed load management. Each user is assumed to know his total electric demand on a daily basis and decides the exact demand distribution over 24 time-slots through game method, with the aim of minimizing the cost paid to energy provider while taking into account its own preference.

The power grid is composed of one energy provider and M users. Denote $d_i = [d_1^i, \cdots, d_N^i]$, $N = 24$, as the demand distribution vector of user i. Firstly, model such system to a directed graph (V, E), as shown in Fig. 11.3. The edge e^j represents time slot j, i.e., the energy demanded by user on time slot j flows through edge e^j, so the total load in this edge, denoted by x_j, equals $\sum_{i=1}^M d_j^i$. Then the energy price in time slot j is $c(x_j)$. Next, map the demand management to a congestion game $\{P, E, \{s_i\}_{i \in M}, \{c_e\}_{e \in E}\}$, where P is the set of players, E is the set of resources, s_i is the strategy space of player i and c_e is a cost function associated with resource $e \in E$. More specifically, the correspondence between demand management and congestion game is listed: 1) Players: M users; 2) Resources: N edges, i.e., N time slots; 3) Strategy space of player i: $d_i = [d_1^i, \cdots, d_N^i]$; 4) Cost of each resource j: energy price in each time slot j, i.e., $c(x_j)$.

In the game process, each user updates its strategy in order to minimize its weighted cost, under given strategies of other users. The cost function of user i is $c_i(s_i, s_{-i}) = \sum_{e^j \in s_i} w_j^i c(x_j)$, where s_{-i} indicates the current strategy of other users, and w_j^i denotes the cost weight in time slot j of user i according to its preference. The Nash equilibrium can be reached after several such repeated update processes. Once convergence is achieved, prices per hour and

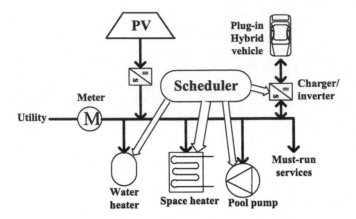

Figure 11.4: A case study in [28].

users' demand distribution are fixed, which are executed accordingly during the day.

In congestion game theory, it has been demonstrated that the Nash equilibrium point is not only a local optimum for the selfish user, but also is a global solution. In this demand management problem, simulation results have shown that it is possible to obtain a smoother demand curve. In addition, the problem is solved in a distributed manner, i.e., each user obtains its demand distribution vector locally. However, each user requires strategies of all other players, so a complex communication protocol is needed to support the information exchange among users.

Instead of applying weighted cost, Michael et al. in [28] consider monetary profit which is equal to monetary benefit that derives from energy services minus charged cost of energy consumption. They aim at maximizing this profit through scheduling hourly energy consumption of various appliances. Fig. 11.4 describes a case study in that paper [28]. The Photo Voltage (PV) system generates energy for local use. In case of shortage, energy is purchased from the wholesale market. Energy services include charging Plug-in hybrid vehicle (PHEV), running space heater, heating storage water heater, operating pool pump, as well as must-run service that contains all other energy services except the four kinds mentioned above. The must-run service requirements are assumed to be fixed over the day. The scheduler determines the hourly charging rate of the PHEV battery, the hourly heating power of the space heater, when the water heater is switched on, and when the pool pump shall run. Denote operation schedules by $x = \{x_i | i = car, heat, water, pool\}$, the scheduling is described as the following mathematical optimization problem:

$$Max \sum_{t=1}^{T} \left(\begin{array}{l} \lambda_{ES,must-run}(t) \times U_{ES,must-run}(t) + \\ \sum_{i} (\lambda_{ES,i}(t) \times U_{ES,i}(t,x_i)) - \lambda(t) \times P(t,x) \end{array} \right) \quad (11.1)$$

where $U_{ES,i}(t, x_i)$ represents the "energy equivalent," i.e., the service provided by appliance i of consuming x_i units energy. The relationship between them can refer to appendix in [28]. $\lambda_{ES,i}(t)$ is the monetary value assigned to each unit of "energy equivalent" at time slot t. $\lambda(t)$ is energy price given by wholesale market. $P(t, x)$ is the amount of hourly energy purchased from wholesale market when local PV generation is short. This mathematical optimization problem is solved and compared by Particle Swarm Optimization (PSO) [31] and its variations.

These results can be readily extended to situations with more appliances. However, the maximum available energy capacity allocated to a household should be taken into account, as discussed in the following DR method.

Shalinee and Lawrence [23] focus on control mechanisms for residential electricity demand in smart grid. They first analyze in-home scheduling, and then a distributed approach to support neighborhood-level scheduling is considered.

First, in the in-home scheduling problem, they present a simple optimization model to determine the optimal operation timing of various appliances. The planning horizon is discretized into T time periods. When the user requests appliance n in period t, the decision of when to turn it on is to find s that solves:

$$\min_{t \leq s \leq t+d_n} (s - t)\psi_n^1 + \sum_{r=s}^{T} \left(\prod_{i=s}^{r-1}(1 - \mu_{ni}) \right) \pi_r c_n \qquad (11.2)$$

where d_n and ψ_n^1 represent maximum allowable delay and the inconvenience to the user incurred by each period of delay, respectively; c_n is the amount of power consumed by appliance n when it is on. The electricity price in period t is denoted by π_t, and it is determined by the wholesale market. In addition, if appliance n is on in period t, then the probability that it completes operation in period $t + 1$ is given by μ_{nt}; hence the product term in the second term calculates the probability that the appliance is still on in period r. The first term represents the delay cost, while the second one represents the expected energy cost while the appliance is on. Because users are selfish, they do not consider power constraints. Each appliance can be optimized individually through formulation (11.2). It is good for each user, but it fails to reduce the load peak, and may create a worse peak.

Next, they propose a distributed scheduling mechanism to reduce peak demand across a collection of local homes. Suppose the available maximum power for the neighborhood is denoted by $P_{max,t}$ for period t. It also assumes that all Energy Management Controllers (EMCs) installed in households transmit/receive information with each other over a common control channel. Firstly, they design a channel competition mechanism to compete with other EMCs for the available power when a new demand request is generated, without considering minimum demand of others. It may result in some consumers receiving little or no power at certain periods. To overcome this problem, each home is allocated a base power level P_b. If the requested demand for a household is less than P_b, then its EMC is on standby. Otherwise its EMC uses its base power as much as possible and then competes with other EMCs to gain additional power. Finally, they introduce a dynamic programming (DP) [32] algorithm to optimize the timing of appliance operation, subject to the available power constraints for a household.

Under this mechanism, it is possible to optimize electricity consumption within a home and also to reduce peak demand within a neighborhood of homes. But DP suffers computational challenges; it is only suitable for small-sized problems.

Safar and Massoud in [24] study a scheduling algorithm for appliance operation to minimize the electricity bill while meeting the scheduling constraints and available power capacity. This algorithm relies on a quasi-dynamic pricing model, in which the energy price consists of TOU dependent base price and a penalty term that penalizes the users when their peak consumption over some recent time windows goes beyond a predetermined threshold. Both non-interruptible and interruptible situations are considered.

Suppose there are K appliances with two sequential power modes. Let $I_{k,m}$ and $P_{k,m}$ represent required operation time and power consumption of appliance k when it is in power mode m, respectively. Total power consumption at time t is denoted by $p(t)$. In the non-interruptible satiation, $p(t)$ is written as:

$$p(t) = \sum_{k=1}^{K} \left(\begin{array}{l} P_{k,0} f(t, a_k, a_k + I_{k,0}) + \\ P_{k,1} f(t, a_k + I_{k,0}, a_k + I_{k,0} + I_{K,1}) \end{array} \right) \qquad (11.3)$$

where a_k is the start time of appliance k, and $f(t, a, b)$ is a pulse function, i.e., $f(t, a, b) = 1$ if appliance is on during time interval $a \le t \le b$, otherwise $f(t, a, b) = 0$. The quasi-dynamic energy price is defined as $C(t)R(p(t))$, where $C(t)$ is TOU-based price, and $R(p(t))$ is the penalty function, with α and p_0 being predetermined constants:

$$R(p(t)) = \left\{ \begin{array}{ll} 1 + \alpha, & p(t) > p_0 \\ 1, & p(t) \le p_0 \end{array} \right. \qquad (11.4)$$

Therefore, the task is to schedule start time for various appliances to achieve:

$$Min \left\{ \sum_{k=1}^{K} \left(\begin{array}{l} \int_{a_k}^{a_k + I_{k,0}} P_{k,0} C(t) R(p(t)) dt + \\ \int_{a_k + I_{k,0}}^{a_k + I_{k,0} + I_{k,1}} P_{k,1} C(t) R(p(t)) dt \end{array} \right) \right\} \qquad (11.5)$$

subject to

$$s_k \le a_k, \quad a_k + I_{k,0} + I_{k,1} \le e_k, \quad p(t) < P_{max} \qquad (11.6)$$

where s_k and e_k denote allowed starting-time and deadline of appliance k, respectively, and P_{max} is the power capacity. This nonlinear optimization problem is solved by Sequential Quadratic Programming (SQP) [33].

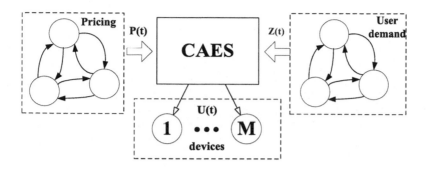

Figure 11.5: CAES energy management system.

In the interruptible case, appliance k can complete its task of power mode 1 in L_k distinct, non-overlapping task fragments. When appliance K resumes working on its task, it takes $I_{k,2} \leq I_{k,0}$ time to restart. From another angle, appliance k is divided into L_k appliances. It equates to increase the number of appliances from K to $\sum_{k=1}^{K} L_k$, with additional constraint on internal order of L_k sub-tasks for appliance k. It is also solved by SQP. Furthermore, the authors discretize this optimization problem considering only a single power mode for each appliance. Dynamic programming is used to solve it.

This solution possesses two advantages. Firstly, the penalty term in the energy price contributes to flattening the demand curve. Second, this scheduling algorithm is readily extended to the situation where appliances have more power modes. Nevertheless, it requires to know characteristics of all appliances that request to be scheduled in advance. It is hard to learn which appliances would be used, because the user selects an appliance randomly. The algorithm elaborated below studies this issue.

A Consumer Automated Energy Management System (CAES) is proposed in [34], inspired by a fact that few users are willing to continuously make a sequence of decisions to defer or advance using a device, especially when it has limited financial impact on them. Users just select devices indicating their desires to run them, then CAES schedules when to run the devices and how much energy will be allocated, with the objective of minimizing the sum of time average financial cost of consuming energy and time average dis-utility to the user for delaying operations of the selected devices. Fig. 11.5 shows CAES. Time is discrete. The user has M devices. The demand request vector $z(t) \in R_+^M$ and energy price $p(t) \in R_+$ are taken as inputs. Both of them are modeled as Markov chains with unknown transition probability distributions. At time t, if device m is selected, then $z_m(t) = \gamma_m$, where γ_m is the required energy to operate it. Otherwise $z_m(t) = 0$. The output is $u(t) \in R_+^M$. The value of $u_m(t)$ is the energy allocated to device m at time t and it may be less than its pending energy requirement. So the pending energy backlog $x(t) \in R_+^M$ is created according to:

$$x(t+1) = x(t) + z(t) - u(t) \tag{11.7}$$

Next, define an auxiliary vector satisfying $y(t+1) = \theta y(t) + (I-\theta)x(t)$, where I is identity matrix and $0 \le \theta \le I$ is a diagonal matrix. Further, define a dis-utility function $\overline{U}_m(y_m(t)) \in R_+$, reflecting user's dissatisfaction with device m of waiting. So θ parameterizes the dis-utility function, i.e., when $\theta \approx I$ user cares about the average delay of completing a device, while it minds the delay associated with the device currently selected if $\theta = 0$. Denoting the state vector by $\Omega(t) = [x(t); y(t); z(t); p(t)]$, the goal is to find optimal $u(t)$ that minimizes infinite time cost, with initial state Ω_0:

$$V_u(\Omega_0) = \lim_{T \to \infty} E\left[\sum_{t=1}^{T} \gamma^t \left\{ \sum_{m=1}^{M} (p(t)u_m(t) + \lambda \overline{U}_m(y_m(t))) \right\}\right] \tag{11.8}$$

where $\lambda \ge 0$ gives the tradeoff between the financial cost and dis-utility cost, and $0 \le \gamma \le 1$ indicates that the user is more concerned about the immediate cost. Bellman's equation describing optimality condition for such Markov Decision Process [35] provides a solution. However, the Markov transition probabilities for $z(t)$ and $p(t)$ are unknown, the value function in the Bellman's equation is difficult to solve. CAES uses an online learning method termed Q-learning [36] to estimate it.

CAES factors in the statistical impact of future prices and the correlation between devices selected by the user. However, the assumption of requiring $z(t)$ and $p(t)$ to be Markov chains is still an over-strict hypothesis in some applications.

11.2.3   Operation Cost of Electric Utility Reduction

Stephane and George in [37] aim at reducing electric utility's operation cost during T periods by scheduling the start time of user demands. Suppose user n has a demand characterized by (d_n, τ_n, s_n), where d_n and τ_n denote instantaneous power consumption and the duration of completing this demand, and $0 \le s_n \le T - \tau_n$ is the flexible start time. It assumes that once the demand service is started, it cannot be interrupted until it is finished. Thus the total instantaneous load at time t is:

$$\lambda(t) = \sum_{n} d_n 1_{\{s_n \le t \le s_n + \tau_n\}} \tag{11.9}$$

Consider a ramp cost function for the electric utility:

$$C_L(\lambda(t)) = C_0 + C_1(\lambda(t) - L)^+ \tag{11.10}$$

where C_0 and C_1 represent the base cost and the overage rate. It implies that the energy generation cost equals the base cost if the total load is below a threshold L, otherwise it needs extra cost which is linear with the overages. Thus, the overall cost of the electric utility becomes:

$$GC_{ramp} = \int_{t=0}^{T} \lambda(t)C_L(\lambda(t))dt$$
$$= C_0 \sum_n d_n\tau_n + C_1 \int_{t=0}^{T} \lambda(t)(\lambda(t) - L)^+ dt \qquad (11.11)$$

Since the problem of minimizing GC_{ramp} is NP-hard, the authors studied and compared different approximation methods based on how much information shared among users:

1) *Users know demand characteristics of each other and exchange observations in a timely manner.* According to game theory, if the electric utility charges user with b_i which is proportional to both the energy he consumed and the global cost, e.g., $b_i = d_i\tau_i/\sum_n d_n\tau_n \times GC_{ramp}$, user will update its strategy to minimize GC_{ramp}, under given strategies of other players. Authors approve the best strategy for user i is to schedule his job at a time minimizing $\int_{s_i}^{s_i+\tau_i} \sum_{i \neq j} d_j\Phi_j(t)dt$, where $\Phi_j(t)$ is the probability of the job j being active at time t.

2) *Users do not share information with each other for privacy reasons, but they know the instantaneous total load.* Inspired by ALOHA protocol [38], at each period, un-scheduled user first judges whether it is his last possible scheduling slot. If so, it starts the demand immediately. If not, it starts the demand with probability p_i if the total instantaneous load currently adding his instantaneous power consumption is still below the threshold L, otherwise with probability q_i, $0 < q_i < p_i < 1$. Authors also discuss two variations of ALOHA strategy.

3) *There is no communication among users, but it assumes all customers have the same demand characteristics.* In this situation, the best strategy for each user is choosing the start time of its demand uniformly at random.

Simulation results confirm that the strategy with more knowledge setting performs better, exactly as our intuition. This literature offers instructive solutions for addressing such problems. However, it takes fixed threshold L in the cost function, which means integration of renewable energy sources is not considered. Because of uncertainties of renewable energy sources, it would thus be interesting to study the problem with time-varying threshold $L(t)$.

Soumyadip et al. in [29] also aim at reducing operation cost of the electric utility but from another perspective. They consider the electric utility has a basic generation capacity, and energy is purchased from the wholesale market in case of shortage. It encourages users to reduce their demand by paying them rebates. Therefore, the operation cost of the electric utility is comprised of rebates paid to all the users and the cost charged by the wholesale market. Rather than simply providing all users with the same rebate contracts, they design a customized, time-varying rebate plan for each user to achieve a minimal operation cost.

Firstly, a single-period situation is studied. Each user reduces its demand according to demand reduction function $f_i(a_i, r_i)$, where r_i is the per unit

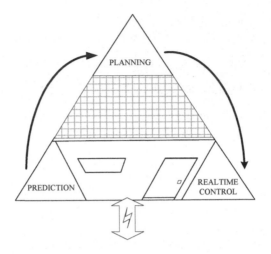

Figure 11.6: A three-step optimization methodology for demand side load management.

rebate of demand reduction for user i , and a_i reflects willingness of user i to reduce demand. Next, they consider a more complex multi-period problem. It is similar to single-period case, but it allows shifting some demand from one period to subsequent periods based on certain rule. Therefore the actual demand level before rebate at time t should take into account the demand shifted from time $1, \cdots, t - 1$.

Indeed, demand reduction equates virtual generation. Besides, the fluctuation of demand corresponds to the uncertainty in renewable energy generator, although they may not be equivalent. This solution provides thought for transaction among users who have installed renewable energy resources.

Albert et al. published a series of results [39–42] on demand side load management using a three-step methodology. The three steps are:

Step 1: Prediction. A system located at each house predicts the energy demand profiles for the upcoming day based on historical consumption pattern and external factors like weather.

Step 2: Global planning of a fleet. Aggregate all the predicted profiles, i.e., schedule the demand distribution of each household to spread the electricity consumption equally over the planning horizon, e.g., one day.

Step 3: Local control. Using steering signals from *Step 2*, a real time control algorithm decides when appliances are switched on/off, when and how much energy flows from or to the buffers, and when and which generators are switched on.

The relationship of the three steps is shown in Fig. 11.6. The main concern here relates to the second step in [41]. To spread the computation and communications, the planning methodology is organized in a tree structure. The root planner decomposes the desired energy profile in subparts to his child planners. Again the sub-desired profile is delegated to lower planners. The planners on the bottom of the tree are directly connected to the controllers located in the house. To achieve the delegated profile, the bottom planner uses a local dynamic programming subject to local constraints and a price vector. The price vector is affected by all the other planners' planning in the current iteration. Then the steering signals are sent to the domestic controllers. Further the domestic controllers generate a planning for the coming day and feed it back to the planner. On every level, the data is aggregated and sent further upward in the tree. Based on the mismatch between the planning and the desired profile, the root planner adjusts the decomposition of the profile, and the process starts again. This iterative process is organized so that after several iterations, the resulting profile falls between the lower and upper bound.

Utilizing tree structure, the complex demand planning problem is divided into smaller, and more easily managed problems that can be computed by dynamic programming in practice.

Michael et al. in [43] investigate the problem of allocating fluctuant renewable energy to delay tolerant demands, using the Lyapunov optimization technique [44] initially developed by themselves for dynamic control of queuing systems in wireless networks.

The time is discrete. During each time-slot t, the renewable resource provides $s(t)$ units of energy, the amount of energy requested is $a(t)$, and the price of purchasing energy from wholesale market is $\gamma(t)$. All the three are assumed time-varying and unpredictable, and bounded by constants $s_{\max}, a_{\max}, \gamma_{\max}$, respectively. It also assumes that no storage is considered, i.e., $s(t)$ must either be used or wasted during period t. Energy demands are buffered in a queue and served in a First-In-First-Out (FIFO) manner. Denoting $x(t)$ as the amount of energy purchased from wholesale market, the total amount of energy requests pending in the queue on timeslot t is updated according to:

$$Q(t+1) = \max[Q(t) - s(t) - x(t), 0] + a(t) \tag{11.12}$$

The goal is to choose $x(t)$ minimizing the expected time average cost experienced by the electric utility, i.e.:

$$
\begin{aligned}
\min \quad & \lim_{t \to \infty} \tfrac{1}{t} \sum_{\tau=0}^{t-1} \mathrm{E}\{\gamma(t)x(t)\} \\
\text{s.t.} \quad & \lim_{t \to \infty} \tfrac{1}{t} \sum_{\tau=0}^{t-1} \mathrm{E}\{Q(\tau)\} < \infty \\
& 0 \le x(t) \le x_{\max} \quad \forall t
\end{aligned}
\tag{11.13}
$$

The first inequality is a constraint on the expected time average queue backlog. Introduce a virtual queue $Z(t)$ with $Z(0) = 0$:

$$Z(t) = \max[Z(t) - s(t) - x(t) + \alpha 1_{\{Q(t) > 0\}}]. \tag{11.14}$$

Following Lyapunov optimization technique [44], define Lyapunov function $L(\Theta(t)) = \tfrac{1}{2}[Z(t)^2 + Q(t)^2]$, where $\Theta(t) = (Z(t), Q(t))$. Then the conditional

1-slot Lyapunov drift is:

$$\Delta(\Theta(t)) = \mathrm{E}\{L(\Theta(t+1)) - L(\Theta(t))|\Theta(t)\} \tag{11.15}$$

The control algorithm is designed to observe $s(t)$, $a(t)$, $\gamma(t)$, $Q(t)$ and $Z(t)$, then to choose $x(t)$ to minimize a bound of the following equation on each timeslot t:

$$\Delta(\Theta(t)) + V\mathrm{E}\{\gamma(t)x(t)|\Theta(t)\} \tag{11.16}$$

where V is a positive parameter that affects tradeoff between time average cost and delay. It is intuitive to compute:

$$\begin{aligned}
\Delta(\Theta(t)) + V\mathrm{E}\{\gamma(t)x(t)|\Theta(t)\} \leq\ & B + V\mathrm{E}\{\gamma(t)x(t)|\Theta(t)\} \\
& + Q(t)\mathrm{E}\{a(t) - s(t) - x(t)|\Theta(t)\} \\
& + Z(t)\mathrm{E}\{\partial - s(t) - x(t)|\Theta(t)\}
\end{aligned} \tag{11.17}$$

where $B = \frac{(s_{\max} + x_{\max})^2 + a_{\max}^2}{2} + \frac{\max[(s_{\max} + x_{\max})^2, \alpha^2]}{2}$. Obviously, minimizing this bound is equivalent to solving the following optimization problem in every timeslot t:

$$\begin{aligned}
\min \quad & x(t)[V\gamma(t) - Q(t) - Z(t)] \\
\text{s. t.} \quad & 0 \leq x(t) \leq x_{\max}
\end{aligned} \tag{11.18}$$

Thus the solution of $x(t)$ is given by:

$$x(t) = \begin{cases} x_{\max}, & Q(t) \geq \max[s(t) + x(t), V\gamma(t) - Z(t)] \\ \min[Q(t) - s(t), 0], & \text{otherwise} \end{cases} \tag{11.19}$$

It has been approved that all energy requests are fulfilled with a maximum delay $D_{\max} = (2V\gamma_{\max} + a_{\max} + \partial)/\partial$, while the expected time average cost of the electric utility satisfies $\frac{1}{t}\sum_{\tau=0}^{t-1} \mathrm{E}\{\gamma(t)x(t)\} \leq c^* + B/V$, c^* is the infimum time average cost without consideration of delay constraint. The parameter V can be tuned to provide average cost arbitrarily close to optimal, with a tradeoff in delay.

The proposed Lyapunov optimization does not need a priori statistical knowledge of the supply, demand and market price. It is very useful in smart grid with unpredictable demand and renewable energy generation. It also provides broad space for the research, since some assumptions in this model could be improved, e.g., energy demand should be served first with highest priority, instead of in a FIFO manner.

11.2.4   Social Welfare Maximization

Pedram et al. in [22] not only consider the profits of users or electric utility, they also focus on social welfare. From a social fairness standpoint, it is desirable to utilize the available energy in such a way that the sum of all user-utilities is maximized and the cost imposed on the energy provider is minimized.

Suppose there are N energy users and one energy provider. The planning time is divided into K timeslots. Denote the energy consumed by user i in timeslot k by x_i^k, and the energy offered by the provider is L^k. Both x_i^k and L^k have to fall into a pre-determined interval. Also, the minimum generation capacity should always cover the minimum requirements of all users, i.e., $L_k^{min} = \sum_{i \in N} m_i^k, \forall k \in K$. The social welfare optimization is stated as:

$$\begin{array}{c} \underset{\substack{m_i^k \le x_i^k \le M_i^k, \\ L_k^{min} \le L^k \le L_k^{max}}}{\text{maximize}} \quad \sum_{k \in K} \left\{ \sum_{i \in N} U(x_i^k, \omega_i^k) - C_k(L^k) \right\} \\ \textit{subject to} \quad \sum_{i \in N} x_i^k \le L^k, \quad \forall k \in K \end{array} \tag{11.20}$$

where $C_k(L^k)$ is a cost function that indicates the cost experienced by the energy provider offering L^k units of energy, and $U(x_i^k, \omega_i^k)$ is a utility function representing user's satisfaction level of consuming x_i^k units of energy while meeting its preference ω_i^k. Quadratic utility function and quadratic cost function are used. Clearly, (11.20) could be solved independently for each timeslot. Let λ^k denote Lagrange multiplier, and the dual optimization problem for timeslot k is written as:

$$\underset{\lambda^k > 0}{\text{minimize}}\, D(\lambda^k) = \sum_{i \in N} B_i^k(\lambda^k) + S_k(\lambda^k) \tag{11.21}$$

$$B_i^k(\lambda^k) = \underset{m_i^k \le x_i^k \le M_i^k}{\text{maximize}}\, U(x_i^k, \omega_i^k) - \lambda^k x_i^k \tag{11.22}$$

$$S_k(\lambda^k) = \underset{L_k^{min} \le L^k \le L_k^{max}}{\text{maximize}}\, \lambda^k L^k - C_k(L^k) \tag{11.23}$$

It is observed that $D(\lambda^k)$ is decomposed into N separable subproblems in form of (11.22) that can be solved by each user and another subproblem in form of (11.23) which could be solved by the energy provider. Using subgradient method [45], during each timeslot k, each user i estimates his power consumption x_i^k through iterative computation, and the energy provider determines generation amount L^k and the Lagrange multiplier λ^k:

Step 1: The energy provider initializes L^k and λ^k randomly and broadcasts λ^k to users.

Step 2: Each user updates his consumption value x_i^k by solving (11.22) based on received value of λ^k, then transmits the estimated value of x_i^k to the energy provider; The energy provider computes the generation amount L^k by solving (11.23) based on the value of λ^k.

Step 3: The energy provider updates the value of λ^k, according to $\lambda^k = [\lambda^k - \gamma \frac{\partial D(\lambda^k)}{\partial \lambda^k}]^+$ upon receiving x_i^k from all users, where γ is pre-determined step size. Again, broadcasts updated λ^k to users.

Step 4: Repeat *Step 2* to *Step 3* until predefined precision is achieved.

Interestingly, (11.22) is the benefit that the user searches for, while (11.23) is the profit that the energy provider attempts to achieve. In fact, if the energy provider charges the users at a price λ^{k*}, i.e., the solution of the dual problem, the computed optimal consumption amount x_i^{k*} maximize the welfare of user i, and L^{k*} maximizes the profit of energy provider.

Mardavij et al. in [25] investigate a similar problem. The difference is that they consider distributed energy suppliers with different retail prices, instead of a single provider. Besides, energy transmission is taken into account.

Suppose the power system is composed of n buses (nodes), r transmission lines and m energy providers. All loads connected to one node are treated as a homogeneous demand. The goal is to find optimum demand vector $d = [d_1, \cdots, d_n]^T$, supply vector $s = [s_1, \cdots, s_m]^T$ and line current flow vector $I = [I_1, \cdots, I_r]^T$ so that social welfare $W(s, d)$ is maximized:

$$\begin{aligned} \text{maximize} \quad & W(s, d) = \sum_{j=1}^{n} u_j(d_j) - \sum_{i=1}^{n} c_i(s_i) \\ \text{subject to} \quad & Ks + EI = d, \ RI = 0 \\ & -I_{\max} \leq I \leq I_{\max}, \ s_{\min} \leq s \leq s_{\max} \end{aligned} \tag{11.24}$$

where $K \in \{0, 1\}^{n \times m}$, $E \in \{-1, 0, 1\}^{n \times r}$, and $R \in \Re^{p \times r}$ are matrix describing the transmission grid. They are matrix aggregating the output of several suppliers connected to one node, graph incidence matrix and loop-impedance matrix. Thus, $Ks + EI = d$ and $RI = 0$ account for Kirchhoff's current and voltage laws (KCL and KVL), respectively. Utilizing dual formulations which are similar to (11.21)-(11.23), we can easily draw the conclusion that a set of Locational Marginal Prices (LMPs) emerge as Lagrange multipliers corresponding to KCL constraints. The Independent System Operator (ISO) is induced to be charge of computing the LMPs, users and producers take actions after receiving LMPs messages to maximize their individual profit. However, ISO may not know user utility functions for privacy reasons. On the other hand, if d is fixed, social welfare function becomes $W(s, d) = \sum_{i=1}^{n} c_i(s_i)$. In this case, ISO can derive LMPs without knowledge of user utility functions. This motivates authors to present an alternative solution:

Step 1: Before each period t, ISO computes the LMPs $\lambda_t = [\lambda_{1,t}, \cdots, \lambda_{n,t}]^T$ based on forecast demand that is of the form $d_{l,t} = \widehat{D}_t(d_{l,t-1}, \cdots, d_{l,t-1-T})$, $l = 1, \cdots, n$. Then ISO announces retail prices $\pi_t = [\pi_{1,t}, \cdots, \pi_{n,t}]^T$, which correspond to the following equations:

$$\begin{aligned} \pi_t &= \prod_t (\tilde{\lambda}_t, \tilde{\pi}_{t-1}) \\ \tilde{\lambda}_t &= [\lambda_t, \cdots, \lambda_{t-T}] \\ \tilde{\pi}_{t-1} &= [\pi_{t-1}, \cdots, \pi_{t-1-T}] \end{aligned} \tag{11.25}$$

Step 2: Upon receiving $\pi_{l,t}$ from ISO, user l adjusts its energy consumption during $[t, t+1]$ according to:

$$d_{l,t} = \arg \max_{x \in \Re^+} u_j(x) - \pi_{l,t}x, \ l = 1, \cdots, n \tag{11.26}$$

Step 3: During $[t, t+1]$, producers match all the demand.

Step 4: Repeat *Step 1* to *Step 3* for next period.

The authors have proved that this method converges to a small neighborhood of solution of (11.24).

Both results in the two literatures above achieve a market equilibrium point which satisfies the following criteria: 1) maximizing welfare for each generating unit; 2) maximizing profit for every individual consumer; 3) maximizing social welfare.

Clearly, market equilibrium is significant. Another research group discusses two analogous market models in [26]. One considers demand shaping by subjecting customers to real-time spot prices and incentivizing them to shift or even reduce their load, which is similar to the technique in [22]. The other designs a DR method to match a deficit supply by shedding users' energy consumption. In this model, the user who sheds its energy consumption is equivalent to an energy generator, while the deficit of supply is viewed as demand. A similar analytical approach of market equilibrium is adopted.

In addition, Arman et al. in [27] focus on perturbation analysis of market equilibrium in the presence of fluctuations in renewable energy resources and demand. They firstly analyze the overall market equilibrium formulation under nominal conditions.

They model the overall electricity market similar to [22, 25] above, including three components:

1) *Generation modeling.* There are N_G generating units, and the production of unit i is divided into N_{G_i} power blocks. Denote the production and associated linear operation cost of power block b in unit i by $P_{G_{ib}}$ and $\lambda^C_{G_{ib}}$, respectively. For generating units, the goal is to maximize the overall profit p_g which is stated as:

$$p_g = \sum_{i=1}^{N_G} \sum_{b=1}^{N_{G_i}} (\rho_{n(i)} - \lambda^C_{G_{ib}}) P_{G_{ib}} \tag{11.27}$$

where $\rho_{n(i)}$ is the LMP of unit i which is located at node n in the power network. The power production should be subject to the maximum available constraints.

2) *Consumption modeling.* There are N_D users owning several consumers for each. Let $P_{D_{jk}}$ and $\lambda^U_{D_{jk}}$ represent the power consumed by consumer k of user j and corresponding linear utility, respectively. The consumption modeling aims at maximizing user welfare u as follows:

$$u = \sum_{j=1}^{N_D} \sum_{k=1}^{N_{D_j}} (\lambda^U_{D_{jk}} - \rho_{n(j)}) P_{D_{jk}} \tag{11.28}$$

subject to both minimum and maximum energy consumption requirements of users.

3) *ISO modeling.* It is responsible for maximizing social welfare s i.e.:

$$s = \sum_{j=1}^{N_D} \sum_{k=1}^{N_{D_j}} \lambda^U_{D_{jk}} P_{D_{jk}} - \sum_{i=1}^{N_G} \sum_{b=1}^{N_{G_i}} \lambda^C_{G_{ib}} P_{G_{ib}} \tag{11.29}$$

subject to several constraints, including power flow balance in each node and power line capacity.

In the energy market model above, the decision variables are $P_{G_{ib}}$, $P_{D_{jk}}$ and $\rho_{n(i)}$. As mentioned previously, LMPs relate to Lagrange multipliers corresponding to power flow balance constraints in ISO model, $\rho_{n(i)}$ can be viewed as fixed values in both generation and consumption modeling. So the three models are linear programming problems. Therefore, the three sets of Karush-Kuhn-Tucker (KKT) optimality conditions are both necessary and sufficient for describing overall market equilibrium. In addition, the three KKT sets result in a Mixed Linear Complementarity Problem (MLCP).

Further, they introduce uncertainties ΔG_{ib} into generators, as $\bar{P}_{G_{ib}} = P_{G_{ib}}(1 - \Delta G_{ib})$. For demand fluctuation, let a control parameter $0 < \kappa_{D_{jk}} < 1$ denote the response of the consumers to change in the Real Time Price so that $\bar{P}_{D_{jk}} = P_{D_{jk}}(1 - \kappa_{D_{jk}})$. Replace $\bar{P}_{G_{ib}}$ and $\bar{P}_{D_{jk}}$ with $P_{G_{ib}}$ and $P_{D_{jk}}$ in the three sets of KKT optimality conditions, respectively. Using properties of MLCP, the authors have proved that these perturbations lead to a limited shift off the equilibrium in nominal conditions.

11.3   Summary of DR Methods and Future Directions

This section summarizes recent DR results in Table 11.1. Each has strengths and weaknesses. We will analyze these strengths and weaknesses, and discuss future research directions.

The advantages in the customer profit optimization group are mainly embodied in two aspects. Firstly, the decision variables under this category are either scheduling the operation timing of requested appliances or computing demand distribution over time. Both imply that power consumption spreads throughout the time, which helps to avoid a demand peak. Secondly, although the goal is to optimize the profit of customer and users are always selfish, some reasonable steps are taken to limit this selfishness, e.g., authors in [24] restrict maximum available energy for user. However, most results only consider the individual user with several appliances. In fact, interactions among customers are very important to both customers and the whole system, especially when distributed renewable energy resources are increasingly integrated. For example, the user with redundant energy can upload its extra energy to the grid to share with others. Authors in [23] design a channel competition mechanism to compete with neighborhood for the power, providing a good example.

We observed that DR methods in operation cost of electric utility reduction group possess the first advantage in customer profit optimization group. Particularly, the result in [43] can deal with a situation where both demand and supply are stochastic, although the model is simple. Because they focus on the cost of electric utility, the profit of customer cannot be guaranteed. As mentioned previously, DR methods in social welfare maximization cate-

Table 11.1: Summarization of Demand Response Algorithms

	Optimization Objective	Scheduling Variables	Energy Price	Constraints	Solution
[7]	weighted electric charge (user)	demand distribution over 24 timeslots	proportion to total load	no	congestion game
[28]	difference of electric charge and utility (user)	hourly energy consumption of appliances	provided by energy market	no	particle swarm optimization
[23]	sum of electric charge and delaying cost (user)	operation timing of requested appliances	provided by energy market	compete for available energy, demand deadline	dynamic programming
[24]	electric charge (user)	operation timing of requested appliances	proportion to total load	available energy, demand deadline	sequential quadratic programming
[34]	sum of infinite horizon electric charge and delaying cost (user)	the amount of energy allocated to pending devices over time	Markov chain with unknown transition probability	no	dynamic programming, Q-learning
[37]	energy generation cost (electric utility)	operation timing of requested appliances	not involved	demand deadline	game method; ALOHA; random selection
[29]	sum of energy purchase cost and rebates to users (electric utility)	time-varying rebate of each user	provided by energy market	no	steepest descent method
[41]	flatten the demand curve (electric utility)	demand distribution over time	proportion to total load	local constraint	tree structure

Table 11.1: *Continued.*

	Optimization Objective	Scheduling Variables	Energy Price	Constraints	Solution
[43]	time average cost of purchasing energy (electric utility)	the amount of energy purchased over time	unpredictable	bounds of demand, renewable supply, and energy price	Lyapunov otimization
[22]	social welfare	amount of energy consumption and generation	Lagrange multiplier	generating capacity; demand bound	sub-gradient method
[25]	social welfare	amount of energy consumption and generation	Lagrange multipliers	power line capacity; generating capacity; KCL and KVL	alternative solution based on sub-gradient method
[26]	social welfare	amount of demand reduction	Lagrange multipliers	fixed total demand reduction	sub-gradient method
[27]	social welfare	perturbation analysis with fluctuations in suppliers and demands	Lagrange multipliers	generating capacity; demand bound; KCL and KVL; power line capacity	KKT optimal conditions

gory can achieve market equilibrium. They also consider interaction among users and generators. Unfortunately, only social welfare in a single period is discussed. Indeed, social welfare in the long term makes more sense.

According to the analysis above, an ideal DR method should possess the following properties: 1) Achieving electricity transaction among customers; 2) spreading energy consumption throughout time; 3) balancing the welfare for both users and electric utility. Thus, further research could focus on two aspects:

1. *Fast distributed DR solution.* Because of privacy reasons and large information flow of centralized solutions, the user with redundant energy should locally decide whether to upload its extra energy to the grid to

share with others or satisfy its other demand in advance. Similarly, the users who are short of energy compute when and where to purchase the energy. It is also related to distributed energy routing problems in the transmission grid. Besides, the fluctuations of both demand and renewable supplies require that the distributed DR method possess the property of fast convergence. This also helps to stabilize the power system quickly to avoid cascaded failure when there is a failure in one of the transmission lines.

2. *Long-time average social welfare optimization.* The ultimate goal of DR method is to flatten the demand curve over a long time. So it should consider demand shifting on the time axis, i.e., delay or advance starting the appliances. On the other hand, the social welfare introduced in a previous section should be the key point of DR research. Only in this way do both energy producers and energy customers become participants in DR. Therefore, the DR method that focuses on long-time average social welfare optimization will be an interesting topic.

11.4 Conclusion

In this chapter, we have surveyed the state-of-the-art of DR in smart grid. According to the different taxonomic approaches, the DR algorithms can be classified into several groups. This chapter focuses on a classification that is based on the optimization objective. A representative number of DR methods have been stated, which belong to the customer profit optimization category, operation cost of electric utility reduction category and social welfare maximization category. Finally, according to the analysis on strengths and weaknesses of each DR category, we believe that fast distributed DR solutions and long-time average social welfare optimization problems are the two research directions for DR in smart grid.

Bibliography

[1] G. Barbose, C. Goldman, and B. Neenan, "A survey of utility experience with real time pricing," [Online]. Available: http://escholarship.org/uc/item/8685983c.

[2] K. Herter, "Residential implementation of critical-peak pricing of electricity," *Energy Policy*, vol. 35, no. 4, pp. 2121-2130, April 2007.

[3] S. J. Rassenti, V. L. Smith, and B. J.Wilson, "Controlling market power and price spikes in electricity networks: Demand-side bidding," in *Proc. National Academy Sciences*, pp. 29983003, Mar. 2003.

[4] How Does DR Work, [Online]. Available: http://www.rtpcontrols.com/public/dema1.html.

[5] Demand Response, [Online]. Available: http://www.pjm.com/markets-and-operations/demand-response.aspx.

[6] US Department of Energy, "Benefits of demand response in electricity markets and recommendations for achieving them," A report to the United States Congress pursuant to section 1252 of the Energy Policy Act of 2005, [Online]. Available: http://eetd.lbl.gov/ea/ems/reports/congress-1252d.pdf.

[7] C. Ibars, M. Navarro, and L. Giupponi, "Distributed Demand Management in Smart Grid with a Congestion Game," in *Proc. of the First IEEE Int'l. Conf. on Smart Grid Communications*, Gaithersburg, MD, pp. 495-500, Oct. 2010.

[8] S. Gormus, P. Kulkarni, and Z. Fan, "The POWER of Networking: How Networking Can Help Power Management," in *Proc. of the First IEEE Int'l. Conf. on Smart Grid Communications*, Gaithersburg, MD, pp. 561-566, Oct. 2010.

[9] J. D. Kueck, A. F. Snyder, F. Li, and I. B. Snyder, "Use of Responsive Load to Supply Ancillary Services in the Smart Grid: Challenges and Approach," in *Proc. of the First IEEE Int'l. Conf. on Smart Grid Communications*, Gaithersburg, MD, pp. 507-512, Oct. 2010.

[10] M. H. Albadi, and E. F. El-Saadany, "Demand Response in Electricity Markets: An Overview," in *IEEE Power Engineering Society General Meeting*, Tampa, FL, pp. 1-5, June 2007.

[11] Demand Response Guide for Small to Midsize Business Customers, [Online]. Available: http://www.ceati.com/freepublications/7047_Guide_Web.pdf.

[12] Time-based pricing, [Online]. Available: http://en.wikipedia.org/wiki/Time-based_pricing.

[13] M. H. Albadi, and E. F. El-Saadany, "A summary of demand response in electricity markets," *Electric Power Systems Research*, vol. 78, no. 11, pp. 1989-1996, Nov. 2008.

[14] Y. Q. He, and A. K. David, "Time-of-use electricity pricing based on global optimization for generation expansion planning," in *the Fourth International Conference on Advances in Power System Control, Operation and Management*, vol. 2, pp. 668-673, Nov. 1997.

[15] K. H. Ng, and G. B. Sheble, "Direct load control—A profit-based load management using linear programming," *IEEE Trans. on Power Systems*, vol. 13, no. 2, pp. 688-694, May 1998.

[16] C. S. Chen, and J. T. Leu, "Interruptible load control for Taiwan Power Company," *IEEE Trans. on Power Systems*, vol. 5, no. 2, pp. 460-465, May 1990.

[17] H. A. Aalami, M. P. Moghaddam, and G. R. Yousefi, "Demand response modeling considering Interruptible/Curtailable loads and capacity market programs," *Applied Energy*, vol. 87, no. 1, pp. 243-250, Jan. 2010.

[18] J. Bushnell, B. Hobbs, and F. A. Wolak, "When it comes to demand response, is FERC its own worst enemy," CSEM WP 191, Center for the Study of Energy Markets, August 2009, [Online]. Available: http://www.ucei.berkeley.edu/PDF/csemwp191.pdf.

[19] California's Renewable Energy Programs, [Online]. Available: http://www.energy.ca.gov/renewables/index.html.

[20] 20% of Renewable Energy by 2020, [Online]. Available: http: //www.our-energy.com/videos/eu_20_percent_of_renewable_energy_by_2020.html.

[21] G. Lu, D. De, and W. Z. Song, "SmartGridLab: A Laboratory-Based Smart Grid Testbed," in *Proc. of the First IEEE Int'l. Conf. on Smart Grid Communications*, Gaithersburg, MD, pp. 143-148, Oct. 2010.

[22] P. Samadi, A. Mohsenian-Rad, R. Schober, V. W.S. Wong, and J. Jatskevich, "Optimal Real-time Pricing Algorithm Based on Utility Maximization for Smart Grid," in *Proc. of the First IEEE Int'l. Conf. on Smart Grid Communications*, Gaithersburg, MD, pp. 415-420, Oct. 2010.

[23] S. Kishore, and L. V. Snyder, "Control Mechanisms for Residential Electricity Demand in SmartGrids," in *Proc. of the First IEEE Int'l. Conf. on Smart Grid Communications*, Gaithersburg, MD, pp. 443-448, Oct. 2010.

[24] S. Hatami, and M. Pedram, "Minimizing the Electricity Bill of Cooperative Users under a Quasi-Dynamic Pricing Model," in *Proc. of the First IEEE Int'l. Conf. on Smart Grid Communications*, Gaithersburg, MD, pp. 421-426, Oct. 2010.

[25] M. Rooabehani, M. Dahleh, and S. Mitter, "Dynamic Pricing and Stabilization of Supply and Demand in Modern Electric Power Grids," in *Proc. of the First IEEE Int'l. Conf. on Smart Grid Communications*, Gaithersburg, MD, pp. 543-548, Oct. 2010.

[26] L. Chen, N. Li, S. H. Low, and J. C. Doyle, "Two Market Models for Demand Response in Power Networks," in *Proc. of the First IEEE Int'l. Conf. on Smart Grid Communications*, Gaithersburg, MD, pp. 397-402, Oct. 2010.

[27] A. Kiani, and A. Annaswamy, "Perturbation analysis of market equilibrium in the presence of renewable energy resources and demand response," *IEEE Innovative Smart Grid Technologies Conference Europe*, Gothenburg, pp.1-8, Oct. 2010.

[28] M. A. A. Pedrasa, T. D. Spooner, and I. F. MacGill, "Coordinated Scheduling of Residential Distributed Energy Resources to Optimize Smart Home Energy Services," *IEEE Trans. Smart Grid*, vol. 1, no. 2, pp. 134-143, Sept. 2010.

[29] S. Ghosh, J. Kalagnanam, D. Katz, M. Squillante, X. Zhang, and E. Feinberg, "Incentive Design for Lowest Cost Aggregate Energy Demand Reduction," in *Proc. of the First IEEE Int'l. Conf. on Smart Grid Communications*, Gaithersburg, MD, pp. 519-524, Oct. 2010.

[30] D. Fuderberg, and J. Tirole, *Game Theory*, MIT Press, 1991.

[31] J. Kennedy, and R. Eberhart, "Particle swarm optimization," in *Proce. of IEEE International Conference on Neural Networks*, Perth, WA, Australia, vol. 4, pp. 1942-1948, Nov./Dec. 1995.

[32] R. A. Howard, "Dynamic Programming," *Management Science*, vol. 12, no. 5, pp. 317-348, January 1966.

[33] M. Bartholomew-Biggs, and M. Bartholomew Biggs, "Sequential quadratic programming," in *Nonlinear Optimization with Engineering Applications*, ser. Springer Optimization and Its Applications. Springer US, vol. 19, pp. 114, 2008.

[34] D. O'Neill, M. Levorato, A. Goldsmith, and U. Mitra, "Residential Demand Response Using Reinforcement Learning," in *Proc. of the First IEEE Int'l. Conf. on Smart Grid Communications*, Gaithersburg, MD, pp. 409-414, Oct. 2010.

[35] D. Bertsekas, *Dynamic Programming and Optimal Control*, Massachusetts: Athena Scientific, 2005.

[36] R. Sutton, and A. Barto, *Reinforcement Learning*, MIT Press, 1998.

[37] S. Caron, and G. Kesisdis, "Incentive-based Energy Consumption Scheduling Algorithms for the Smart Grid," in *Proc. of the First IEEE Int'l. Conf. on Smart Grid Communications*, Gaithersburg, MD, pp. 391-396, Oct. 2010.

[38] N. Abramson, "The ALOHA System: another alternative for computer communications," in *Proc. of the Fall Joint Computer Conference*, pp. 281-285, Nov. 1970.

[39] A. Molderink, V. Bakker, M. Bosman, J. Hurink, and G. Smit, "A three-step methodology to improve domestic energy efficiency," in *IEEE PES Conference on Innovative Smart Grid Technologies*, Gaithersburg, MD, pp. 1-8, Jan. 2010.

[40] V. Bakker, M. Bosman, A. Molderink, J. Hurink, and G. Smit, "Improve heat demand prediction of individual households," in *Conference on Control Methodologies and Technology for Energy Efficiency*, Mar. 2010.

[41] V. Bakker, M. G. C. Bosman, A. Molderink, J. L. Hurink, and G. J. M. Smit, "Demand side load management using a three step optimization methodology," in *Proc. of the First IEEE Int'l. Conf. on Smart Grid Communications*, Gaithersburg, MD, pp. 431-436, Oct. 2010.

[42] A. Molderink, V. Bakker, M. Bosman, J. Hurink, and G. Smit, "Domestic energy management methodology for optimizing efficiency in smart grids," in *Proc. of IEEE Conference on Power Technology, Bucharest*, pp. 1-7, June/July 2009.

[43] M. J. Neely, A. S. Tehrani, and A. G. Dimakis, "Efficient Algorithms for Renewable Energy Allocation to Delay Tolerant Consumers," in *Proc. of the First IEEE Int'l. Conf. on Smart Grid Communications*, Gaithersburg, MD, pp. 549-554, Oct. 2010.

[44] M. J. Neely, "Energy optimal control for time varying wireless networks," *IEEE Trans. on Information Theory*, vol. 52, no. 7, pp. 2915-2934, July 2006.

[45] L. T. Dos Santos, "A parallel subgradient method for the convex feasibility problem," *Journal of Computational and Applied Mathematics* vol. 18, pp. 307–320, 1987.

Chapter 12

Green Wireless Cellular Networks in the Smart Grid Environment

Shengrong Bu, F. Richard Yu and Peter X. Liu

Department of Systems and Computer Engineering, Carleton University, Ottawa, ON, Canada

Email: shengrbu@sce.carleton.ca; richard_yu@carleton.ca; xpliu@sce.carleton.ca

Recently, there is great interest in considering the energy efficiency aspect of cellular networks. On the other hand, the power grid infrastructure, which provides electricity to cellular networks, is experiencing a significant shift from the traditional electricity grid to the smart grid. When a cellular network is powered by the smart grid, considering only energy efficiency in the cellular network may not be enough. In this chapter, we consider not only energy-efficient communications but also the dynamics of the smart grid in designing green wireless cellular networks. Specifically, the dynamic operation of cellular base stations depends on the traffic, real-time electricity price, and the pollutant level associated with electricity generation. Coordinated multipoint (CoMP) is used to ensure acceptable service quality in the cells whose base stations have been shut down. The active base stations decide on which retailers to procure electricity from and how much electricity to procure. We formulate the system as a Stackelberg game, which has two levels: a cellular network level and a smart grid level. Simulation results show that the smart grid has significant impacts on green wireless cellular networks, and our proposed scheme can significantly reduce operational expenditure and CO_2 emissions in green wireless cellular networks.

12.1   Introduction

The continuously growing demand for ubiquitous network access leads to the rapid development of wireless cellular networks. Recently, increasingly rigid environmental standards and rapidly rising energy price have led to a trend of considering the *energy efficiency* aspect of wireless cellular networks. The electricity bill has become a significant portion of the operational expenditure of cellular operators. The CO_2 emissions produced by wireless cellular networks are equivalent to those from more than 8 million cars [1].

In a typical wireless cellular network, base stations (BSs) account for 60–80% of the whole cellular network's energy consumption [2]. Approximately 120,000 new base stations are deployed every year to service 400 million new mobile subscribers around the world [3]. Therefore, an important energy-efficient approach is the dynamic operation of cellular base stations based on the traffic. A base station consumes more than 90% of its peak energy even when there is little or no traffic [1]. To improve energy efficiency, base stations can dynamically coordinate, and switch off redundant base stations during periods of low traffic. Coordinated multipoint (CoMP) communication is a new method that helps with the implementation of dynamic base station coordination in practice. CoMP communication is considered as a key technology for future mobile networks, and is expected to be deployed in the future long-term evolution-advanced (LTE-A) systems to improve the cellular network's performance [4]. CoMP communication can improve energy efficiency, extend the coverage of the active base stations and ensure service quality at an acceptable level for mobile users in nearby cells whose base stations have been shut down during a low activity period.

On the other hand, the power grid infrastructure, which provides electricity to cellular networks, is experiencing a significant shift from the traditional electricity grid to the *smart grid*. The electricity demand of consumers has sharply increased in recent years. Moreover, there is increasing interest in integrating renewable resources into the power grid, in order to decrease greenhouse gas emissions. In addition, demand-side management (DSM), such as dynamic pricing, and demand response programs should be used to improve the reliability of the grid by dynamically changing or shifting the electricity consumption. These new requirements and aging existing grids make the modernization of the grid infrastructure a necessity. The next-generation smart grid can optimize electricity generation, transmission, and distribution, reduce peaks in power usage, and sense and prevent power blackouts, by incorporating network and information technologies with intelligent control algorithms [5].

When a wireless cellular network is powered by the smart grid, only considering energy efficiency in the cellular network may not be enough. Actually, in the smart grid environment, consuming more energy can be better than consuming less energy in some circumstances. This is because large amounts of renewable energy will be integrated in the future smart grid, and these renewable sources are highly intermittent in nature and are often uncontrollable [6]. In addition, shut-down and ramp-up of a power plant can be costly or sometimes not technically viable, and the electricity storage capacities are limited in practice [6–8]. Indeed, power generators may use negative prices (i.e., consumers will be paid by consuming electricity) to encourage consumers to consume more energy [6, 9]. Therefore, the dynamics of the smart grid will have significant impacts on the operation of green wireless cellular networks.

In this work, we consider not only energy-efficient communications but also the dynamics of the smart grid in designing green wireless cellular networks. Several distinct features of this work are as follows:

- The dynamic operation of cellular base stations depends on the traffic, real-time price provided by the smart grid and the pollutant level associated with the generation of the electricity.

- CoMP is used to ensure acceptable service quality in the cells whose base stations have been shut down to save energy.

- The active base stations decide on which retailers to procure electricity from and how much electricity to procure considering the pollutant level of each retailer and the price offered by the retailer.

- We formulate the system as a Stackelberg game, which has two levels: a cellular network level and a smart grid level. The close-form solutions to the proposed scheme are given.

- Simulation results show that the smart grid has significant impacts on green wireless cellular networks, and our proposed scheme can significantly reduce operational expenditure and CO_2 emissions in green wireless cellular networks.

The rest of the chapter is organized as follows. Section 12.2 describes the related background research, which includes green wireless cellular networks and the smart grid. We describe the system models in Section 12.3. The problem is formulated as a two-level Stackelberg game in Section 12.4. The proposed game is analyzed in Section 12.5. Simulation results are presented and discussed in Section 12.6. Finally, we conclude this study in Section 12.7 with future work.

12.2   Background

This section covers two background research topics: green wireless cellular networks and the smart grid.

12.2.1   Green Wireless Cellular Networks

The rising energy costs and carbon footprint of operating cellular networks have led to a trend of improving the networks' energy efficiency. In this section, three important energy-efficient solutions will be presented, and followed by the related work in this area. Finally, four different tradeoffs are described, since they have a significant impact on the energy-efficient solutions.

12.2.1.1   Energy-Efficient Solutions

Three important energy-efficient solutions for cellular networks are as follows:

- Renewable Energy Resources
  Powering BSs with renewable energy resources, especially in off-grid sites, could reduce greenhouse gas emissions and a large amount of expenditures for cellular companies. Renewable energy sources, such as wind and solar power, are environment friendly, since adopting them does not produce any greenhouse gases. It can also reduce the amount

of electricity taken from the power grid. In off-grid sites, where expensive diesel-powered generators are generally used, renewable energy resources can become a viable option to reduce the overall network expenditure [10]. Moreover, using air cooling and cold climates to cool the electronic devices in the BSs can further reduce the electricity consumption [11]. Therefore, a number of solutions, such as the Nokia Siemens Networks Flexi Multiradio base station and the Huawei green base station, are offered by BS equipment manufactures to reduce electricity consumption and support off-grid BSs with renewable energy resources [12, 13]. However, the renewable energy sources cannot be the sole electricity source for a BS, since BSs require high reliability, and any power shortage will disturb the network's service provision [14].

- Heterogeneous Networks
  The deployment of heterogeneous networks based on smaller cells is an important technique to increase the energy efficiency of wireless cellular networks [10]. In recent years, the demand for cellular data traffic has grown significantly with the introduction of mobile devices such as iPhone and iPad. Macrocells have been used to provide large area coverage and to better handle user mobility in cellular networks. However, they are not efficient in providing high data rates. Recently, femtocells have been deployed to provide higher data rates and also enhance in-building coverage. Due to their small coverage area, femtocells require much less transmission power than macrocells, and therefore their BSs are much more energy efficient in providing broadband coverage [10]. However, a large number of femtocell BSs deployed may increase the handoff rates of mobile users among adjacent cells, and also degrade the energy efficiency of the whole network [14]. Therefore, a joint deployment of BSs with different cell sizes is desirable in energy-efficient networks.

- Energy-Aware Cooperative BS Power Management
  The traffic load in a cellular network can have significant spatial and temporal fluctuations due to a number of factors such as user mobility and activities [15]. Therefore, some cells might be under low traffic load, while others may be under heavy traffic load. Since operating a BS consumes a considerable amount of electricity, selectively switching off some BSs or some of the resources of an active BS under low load conditions can save a substantial amount of energy. When some cells are switched off or in a sleep mode, the radio coverage and service provision for these cells can be guaranteed by the remaining active cells. Network-level power management is required where multiple BSs coordinate together. Cell zooming is an important technique through which BSs can adjust their cell sizes according to the network or traffic situation, and therefore reduce the energy consumption of the whole network.

There are a few studies on saving energy using energy-efficient approaches in the operation of base stations. The dynamic energy-efficient management scheme is analyzed in [2, 16] as the traffic varies over time. Authors of [17] propose BS energy-saving algorithms, which can dynamically minimize the number of the active BSs with respect to spatial-time traffic variation. Jardosh et al. propose the adoption of resource on-demand strategies to power on or off WLAN access points (APs) dynamically based on the volume and the location of users' demand [18]. Cao et al. analyze the energy-saving performance of CoMP transmission and wireless relaying with an average outage constraint [19]. To the best of our knowledge, all the previous work does not consider the impact of the smart grid on the operational decisions of cellular networks.

12.2.1.2   Key Tradeoffs in the Network

The following four key tradeoffs need to be considered when designing energy-efficient solutions for cellular networks [20].

The tradeoff between deployment efficiency (DE) and energy efficiency (EE) can be used to balance the deployment cost, throughput and energy consumption in the network as a whole [20]. DE, defined as system throughout per unit of deployment cost, is an important measure of network performance for mobile operators [20]. These two metrics often lead to opposite design criteria for network planning. For example, in order to save expenditures on base station equipment and maintenance, network planning engineers favor making individual cell coverage as wide as possible. However, radio resource management engineers prefer the deployment of small cells in order to minimize energy radiation.

The tradeoff between spectrum efficiency (SE) and EE can be used to balance the achievable rate for a given bandwidth and energy consumption of the network [20]. SE, a measure of the system throughput per unit of bandwidth, is normally used as optimization objective for wireless networks [20]. SE and EE are sometimes in conflict. Therefore, it is important to investigate how to balance these two important metrics.

The tradeoff between bandwidth (BW) and power (PW) can be used to balance the bandwidth utilized and the power needed for transmission for a given data transmission rate [20]. In wireless communications, BW and PW are two most important resources, but they are very limited. The fundamental BW-PW relation shows that in order to increase energy efficiency at a given data transmission rate, the transmit power must decrease, and therefore the signal bandwidth needs to undergo a corresponding increase.

The tradeoff between delay (DL) and PW can be used to balance the average end-to-end service delay and average power consumed in transmission [20]. DL, also known as service latency, is used to measure quality of service (QoS) and user experience in the network. It is related to the upper layer traffic types.

12.2.2 Smart Grid

As concerns about climate change grow, there is increasing interest in obtaining energy from renewable resources, such as solar and wind. Smart grid technologies can facilitate the integration of these renewable energy sources into the power grid, by coordinating and managing dynamically interacting power grid participants [5]. These renewable energy sources might be highly intermittent in nature and often uncontrollable, which produces a significant challenge for the reliability of the grid. In addition, the use of smart meters and smart appliances in the smart grid, which are an emerging class of energy users, can cause uncertainties on the demand side. Therefore, it is a challenging task to guarantee that the power demand load and power generation remain balanced, which is very important for system reliability. A mismatch between the supply and demand could cause a deviation of zonal frequency from nominal value, and power outages and blackouts may occur [21].

12.2.2.1 Demand-Side Management

Demand-side management (DSM), an important mechanism for improving the reliability of the smart grid, is a set of programs implemented by utility companies that allow customers a greater role in dynamically changing or shifting electricity consumption [22]. Multiple DSM programs can be implemented together in real systems to provide a combined improvement in demand management performance. DSM can help utilities operate more efficiently, reduce emission of greenhouse gases and also decrease the cost for electricity consumers. Recently, dynamic pricing programs have attracted much attention as one of the most important DSM strategies to encourage users to consume electricity more wisely and efficiently [23].

Among various dynamic pricing models, time-of-use (TOU) pricing, critical-peak pricing (CPP) and real-time pricing (RTP) are three important ones [21]. In TOU, variable pricing is decided for prescheduled blocks of time [24, 25]. In CPP, the price is decided in advance based on the demand hours. In RTP, the prices offered by retailers change frequently to reflect variations in the cost of the energy supply. Since the renewable energy sources integrated in the smart grid are highly intermittent in nature and often uncontrollable (e.g., the amount of electrical energy they produce varies over time and depends heavily on random factors such as the weather), it is a challenging task to integrate a significant portion of renewable energy resources into the power grid infrastructure, which needs to have means of effectively coordinating energy demand and generation. In addition, grid electricity energy storage, where the surplus energy is stored during times when production exceeds consumption, has limited capacity, and may not be economical in practice [6, 8]. Moreover, shut-down and ramp-up of a power plant can be costly. Therefore, power generators may use negative prices (i.e., consumers will be paid by consuming electricity) to encourage consumers to consume more energy. From

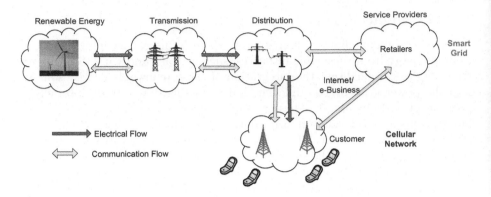

Figure 12.1: A cellular network powered by the smart grid.

an economic perspective, negative prices can be rational, because the costs to shut down and ramp up a power plant may exceed the loss for accepting negative prices. Negative prices can also occur if market actors have to fulfill other contracts, e.g., a heat delivery contract of a combined-heat and power plant (CHP), and therefore the power plant has to be run, although making losses due to negative power prices [7, 9].

12.3   System Models

We consider a cellular network powered by the smart grid, as shown in Fig. 12.1. We first present CoMP communication and service blocking probability model. We then present the electricity consumption model for base stations.

12.3.1   Coordinated Multipoint (CoMP) Communication

In cellular networks, interference exists between inter-cells, which affects spectral efficiency, especially in urban cellular systems. CoMP, which is originally proposed to overcome this limitation, has been selected as a key technology for LTE-A [4]. CoMP can significantly improve average spectral efficiency and also increase the cell edge and average data rates.

In our system, redundant base stations are turned off when the traffic is low, the real-time price is too high or the pollutant level of the electricity retailer is too high. CoMP is used among the active base stations to guarantee the coverage requirements [4]. In the downlink, the neighboring active base stations transmit cooperatively, and thus their coverage increases. In the uplink, multiple active base stations receive in coordination, which effectively reduces the requirement of received signal power at each individual base station. Together, CoMP can provide coverage for mobile users in nearby cells whose base station has been shut down during low activity period, and ensure the service blocking probability is within certain acceptable levels.

In order to achieve the theoretical maximum capacity, all of the base stations in the network should cooperate with each other in each transmission or reception process. However, the introduced complexity is not acceptable in real systems. Therefore, standards have specified the maximal number of BSs that may cooperate with each other [26]. We assume that there are i BSs in each CoMP cooperation cluster. All the possible combinations of the BSs in the cluster are denoted as the coordination set Θ, whose cardinality is 2^i. The base stations in the cluster can be switched on in any combination, where each combination set Ψ is an element of set Θ.

12.3.2 Service Blocking Probability Model in the Cellular Network

For an arbitrary element $\Psi \in \Theta$, the uplink sum-capacity for the cluster $C(\Psi)$ can be calculated as follows [27]:

$$C(\Psi) = \log_2 \det(\mathbf{I}_{|\Psi|} + P\mathbf{H}\mathbf{H}^\dagger), \tag{12.1}$$

where $\mathbf{I}_{|\Psi|}$ denotes a $|\Psi| \times |\Psi|$ identity matrix, P denotes the transmission power of each user terminal, $\mathbf{H} \in \mathbb{C}^{|\Psi| \times |\Psi|}$ denotes the channel matrix. The number of channels available in the cluster m can then be calculated as follows:

$$m = \left\lfloor \frac{BC(\Psi)}{B_0} \right\rfloor, \tag{12.2}$$

where B denotes the bandwidth allocated for each CoMP cluster, B_0 denotes the effective bandwidth required for a service. We assume that traffic characteristics, the desired packet-level QoS (e.g., packet delay and loss) guarantees, and the scheduling can together be represented by this effective bandwidth. Techniques for computing this effective bandwidth for different QoS requirements and traffic characteristics can be found in [28,29]. Therefore, the service blocking probability in the cluster $\mathcal{P}_b$ can be calculated according to Erlang-B formula [30]:

$$\mathcal{P}_b = \frac{\frac{E^m}{m!}}{\sum_{j=0}^{m} \frac{E^j}{j!}}, \tag{12.3}$$

where $E = \lambda/\mu$. Here, λ denotes the traffic arrival rate in the cluster, and μ denotes the average service rate. We define the relationship between the service blocking probability and a combination set as: $\mathcal{P}_b = \mathcal{D}(\Psi)$, where $\mathcal{D}(\Psi)$ is constructed by combining (12.1), (12.2) and (12.3).

12.3.3 Electricity Consumption Model for Base Stations

The energy consumption of each active cellular base station P_{BS} includes three parts: electricity consumption due to the transmission power, signal processing and backhauling, which can be denoted as follows [31]:

$$P_{BS} = aP_{tx} + bP_{sp} + cP_{bh}, \tag{12.4}$$

where P_{tx}, P_{sp}, P_{bh} denote the radiated power per base station, the signal processing power per base station, and the power due to backhauling per base station, respectively. a, b and c denote the scale factor associated with the corresponding power type, such as cooling and battery backup.

The transmit power of each base station depends on the path loss in the propagation channel. Therefore, the average transmit power per base station can be denoted as follows [31]:

$$\log(P_{tx}) = \log P_{min} - \log K + \phi \log(D/2), \tag{12.5}$$

where P_{min} denotes minimum receiving power required by user terminal, K is a parameter accounting for effects including BS antenna settings, carrier frequency and propagation environment, ϕ is the path loss exponent, and D denotes the inter site distance.

The energy consumption due to signal processing operations is partly determined by the employed air interface as well as the amount of cooperation between base stations. The uplink channel estimation operation consumes around 10% of the overall signal processing power [31]. The amount of power increases linearly with Ψ due to the increase in the number of estimated links. Around 3% of signal processing power is spent on uplink and downlink MIMO processing [31]. Assuming an MMSE filter operation, the total MIMO processing within each CoMP cooperation cluster requires Ψ^3 operations [31]. Therefore, the average MIMO processing per base station only increases quadratically with Ψ, and the energy consumption due to the signal processing for each base station can be calculated as follows [31]:

$$P_{sp} = p_{sp} \left(0.87 + 0.1|\Psi| + 0.03|\Psi|^2 \right), \tag{12.6}$$

where p_{sp} denotes a base line signal processing power per base station.

Backhaul is modeled as a collection of wireless micro wave links, matching the state of the art in most cellular networks. Each link has a capacity of 100 Mbit per second and spends 50W power. Thus for a given average backhaul requirement c_{bh} per base station, the backhaul power can be calculated [31]:

$$P_{bh} = \frac{50c_{bh}}{10^8}. \tag{12.7}$$

Substituting (12.5), (12.6) and (12.7) into (12.4), we can obtain:

$$P_{BS} = \frac{aP_{min}(\frac{D}{2})^{\phi}}{K} + bp_{sp}(0.87 + 0.1|\Psi| + 0.03|\Psi|^2) + \frac{50cc_{bh}}{10^8}. \tag{12.8}$$

Namely,

$$P_{BS} = U + V|\Psi| + W|\Psi|^2, \tag{12.9}$$

with $U = \frac{aP_{min}(\frac{D}{2})^{\phi}}{K} + 0.87bp_{sp} + \frac{50cc_{bh}}{10^8}$, $V = 0.1bp_{sp}$, and $W = 0.03bp_{sp}$.

12.4 Problem Formulation

Each retailer in the smart grid provides the real-time price to the base stations in each cluster, and then each cluster will decide how many base stations need to be turned on, and how much electricity needs to be procured from each retailer in order to provide enough electricity for all of the active base stations in the cluster. The system is illustrated in Fig. 12.1. Each retailer and each cluster of base stations make decisions to maximize respective utility.

We employ a Stackelberg game-theoretical scheme to jointly consider the utility of the retailers and base stations in the cluster. Our proposed Stackelberg game can be considered a generalized Stackelberg game in which there are multiple leaders and one or more followers. According to the behaviors of the base stations in each cluster and those of the retailers, the proposed Stackelberg game is divided into two levels: a *cellular network level* game and a *smart grid level* game. The base stations in one cluster, as a Stackelberg follower, play a cellular network level game. On the other hand, each retailer, as a Stackelberg leader, plays the smart grid level game. The prices and the information about how much electricity to procure need to be exchanged between the retailers and the base stations in the cluster. The proposed two-level game-theoretical approach can be implemented in a distributed manner.

12.4.1 Cellular Network Level Game

The base stations in a cluster aim to obtain the lowest service blocking probability with the least possible costs. There is a tradeoff between the blocking probability and the costs for the base stations. Therefore, it is very important for the base stations to decide how many base stations will be active in a cluster, and how much electricity is procured from each retailer in order to maximize utility.

With the increasing concerns about environmental protection, stricter regulations on pollutant emissions have been introduced, often including financial penalties associated with emissions (e.g., carbon credits), as well as non-financial costs associated with environmental damage, which is represented as financial penalties as well in our model. The appropriately designed pollutant emission costs make it possible for the base stations to procure electricity from retailers with renewable energy, thus reducing CO_2 emissions. We assume this cost is passed to the electricity users.

Therefore, the net utility function of all base stations in the cluster can be defined as:

$$\mathcal{U}_s = \xi \mathcal{F}(\mathcal{P}_b) - \sum_{n \in \mathcal{N}} p_n q_n - \sum_{n \in \mathcal{N}} \mathcal{I}_n(q_n), \qquad (12.10)$$

where ξ is a weight, $\mathcal{F}(\mathcal{P}_b)$ is the utility function of the base stations with regard to the service blocking probability $\mathcal{P}_b$, $\mathcal{N}$ ($\mathcal{N} = \{1, \ldots, N\}$) denotes the retailer set, p_n denotes the price provided by retailer n, q_n denotes the amount of electricity procured from retailer n and the function $\mathcal{I}_n()$ denotes the total pollutant emission cost incurred by the base stations when they procure electricity from retailer n.

The utility of the base station with regard to the service blocking probability is very high as long as the blocking probability is below a threshold value, and very low when the blocking probability passes the threshold. Therefore, the utility function $\mathcal{F}(\mathcal{P}_b)$ can be calculated as follows:

$$\mathcal{F}(\mathcal{P}_b) = U_{const} - \rho \exp\left((\mathcal{P}_b - \delta P_{th} + G_{bk})K_{bk}\right), \qquad (12.11)$$

where P_{th} denotes the blocking probability threshold value, U_{const}, ρ, δ, G_{bk} and K_{bk} are parameters for defining the utility functions for users, which might change with people and time.

The emissions of a generating unit can be described by a quadratic function of the amount of electricity generated [32–34]. Therefore, we use the following function to model the pollutant emission cost for the base stations when procuring an amount of electricity q_n from retailer n:

$$\mathcal{I}_n(q_n) = \alpha_n q_n^2 + \beta_n q_n, \qquad (12.12)$$

where $\alpha_n > 0$, $\beta_n > 0$, and α_n and β_n depends on the pollutant level of retailer n.

Therefore, the optimization problem for the base stations in the cluster can be formulated as:

$$\max_{\Psi, q_n} \mathcal{U}_s = \xi\left(U_{const} - \rho \exp\left((\mathcal{P}_b - \delta P_{th} + G_{bk})K_{bk}\right)\right) - \sum_{n \in \mathcal{N}} p_n q_n$$

$$- \sum_{n \in \mathcal{N}} (\alpha_n q_n^2 + \beta_n q_n),$$

s.t.

$$\sum_{n \in \mathcal{N}} q_n = U|\Psi| + V|\Psi|^2 + W|\Psi|^3. \qquad (12.13)$$

12.4.2   Smart Grid Level Game

We assume that each retailer is independent and acts selfishly, and aims to gain as much extra profit as possible. Retailer n's utility function can be defined as:

$$\mathcal{U}_n = (p_n - c_n)q_n, \qquad (12.14)$$

where c_n denotes the cost of electricity for retailer n. The optimization problem for retailer n is

$$\max_{p_n} \mathcal{U}_n = (p_n - c_n)q_n, \forall n. \qquad (12.15)$$

The choice of the optimal price for each retailer is affected by the other retailers' prices and its own pollutant level, since the retailers compete to get selected by the base stations. If a certain retailer asks such a high price that makes it less beneficial than the other retailers to the base stations, then the base stations will procure less from that retailer. On the other hand, if the provided price is too low, the profit of that retailer will be unnecessarily low.

12.5 Analysis of the Proposed Two-Level Game

For a Stackelberg game, existence and uniqueness of a Stackelberg equilibrium are two desirable properties. If we know there exists exactly one equilibrium, we can predict the equilibrium strategy of the players and resulting performance of the system [35, 36]. In this section, we first obtain the close-form solutions to the proposed game. Then, we prove that the solutions are the Stackelberg equilibrium for the proposed game. Finally, we prove that the Stackelberg equilibrium of the proposed game is unique.

12.5.1 Analysis of the Cellular Network Level Game

For an arbitrary combination set Ψ, we can represent the Lagrangian for (12.13) as follows:

$$L_{bs}(q_n, \nu) = U_s + \nu \left(\sum_{n \in \mathcal{N}} q_n - U|\Psi| - V|\Psi|^2 - W|\Psi|^3 \right), \qquad (12.16)$$

where ν is the Lagrangian multiplier. Set the derivative $dL_{bs} = 0$, which yields the system of equations:

$$\frac{\partial L_{bs}}{\partial q_1} = -p_1 - 2\alpha_1 q_1 - \beta_1 + \nu = 0, \qquad (12.17)$$

$$\vdots$$

$$\frac{\partial L_{bs}}{\partial q_N} = -p_N - 2\alpha_N q_N - \beta_N + \nu = 0, \qquad (12.18)$$

$$\frac{\partial L_{bs}}{\partial \nu} = \sum_{n \in \mathcal{N}} q_n - U|\Psi| - V|\Psi|^2 - W|\Psi|^3 = 0. \qquad (12.19)$$

Therefore, for set Ψ, the optimal quantity of electricity $q_{n,\Psi}^*$ procured from retailer n can be obtained by solving the above equations. Among all elements of the coordination set Θ, the optimal combination set Ψ^* is an element that maximizes the net utility of the base stations in the cluster. For set Ψ^*, the optimal quantity of electricity q_{n,Ψ^*}^* procured from retailer n can be calculated as follows:

$$\begin{aligned} q_{n,\Psi^*}^* &= \frac{1 - 2\alpha_n X}{4\alpha_n^2 X} p_n + \sum_{j \neq n} \frac{1}{4\alpha_n \alpha_j X} p_j + \frac{(1 - 2\alpha_n X)\beta_n}{4\alpha_n^2 X} + \\ &\quad \sum_{j \neq n} \frac{\beta_j}{4\alpha_n \alpha_j X} + \frac{U|\Psi^*| + V|\Psi^*|^2 + W|\Psi^*|^3}{2\alpha_n X}, \end{aligned} \qquad (12.20)$$

where $X = \sum_{n \in \mathcal{N}} \frac{1}{2\alpha_n}$, and $n, j \in \mathcal{N}$.

Property 1. *The optimal amount of electricity* q^*_{n,Ψ^*} *procured from retailer n decreases with its price* p_n, *when other retailers' prices are fixed.*

Proof. Taking the first-order derivative of q^*_{n,Ψ^*}, we have

$$\frac{\partial q^*_{n,\Psi^*}}{\partial p_n} = \frac{1 - 2\alpha_n X}{4\alpha_n^2 X}. \tag{12.21}$$

Since $X > 0$, and $1 - 2\alpha_n X < 0$, $\frac{\partial q^*_{n,\Psi^*}}{\partial p_n}$ is less than 0. Therefore, q^*_{n,Ψ^*} is decreasing with p_n. □

12.5.2 Analysis of the Smart Grid Level Game

Substituting (12.20) into (12.15), we have

$$\max_{p_n} \mathcal{U}_n = (p_n - c_n) q^*_{n,\Psi^*}. \tag{12.22}$$

Taking the derivative of $\mathcal{U}_n$ to p_n and equating it to zero, we have

$$\frac{\partial \mathcal{U}_n}{\partial p_n} = q^*_{n,\Psi^*} + (p_n - c_n)\frac{\partial q^*_{n,\Psi^*}}{\partial p_n} = 0. \tag{12.23}$$

Solving the above equations of p_n, we denote the solutions as p^*_n.

Property 2. *The utility function* $\mathcal{U}_n$ *of retailer n is concave in its own price* p_n, *when the other retailers' prices are fixed and its optimal amount of electricity provided can be calculated in (12.20).*

Proof. Taking the derivatives of $\mathcal{U}_n$ to p_n results in

$$\begin{aligned}
\frac{\partial \mathcal{U}_n}{\partial p_n} &= q^*_{n,\Psi^*} + (p_n - c_n)\frac{\partial q^*_{n,\Psi^*}}{\partial p_n} \\
&= \frac{1 - 2\alpha_n X}{2\alpha_n^2 X}p_n + \sum_{j\neq n}\frac{1}{4\alpha_n\alpha_j X}p_j + \frac{(1 - 2\alpha_n X)(\beta_n - c_n)}{4\alpha_n^2 X} \\
&\quad + \sum_{j\neq n}\frac{\beta_j}{4\alpha_n\alpha_j X} + \frac{U|\Psi^*| + V|\Psi^*|^2 + W|\Psi^*|^3}{2\alpha_n X},
\end{aligned} \tag{12.24}$$

and,

$$\frac{\partial^2 \mathcal{U}_n}{\partial p_n{}^2} = \frac{1 - 2\alpha_n X}{2\alpha_n^2 X}. \tag{12.25}$$

Since $1 - 2\alpha_n X < 0$ and $X > 0$, we have $\frac{\partial^2 \mathcal{U}_n}{\partial p_n{}^2} < 0$. Therefore, $\mathcal{U}_n$ is concave with respect to p_n. □

12.5.3   Existence of Stackelberg Equilibrium for the Proposed Two-Level Game

In this subsection, we will prove that the solutions $q^*_{n,\Psi*}$ and $p^*_n (n \in \mathcal{N})$ are the Stackelberg Equilibrium for the proposed game.

Definition 12.5.1. q^{SE}_n and p^{SE}_n are the Stackelberg Equilibrium of the proposed two-level game if for every retailer n, when p_n is fixed

$$\mathcal{U}_s \left(\{q^{SE}_n\} \right) = \sup_{\{q_n\}} \mathcal{U}_s (\{q_n\}), \qquad (12.26)$$

and when q_n is fixed

$$\mathcal{U}_n \left(p^{SE}_n \right) = \sup_{p_n} \mathcal{U}_n (p_n). \qquad (12.27)$$

In the following, we prove that the solution $q^*_{n,\Psi*}$ in (12.20) is the global optimum that maximizes the base stations' utility $\mathcal{U}_s$. Namely, we verify that $q^*_{n,\Psi*}$ in (12.20) meets Karush-Kuhn-Tucer (KKT) conditions. Based on (12.16), we get

$$\nabla L_{bs}(q^*_{n,\Psi*}) = -p_n - 2\alpha_n q^*_{n,\Psi*} - \beta_n + \nu = 0, \qquad (12.28)$$

and

$$\nabla^2 L_{bs}(q^*_{n,\Psi*}) = -2\alpha_n < 0. \qquad (12.29)$$

Therefore, $q^*_{n,\Psi*}$ in (12.20) is the global optimum that maximizes the base station's utility $\mathcal{U}_s$ [37]. $q^*_{n,\Psi*}$ satisfies (12.26) and is the Stackelberg Equilibrium q^{SE}_n. Due to the concavity of $\mathcal{U}_n$, retailer n can always find its optimal price p^*_n. Together, we construct the following theorem:

Theorem 1. *The pair of $q^*_{n,\Psi*}$ and $\{p^*_n\}$ is the Stackelberg Equilibrium for the proposed two-level game, where the Stackelberg Equilibrium is defined in (12.26) and (12.27).*

12.5.4   Uniqueness of the Stackelberg Equilibrium for the Proposed Two-Level Game

We can show that the Stackelberg Equilibrium of the proposed game is unique. Namely, we can prove that the smart grid level game has the unique Nash Equilibrium. The Nash Equilibrium gives the set of prices such that none of the retailers can increase its individual utility by choosing a different price given the prices offered by the other retailers.

Because the retailers are independent and rational, and have the goal of maximizing their profits, the best response functions can be defined and used to obtain the solution of the pricing problem in the smart grid level game. When the price strategies $\mathbf{p}_{-n}$ offered by the retailers other than retailer n

are given, the best response function $\mathcal{B}_n(\mathbf{p}_{-n})$ of retailer n can be defined as follows:

$$\mathcal{B}_n(\mathbf{p}_{-n}) = \arg\max_{p_n} \mathcal{U}_n(p_n, \mathbf{p}_{-n}). \tag{12.30}$$

Therefore, substituting (12.20) into (12.23), the best response function $\mathcal{B}_n(\mathbf{p}_{-n})$ of retailer n is as follows:

$$\mathcal{B}_n(\mathbf{p}_{-n}) = \frac{\alpha_n}{2(2\alpha_n X - 1)}\left(\sum_{j\neq n}\frac{p_j}{\alpha_j}\right) + Y_n, \tag{12.31}$$

where

$$Y_n = \frac{\alpha_n}{2(2\alpha_n X - 1)}\left(\sum_{j\neq n}\frac{\beta_j}{\alpha_j}\right) + \frac{c_n - \beta_n}{2}$$

$$+ \frac{\alpha_n}{2\alpha_n X - 1}\left(U|\Psi^*| + V|\Psi^*|^2 + W|\Psi^*|^3\right). \tag{12.32}$$

In the following, we will show that an arbitrary retailer n's best response function $\mathcal{B}_n(\mathbf{p}_{-n})$ is a *standard* function.

Definition 12.5.2. *A function $\mathcal{B}_n(\mathbf{p}_{-n})$ is standard if for all $\mathbf{p}_{-n} \geq 0$, the following properties are satisfied [38]:*

- *Positivity: $\mathcal{B}_n(\mathbf{p}_{-n}) > 0$.*

- *Monotonicity: If $\mathbf{p}_{-n} \geq \mathbf{p}'_{-n}$, then $\mathcal{B}_n(\mathbf{p}_{-n}) \geq \mathcal{B}_n(\mathbf{p}'_{-n})$.*

- *Scalability: For all $\omega > 1$, $\omega\mathcal{B}_n(\mathbf{p}_{-n}) > \mathcal{B}_n(\omega\mathbf{p}_{-n})$.*

Proposition 12.5.1. *An arbitrary retailer n's best response function $\mathcal{B}_n(\mathbf{p}_{-n})$ is standard.*

1) Positivity.
In (12.31),

$$\mathcal{B}_n(\mathbf{p}_{-n}) = \frac{\alpha_n}{2(2\alpha_n X - 1)}\left(\sum_{j\neq n}\frac{p_j}{\alpha_j}\right) + Y_n.$$

Since $\alpha_n > 0$, $2\alpha_n X - 1 > 0$, $\alpha_j > 0$ and $p_j > 0$, the first item of the RHS of the above equation is greater than 0. Since $Y_n > 0$, the best response function $\mathcal{B}_n(\mathbf{p}_{-n})$ is positive.
2) Monotonicity.
Suppose $\mathbf{p}_{-n}$ and $\mathbf{p}'_{-n}$ are different price vectors, and the vector inequality $\mathbf{p}_{-n} \geq \mathbf{p}'_{-n}$ means that $p_j \geq p'_j$, $\forall j \in \{1,\ldots,N\}, j \neq n$. If $\mathcal{B}_n([p_1,\ldots,p_j,\ldots,p_N]) \geq \mathcal{B}_n([p_1,\ldots,p'_j,\ldots,p_N])$, then monotonicity can be

shown to hold. Therefore, the problem reduces to provide $\frac{\partial \mathcal{B}_n(\mathbf{p}_{-n})}{\partial p_j} \geq 0$. Taking the derivitive of the best response function $\mathcal{B}_n(\mathbf{p}_{-n})$ to p_j, we obtain

$$\frac{\partial \mathcal{B}_n(\mathbf{p}_{-n})}{\partial p_j} = \frac{\alpha_n}{2\alpha_j(2\alpha_n X - 1)} > 0. \tag{12.33}$$

3) Scalability.

Based on (12.31), we can obtain

$$\omega \mathcal{B}_n(\mathbf{p}_{-n}) = \frac{\alpha_n \omega}{2(2\alpha_n X - 1)} \left(\sum_{j \neq n} \frac{p_j}{\alpha_j} \right) + \omega Y_n. \tag{12.34}$$

$$\mathcal{B}_n(\omega \mathbf{p}_{-n}) = \frac{\alpha_n \omega}{2(2\alpha_n X - 1)} \left(\sum_{j \neq n} \frac{p_j}{\alpha_j} \right) + Y_n. \tag{12.35}$$

Based on (12.34) and (12.35), we can obtain

$$\omega \mathcal{B}_n(\mathbf{p}_{-n}) - \mathcal{B}_n(\omega \mathbf{p}_{-n}) = (\omega - 1)Y_n > 0. \tag{12.36}$$

Therefore, for all $\omega > 1$, $\omega \mathcal{B}_n(\mathbf{p}_{-n}) > \mathcal{B}_n(\omega \mathbf{p}_{-n})$.

Theorem 2. *The pair of q^*_{n,ψ^*} and $\{p^*_n\}$ is the unique Stackelberg Equilibrium for the proposed two-level game, where the Stackelberg Equilibrium is defined in (12.26) and (12.27).*

Proof. Since an arbitrary retailer n's best response function $\mathcal{B}_n(\mathbf{p}_{-n})$ is standard, the smart grid level game has the unique Nash Equilibrium [39]. Therefore, the proposed game has the unique Stackelberg Equilibrium. □

12.6   Simulation Results and Discussions

In this section, we use computer simulations to evaluate the performance of the proposed dynamic base station operation scheme (DBSOS). In DBSOS, according to Subsection 12.5.3, we derive the optimal solution for the cellular network to decide how many base stations will be active in a cluster, and how much electricity is procured from each retailer in order to maximize utility. We also derive the optimal solution for retailers to decide the electricity price. In the simulations, we assume that there are three base stations in a cluster. These base stations can procure electricity from two retailers. We set the parameters as follows: a = 7.35, b = 2.9, c = 1, p_{sp} = 58 W, $10 \log(P_{min})$= -120, $K = 0.0001$, $\phi = 4.0$, $D = 800$ m, $c_{bh} = 10^9$ bps [31]. We set λ and μ as 1.1 and 1/180, respectively. We choose $P_{th} = 0.1$ [40,41], $c_1 = 0.3$ cents/kWh, $c_2 = 0.4$ cents/kWh, $\beta = 0.1$ cents/kWh, $\alpha_1 = 2 \times 10^{-2}$ and $\alpha_2 = 5 \times 10^{-2}$. We compare the proposed DBSOS with the traditional method, which does

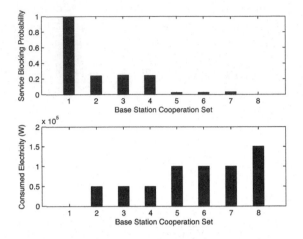

Figure 12.2: Service blocking probability and amount of electricity consumed as the combination of the base stations in a cluster varies, $\lambda = 1.1$ and $\mu = 1/180$. The numbers on the x axis represent the indices of different elements of the cooperation set Θ. In base station cooperation sets 2, 3, 4, there is one active base station. In cooperation sets 5, 6 and 7, there are two active base stations. In cooperation set 8, all three base stations are active.

not consider the dynamics of smart grid and all of the base stations are active, in terms of operational expenditure and CO_2 emissions.

We first investigate how different coordinations among active base stations in DBSOS with CoMP affect the service blocking probability in the cluster and the total amount of electricity consumed. Fig. 12.2 shows that the service blocking probability starts at 1.0 when there are no active base stations in the cluster, gradually decreases, and finally reaches 0.0 when three active base stations coordinate. Each base station consumes approximately 5×10^3 W electricity when it is active. The figure also shows that the amount of electricity consumed increases with the number of active base stations in the cluster. In this scenario, if one base station in the cluster is shut down, the service blocking probability is still less than the threshold 0.1. The two active base stations in the cluster can extend their coverage using CoMP transmissions and receptions.

Fig. 12.3 shows that using the traditional method, these three base stations spend a constant amount of money to procure a constant amount of electricity, and therefore produce a constant amount of CO_2, even though the traffic arrival rate λ varies. The figure also shows that for DBSOS, the number of base stations that need to be activated depends on the traffic arrival rate. For example, when the traffic arrival rate is less than 0.8, one active base station is

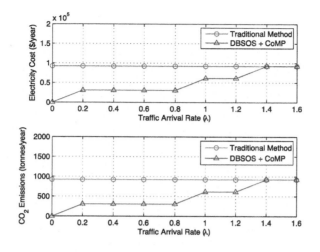

Figure 12.3: Compare the traditional method with the proposed DBSOS with CoMP, in terms of electricity cost and CO_2 emission as the traffic arrival rate λ varies.

enough to meet the service blocking probability requirement. CO_2 emissions depend on how many base stations are active during that period. DBSOS has a much better performance compared to the traditional method in terms of electricity costs and CO_2 emissions, especially when the traffic arrival rate is low. For example, DBSOS leads to a 66.7% decrease in electricity costs and CO_2 emissions when the traffic arrival rate is between 0.2 and 0.8 compared to the traditional method.

We compare the traditional method, DBSOS with CoMP, and DBSOS with CoMP and demand response (i.e., in the smart grid). This comparison is in terms of the operational expenditure of the base stations and the service blocking probability when negative electricity price is sometimes offered in the smart grid. The electricity price is set to 0.0464 $/kWh, the average negative electricity price is set to -0.2356 $/kWh [6, 8, 9]. Negative price is offered 5% of the time, and the value can increase with the higher amount of renewable energy intergrated into the smart grid. Fig. 12.4 shows that the traditional method has the highest operational expenditure. Compared to DBSOS with CoMP, DBSOS with CoMP in the smart grid decreases the operational expenditure by 27.06%, even though the negative price is offered only rarely. The reason for this is that in the BSSOS with CoMP in the smart grid, the operation of the base stations depends not only on the service blocking probability, but also on the price offered by the smart grid. Therefore, all of the base stations might be turned on when the electricity price is low or negative.

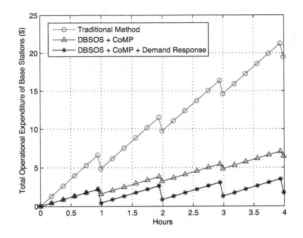

Figure 12.4: Total operational expenditure of base stations.

12.7   Conclusions

In this chapter, we have discussed how the dynamic operation of cellular base stations based on the traffic arrival rate, real-time price provided by the smart grid and pollutant levels of electricity retailers can reduce operational expenditure and CO_2 emissions in green wireless cellular networks. In the proposed scheme, the base stations can be switched off to save energy. CoMP is used to extend the coverage of the active BSs, and thus the service quality of mobile users can still be ensured. The active base stations also need to decide which retailers to procure electricity from and the amount to procure from the smart grid. The system's decision problem has been modeled as a two-level Stackelberg game, where the cellular network is the Stackelberg follower, and the electricity retailers are the Stackelberg leaders. Simulation results have been presented to show that the smart grid has significant impacts on green wireless cellular networks, and our proposed scheme can significantly reduce operational expenditure and CO_2 emissions in green wireless cellular networks. Future work is in progress to consider heterogeneous wireless cellular networks with femtocells in the proposed framework.

Bibliography

[1] E. Oh, B. Krishnamachari, X. Liu, and Z. Niu, "Toward dynamic energy-efficient operation of cellular network infrastructure," *IEEE Comm. Mag.*, vol. 49, pp. 56–61, Jun. 2011.

[2] M. A. Marsan, L. Chiaraviglio, D. Ciullo, and M. Meo, "Optimal energy savings in cellular access networks," in *Proc. IEEE ICC'09 Workshops*, (Dresden, Germany), Jun. 2009.

[3] H. Sistek, "Green-tech base stations cut diesel usage by 80 percent," Apr. 2008. http://news.cnet.com/8301-11128_3-9912124-54.html.

[4] R. Irmer, H. Droste, P. Marsch, M. Grieger, G. Fettweis, S. Brueck, H.-P. Mayer, L. Thiele, and V. Jungnickel, "Coordinated multipoint: concepts, performance, and field trial results," *IEEE Comm. Mag.*, vol. 49, pp. 102–111, Feb. 2011.

[5] C. Feisst, D. Schlesinger, and W. Frye, "Smart grid, the role of electricity infrastructure in reducing greenhouse gas emissions," tech. rep., Cisco Internet Business Solution Group, Oct. 2008.

[6] F. Genoese, M. Genoese, and M. Wietschel, "Occurrence of negative prices on the German spot market for electricity and their influence on balancing power markets," in *Proc. Int'l Conf. on the European Energy Market*, (Madrid, Spain), Jun. 2010.

[7] D. Keles, M. Genoese, D. Most, and W. Fichtner, "Comparison of extended mean-reversion and time series models for electricity spot price simulation considering negative prices," *Energy Economics*, Aug. 2011. online, doi:10.1016/j.physletb.2003.10.071.

[8] M. Nicolosi, "Wind power integration and power system flexibility - an empirical analysis of extreme events in Germany under the new negative price regime," *Energy Policy*, vol. 38, pp. 7257–7268, Nov. 2010.

[9] C. Brandstatt, G. Brunekreeft, and K. Jahnke, "How to deal with negative power price spikes? - Flexible voluntary curtailment agreements for large-scale integration of wind," *Energy Policy*, vol. 39, pp. 3732–3740, Jun. 2011.

[10] Z. Hasan, H. Boostanimehr, and V. K. Bhargava, "Green cellular networks: a survey, some research issues and challenges," *IEEE Communications Surveys & Tutorials*, vol. 13, Fourth Quarter 2011.

[11] A. Bianzino, C. Chaudet, D. Rossi, and J. Rougier, "A survey of green networking research," *IEEE Communications Surveys and Tutorials*, vol. 14, pp. 3–20, First Quarter 2012.

[12] "Multiradio base station makes network evolution easier and greener than ever," http://www.nokiasiemensnetworks.com/sites/default/files/document/NokiaSiemensNetworks_2009_02_05_enFlexiMultiradioBTS.pdf, Press Release Feb. 2009, Nokia Siemens Networks [online access, 2 Feb. 2012].

[13] L. Silu, "The green cdma base station," Huawei Communicate, pp.42–42, Dec. 2008.

[14] M. Ismail and W. Zhuang, "Network cooperation for energy saving in green radio communications," *IEEE Wireless Communications*, vol. 18, pp. 76–81, Oct. 2011.

[15] Z. Niu, Y. Wu, J. Gong, and Z. Yang, "Cell zooming for cost-efficient green cellular networks," *IEEE Communications Magazine*, vol. 48, pp. 74–79, Nov. 2010.

[16] L. Chiaraviglio, D. Ciullo, M. Meo, and M. A. Marsan, "Energy-aware UMTS access networks," in *Proc. Int'l Symp. on Wireless Personal Multimedia Comm.*, (Lapland, Finland), Sept. 2008.

[17] S. Zhou, J. Gong, Z. Yang, Z. Niu, and P. Yang, "Green mobile access network with dynamic base station energy saving," in *Proc. Mobicom Poster*, (Beijing, China), Sept. 2009.

[18] A. P. Jardosh, K. Papagiannaki, E. M. Belding, K. C. Almeroth, G. Iannaccone, and B. Vinnakota, "Green WLANs: on-demand WLAN infrastructures," *Mobile Networks and Applications*, vol. 14, pp. 798–814, Dec. 2009.

[19] D. Cao, S. Zhou, C. Zhang, and Z. Niu, "Energy saving performance comparison of coordinated multi-point transmission and wireless relaying," in *Proc. IEEE Globecom'10*, (Miami, USA), Dec. 2010.

[20] Y. Chen, S. Zhang, S. Xu, and G. Y. Li, "Fundamental trade-offs on green wireless networks," *IEEE Communication Magazine*, vol. 49, pp. 30–37, Jun. 2011.

[21] M. Parvania and M. Fotuhi-Firuzabad, "Demand response scheduling by stochastic SCUC," *IEEE Trans. Smart Grid*, vol. 1, pp. 89–98, Jun. 2010.

[22] G. M. Masters, *Renewable and Efficient Electric Power Systems*. New York: Wiley, July 2004.

[23] P. Samadi, A. Mohsenian-Rad, R. Schober, V. W. S. Wong, and J. Jatskevich, "Optimal real-time pricing algorithm based on utility maximization for smart grid," in *Proc. First IEEE Conf. on Smart Grid Comm.*, (Gaithersburg, MD), Oct. 2010.

[24] M. Erol-Kantarci and H. T. Mouftah, "Tou-aware energy management and wireless sensor networks for reducing peak load in smart grids," in *Proc. IEEE VTC'10F*, (Ottawa, Canada), Sep. 2010.

[25] S. Shao, T. Zhang, M. Pipattanasomporn, and S. Rahman, "Impact of tou rates on distribution load shapes in a smart grid with PHEV penetration," in *Proc. IEEE PES Transmission and Distribution Conference and Exposition*, (New Orleans, USA), Apr. 2010.

[26] L. Wang and C. Yeh, "3-cell network MIMO architectures with sectorization and fractional frequency reuse," *IEEE J. Sel. Areas Commun.*, vol. 29, pp. 1185–1199, Jun. 2011.

[27] D. Gesbert, S. Hanly, H. Huang, S. S. Shitz, O. Simeone, and W. Yu, "Multi-cell MIMO cooperative networks: A new look at interference," *IEEE J. Sel. Areas Commun.*, vol. 28, pp. 1380–1408, Dec. 2010.

[28] A. I. Elwalid and D. Mitra, "Effective bandwidth of general Markovian traffic sources and admission control of high speed networks," *IEEE/ACM Trans. Netw.*, vol. 1, pp. 329–343, Jun. 1993.

[29] F. Yu and V. Krishnamurthy, "Effective bandwidth of multimedia traffic in packet wireless CDMA networks with LMMSE receivers – a cross-layer perspective," *IEEE Trans. Wireless Commun.*, vol. 5, pp. 525–530, Mar. 2006.

[30] L. Kleinrock, *Queueing Systems, Volume I: Theory.* New York: Wiley Interscience, 1975.

[31] A. J. Fehske, P. Marsch, and G. P. Fettweis, "Bit per joule efficiency of cooperating base stations in cellular networks," in *Proc. IEEE Globecom'10 Workshops*, (Miami, USA), Dec. 2010.

[32] K. Senthil, "Combined economic emission dispatch using evolutionary programming technique," *IJCA Special Issue on Evolutionary Computation for Optimization Techniques*, no. 2, pp. 62–66, 2010.

[33] K. Senthil and K. Manikandan, "Improved tabu search algorithm to economic emission dispatch with transmission line constraint," *Int'l J. of Computer Science and Comm.*, vol. 1, pp. 145–149, Jul.-Dec. 2010.

[34] S. M. V. Pandian et al., "An efficient particle swarm optimization technique to solve combined economic emission dispatch problem," *European Journal of Scientific Research*, vol. 54, no. 2, pp. 187–192, 2011.

[35] G. He, S. Lasaulce, and Y. Hayel, "Stackelberg games for energy-efficient power control in wireless networks," in *Proc. IEEE INFOCOM Mini-Conference*, (Shanghai, China), Apr. 2011.

[36] B. Wang, Y. Wu, and K. J. R. Liu, "Game theory for cognitive radio networks: an overview," *Computer Networks*, vol. 54, pp. 2537–2561, Oct. 2010.

[37] M. S. Barzaraa, *Nonlinear Programming: Theory and Algorithms*. John Wiley & Sons, 1993.

[38] R. Yates, "A framework for uplink power control in cellular radio systems," *IEEE J. Sel. Areas Commun.*, vol. 13, pp. 1341–1347, Sept. 1995.

[39] C. Saraydar, N. Mandayam, and R. Goodman, "Efficient power control via pricing in wireless data networks," *IEEE Trans. Commun.*, vol. 50, pp. 291–303, Feb. 2002.

[40] R. G. Akl, M. V. Hegde, and M. Naraghi-Pour, "Mobility-based CAC algorithm for arbitrary call-arrival rates in CDMA cellular systems," *IEEE Trans. Veh. Tech.*, vol. 54, pp. 639–651, Mar. 2005.

[41] D. Niyato and E. Hossain, "A novel analytical framework for integrated cross-layer study of call-level and packet-level qos in wireless mobile multimedia networks," *IEEE Trans. Mobile Comput.*, vol. 6, pp. 322–335, Mar. 2007.

Index